Bruno Daniotti · Marco Gianinetto ·
Stefano Della Torre
Editors

Digital Transformation of the Design, Construction and Management Processes of the Built Environment

Editors
Bruno Daniotti
Architecture, Built Environment
and Construction Engineering—ABC
Department
Politecnico di Milano
Milan, Italy

Marco Gianinetto
Architecture, Built Environment
and Construction Engineering—ABC
Department
Politecnico di Milano
Milan, Italy

Stefano Della Torre
Architecture, Built Environment
and Construction Engineering—ABC
Department
Politecnico di Milano
Milan, Italy

https://doi.org/10.1007/978-3-030-33570-0

Preface

The chapters included in this book deal with the digital transformation of the built environment.

Digitalization is perhaps the outstanding trend in all the sectors of the life, all around the world: an Italian perspective on what is happening in the industry of the built environment may be especially interesting because of the peculiarities of the construction sector, and because criticalities and barriers are in Italy more challenging than in many other countries.

The book belongs to a series designed to present a structured vision of the many possible approaches—within the field of architecture and civil engineering—to the development of researches dealing with the processes of planning, design, construction, management and transformation of the built environment. Each book contains a selection of essays reporting researches and projects, developed during the last six years within the ABC Department (Architecture, Built environment and Construction engineering) of Politecnico di Milano, concerning a cutting-edge field in the international scenario of the construction sector. Following the concept that innovation happens as different researches stimulate each other, skills and integrated disciplines are brought together within the department, generating a diversity of theoretical and applied studies.

Each part of the book was designed to include relevant research outcomes in all the different stages of the transformation process (i.e. Interoperable Management of the Process, Design Stage, Executive Stage and Management Stage) with a common thread (e.g. handling of the information, support of designers and clients in the first stage, optimization of production and construction activities, efficient management of buildings). Moreover, the volume includes a final part on Digital Technologies for Multi-Scale Survey and Analysis, which embraces all the previously mentioned stages and also presents state-of-the-art case histories on Italian monuments well known worldwide, transferring to historic preservation the processual vision matured through the multidisciplinary approach to the built environment as a whole.

The papers have been selected based on their capability to describe the outputs and the potentialities of carried out researches, giving at the same time a report on the reality as well as on the perspectives for the future.

Stefano Della Torre
Head of the Department Architecture
Built Environment and Construction Engineering
Politecnico di Milano
Milan, Italy
e-mail: stefano.dellatorre@polimi.it

Introduction

The construction sector is facing demands for structural innovation in terms of digital dematerialization in managing transformation processes from the building to the territory.

Demands for simplification and transparency in process information management and the rationalization and optimization of a very fragmented and splintered process are a key driver for digitization. In this regard, there are a number of elements that highlight the need to focus on digitization.

The digital transformation of EU business and society presents enormous growth potential for Europe. European industry can build on its strengths in advanced digital technologies and its strong presence in traditional sectors to seize the range of opportunities that technologies such as the Internet of things, big data, advanced manufacturing, robotics, 3D printing, blockchain technologies and artificial intelligence offer. This will enable our industry to capture a share in the emerging markets for the products and services of the future.

The European Directives for Public Procurement are pushing the entire construction supply chain towards radical transformation over the next few years, requiring research, development and training for digitalization.

The private sector is also being greatly affected by demands from purchasing at an international level, which very often requires the use of BIM, particularly for large-scale operations.

ICT standardization and interoperability are a precondition for the uptake of digital innovations. The challenge is to develop and ensure the adoption of European standards that ensure compatibility between systems and guarantee the competitiveness of the European industry and the openness of ICT markets; then, standardization activity is very important in order to transfer research results to actual application.

At the start of the twenty-first century, the acronym BIM was not well known as nowadays. Though, while the academy and the researchers were talking in the wider terms of ICTC—Information and Communication Technologies for Construction—the industry started introducing the new terms of BIM—Building Information Model or Modelling—to progressively substitute the words CAD, computer-aided design, that, despite being already exhaustive of a decisional and management process supported by computers (and this is evident when referring to the way the mechanic industry implemented CAD to support its production process management), was still interpreted—by the majority of the players in the building industry—in a reductive way as CAd, or computer-aided drafting: the building constructions field, at that time is not mature yet to understand the potential of the computer-aided design as a decisional tool to support the whole construction process. The roots of the modern interpretation of the CAD and BIM systems can actually be traced back to the early '60s and to the activities developed inside the MIT—Labs of the Boston University: many of the concepts still at the base of the modern CAD-BIM software, such as software–user interactivity, modular design, object-oriented modelling.

Similar to other sectors, construction is now seeing its own "Digital Revolution", having previously benefitted from only modest productivity improvements. Building Information Modelling ("BIM") is being adopted rapidly by different parts of the value chain as a strategic tool to deliver cost savings, productivity and operations efficiencies, improved infrastructure quality and better environmental performance.

The use of BIM software has allowed users to simplify the entire project development in its entirety, especially for new buildings. This great change in the building construction is characterized by the opportunity to manage the whole process with a single tool in a clear and transparent manner. The data can be modified in real time by anyone who is licensed, appointing party and appointed party.

Therefore, research in this specific area aims to develop interoperable and integrated methods and ICT tools to manage the various phases in the process of transforming the environment.

In the different parts, the most relevant research activities in this sector are presented, focusing on the development of methods and interoperable and integrated ICT tools for the management of the different stages of the process for the transformation of the built environment:

- Interoperable Management of the Process
- Design Stage
- Executive Stage
- Management Stage
- Methods and Tools for the Condition Assessment Survey.

The part I presents a set of research activities, focused on the development of methods and tools, for the management of Information and Communication Technologies through the whole process.

The part II presents an overview of various research activities developed in order to support designers and clients in the first design stages, in the framework of progressive digital transformation and BIM introduction.

A new design approach is changing the traditional one with the introduction of Building Information Modelling, allowing then to anticipate choices, preventing mistakes, pathologies and useless loss of time and economic troubles.

Part III introduces the main research results useful for the optimization of production and construction activities, where the supply chain is involved.

The part IV is aimed to present the most relevant research results of the digital transformation about the management of existing buildings, in general and for historical heritage.

The Facility Management (FM) sector is undergoing a profound transformation of practices, processes, tools and references due to the adoption of novel Information and Communication Technology (ICT) solutions which nowadays promise to improve the traditionally conceived FM streamlining processes, making new knowledge bases available to support data-driven decision-making processes and embracing a network approach to stakeholder management.

The part V describes the outcomes of some relevant research activities related to digital survey technologies of the built environment at different territorial scales, from that of large historical monuments to that of their surroundings and landscape. Specifically, the case histories of the Cathedral of Milan (Duomo di Milano) and Saint Mark's Basilica (Basilica di San Marco) in Venice describe the technical development of the digital survey techniques for cultural and historical heritage during the last decade.

The synergy of traditional and cutting-edge survey approaches made possible to build a digital replica of two of the most famous churches in the world, which are today used for both documentation and conservation of these exceptional monuments.

Bruno Daniotti
Marco Gianinetto

Contents

Digital Technologies for Multi-Scale Survey and Analysis

About the Editors

Stefano Della Torre Graduated in Civil Engineering and in Architecture, he is a full professor in restoration at the Politecnico di Milano. He is director of the ABC Department - Architecture, Built environment and Construction engineering. He is the author of more than 250 publications. He serves as advisor for: CARIPLO Foundation (Cultural districts) Province of Como and Lombardy Region, (policies of programmed conservation of historical-architectural heritage). He has been President of BuildingSMART Italia - national chapter of association BuildingSMART international.

Bruno Daniotti is a full professor of Building Production in ABC Department. PhD in Building Engineering, he has more than thirty years of experience in various fields of research in construction at national and international level, working in Universities, Research Centres such as ICITE of C.N.R. and EC JRC (Ispra (VA)) and collaborating with Research Organizations, public and regulatory bodies, as ISO, EOTA and UNI. He was the Italian Coordinator of STANDINN Project, PM of INNOVance Project and representative of Politecnico di Milano in BuildingSmart Italia (2005-2015). He is coordinator of the Focus Area on ICT and processes of the Italian Construction Technology Platform, of the BIMREL project, developing the Regione Lombardia Platform and BIM library for buildings' products, and of BIM4EEB EU research project "BIM based fast toolkit for Efficient rEnovation in Buildings". Director of II level Master "BIM Manager", he is a member of ISTeA Steering Committee, and coordinator of the research line "Digitization" within the triennial scientific project 2017/2019 of ABC Department.

Marco Gianinetto (MS in Environmental and Land Management Engineering, PhD in Geodesy and Geomatics) is Associate Professor of Geomatics at Politecnico di Milano. His research interests are in the field of Remote Sensing, Earth Observation and Spatial Information Technologies for environmental monitoring. Today, Prof. Gianinetto is a member of the Governing Council of Italian Remote Sensing Society, Co-Editor-in-Chief for European Journal of Remote Sensing (UK), Associate Editor for International Journal of Remote Sensing (UK) and Associate

Editor for Journal of Applied Remote Sensing (USA). Furthermore, from 2007 to 2018 he was Associate Editor for International Journal of Navigation and Observation (Egypt/USA). Marco Gianinetto has been a member of international advisory/evaluator panels, including the Executive Agency for SMEs of European Commission, the Israel Science Foundation, the National Research Council of Romania, the Belgian Federal Science Policy Office, the French National Research Agency. Prof. Gianinetto has been scientific advisor for the study 'Space Market Uptake in Europe' commissioned for the European Parliament by the Directorate General for Internal Policies - Policy Department A.

Interoperable Management of the Process

Introduction

Bruno Daniotti, Marco Gianinetto

The part I presents a set of research activities, focused on the development of methods and tools, for the management of Information and Communication Technologies through the whole process.

In the case of INNOVANCE and BIMREL research projects the goal is to develop BIM based platforms to support information exchange among the different actors of building process; other examples are related with platforms for specific production sectors, as for chemical and electrical products.

The purpose of the Da.Ma.Trà research project is to build a web-based digital platform prototype that can handle the constructive traceability of materials for civil buildings. The project's attention to new materials has a specific focus on the use of bio base materials and agricultural waste in construction.

Another set of research activities deals with specific methods and tools as for Blockchain and the natural language processing for information and project management.

Blockchain is an innovative methodology capable of managing contracts and transactions through which assets are organized and guarded, social actions are governed and relations between nations, institutions and individuals are guided.

Natural language processing (NLP) can be applied to translate text into numerical data, with a possible application in the field of information modeling and project management. Where Information modelling requires a precise and comprehensive definition of the initial requirements.

For updating the construction sector in line with current trends, Life Cycle Thinking (LCT) has to be integrated into the building process from the beginning. In this perspective, the digitalization increasingly assists practitioners in the task, taking advantage in particular of the now widespread Building Information Modeling (BIM).

Then other research activity aims at the definition of methods for managing new digital tendering in the public works, with a focus on existing public assets. The new methodologies of surveying and modelling allow for important advantages in the whole life cycle of a building.

Information Specifications are an important tool that allow the Customer or the Commissioning Body to define both the general and the specific rules and strategic information requirements, as well as the general guidelines for the formulation first of the Pre-Contract BIM Execution Plan, by the Competitors, then of the BIM Execution Plan, by the Contractor.

National BIM Digital Platform for Construction (INNOVance Project)

Alberto Pavan, Sonia Lupica Spagnolo, Vittorio Caffi, Claudio Mirarchi and Bruno Daniotti

Abstract INNOVance represents the first digital platform in the construction sector with BIM methodology on a national level. It's also a BIM library, a Common Data Environment (CDE) of BIM projects in a contract, for sharing work information, and a data exchange platform for the entire construction sector. The platform operates in accordance with the UNI 11337: 2009 standard and it is the origin of the UNI 11337: 2017 standard group. The project was formed within the research relating to the Competitive Call: Energy Efficiency, Industry 2015, promoted by the Ministry of Economic Development (MISE).

Keywords Digital construction platform · Common data environment · BIM library

1 Introduction

At the start of the twenty-first century, the acronym BIM was not well known as nowadays. Though, while the academy and the researchers were talking in the wider terms of ICTC—Information and Communication Technologies for Construction—the industry started introducing the new terms of BIM—Building Information Model or Modeling—to progressively substitute the words CAD, Computer-Aided Design, that, despite being already exhaustive of a decisional and management process supported by computers (and this is evident when referring to the way the mechanic industry implemented CAD to support its production process management) was still interpreted—by the majority of the players in the building industry—in a reductive way as CAd, or Computer-Aided drafting: the building constructions field, at that time, is not mature yet to understand the potential of the Computer-Aided Design as a decisional tool to support the whole construction process. The roots of the modern interpretation of the CAD and BIM systems can actually be traced back to the early

A. Pavan (✉) · S. Lupica Spagnolo · V. Caffi · C. Mirarchi · B. Daniotti
Architecture, Built Environment and Construction Engineering—ABC Department, Politecnico di Milano, Milan, Italy
e-mail: alberto.pavan@polimi.it

B. Daniotti et al. (eds.), *Digital Transformation of the Design, Construction and Management Processes of the Built Environment*, Research for Development,
https://doi.org/10.1007/978-3-030-33570-0_1

60s and to the activities developed inside the MIT—Labs of the Boston University: many of the concepts are still at the base of the modern CAD-BIM software, such as software-user interactivity, modular design, Object-Oriented modeling, are already evident in the Ph.D. thesis work of Ivan Sutherland, the software Sketchpad (Sutherland 1964). In the same period, the concept of a decisional design and management process supported by computers is developed inside the Architecture Machine Group, founded by Nicholas Negroponte, at the MIT laboratories. Negroponte, with the system Urban5 (Negroponte 1970), envisions the idea of expert systems supporting architects and designers in general in the project decision process. During the same period, devices prefigurating the tools nowadays used to support Virtual Reality and Augmented Perceived Reality are already experimented by the Academy (Sutherland 1968). The acronym BDS—Building Description System (Eastman et al. 1974) appears in 1974 to emphasize the potential of CAD tools to virtually describe the complexity of a building in terms of its geometry and the other data necessary to fully represent it. The first releases of 3D CAAD OO—3-Dimensional Computer Architectural Aided Design Object Oriented—commercial software, appeared on the market during the 80s. Graphisoft developed, in 1982, Radar-Ch also known as ArchiCAD, since 1995 that is be the base of the Virtual Building Environment—VBE—concept developed by the Hungarian company. In 1984, in Germany, Nemetscheck launched the first release of its product Allplan. In France, Gimeor developed Architrion, another tool that had a wide diffusion in the 90s. The biggest software companies, AutoDesk and Bentley, in between the 80s and the 90s developed solutions based on their multipurpose products: AutoDesk has Architectural Desktop built on top of AutoCAD (now AutoCAD Architecture), and Bentley, on top of MicroStation TriForma, launched Bentley Architecture, evolved nowadays into AECOsim Building designer. At the same time, other companies started developing solutions for the building industry, based on the most advanced tools developed for the mechanical industry, creating the necessary conditions for a radical step forward of the building industry, in comparison to the former technologies: it is the case of Digital Project, (Gehry Technologies) based on CATIA (Dassault Systems) and Revit (Charles River Software 97, later Revit Technologies, 2000). In 2002 AutoDesk buyed Revit Technologies, and starts pushing, with a heavy marketing campaign, the BIM acronym, making it the preferred buzzword in the building industry.

Despite BIM having now substituted CAD (that anyway, in the proper acceptation, as already highlighted, is as exhaustive as BIM) as a common terminology, this did not correspond to a wider diffusion of it, as an operational tool in the sector, which remained a prerogative that is strongly limited to the US market. It was then, with the publication of the British PAS 1192 (parts 2 and 3), in 2013, and with the optional introduction of BIM in European Community public procurement in 2014 (D. 2014/24/EU), that, in Europe and across the world, the slow but continuous spread of the new BIM technology and methodology began throughout the entire sector. In this evolutionary scenario of BIM, Italy has played an important role, albeit little known. The creation of a digital BIM platform for the construction industry was theorized in 2007 (cit.). In 2009 (4 years before the PAS 1192), the first standard was published for the Italian market, the UNI 11337:2017: "*Building and civil engineering works. Codification criteria for construction products and activities and resources. Identification, description and interoperability. (6.2.1—Object-oriented Planning and design).*" In 2009, a consortium of private entities and research bodies, led by the ANCEenergia consortium, won the Industry 2015 Energy Efficiency Call promoted by the MISE (57 points out of 60), with a supply chain project named INNOVance: a digital platform for the construction sector. A co-financed project which started in 2011, for a total duration of 3 years.

2 The Project

The INNOVance project (product/process innovations and integration of the building construction chain for energy efficiency and sustainable development: action on a national level of the reed system with integrated strategic vision and life cycle) aims at the construction of the first "... *free access data base containing the information, all of the information, whether of a technical, scientific, economic legal nature, or more, that is useful for the entire building process. The program concerns the creation of the first national database for the construction industry...*"

A DB for the dissemination of knowledge in construction, indispensable for a truly innovative transition toward new technologies that support energy efficiency, sustainability, etc.

> *...The energy problems and, in general, the environmental sustainability of production and the products, require a radical internal reorganization of the construction sector that will enable access to growing sophisticated technologies and high standard production systems. Today's construction system, rightfully defined as "traditional", does not adapt to the ever more requiring needs of performance in terms of energy and acoustics containment imposed by the recent national and European regulations...*

> *...In the data base, for each phase of the construction processes, all the procedures and processes (components and results) of the entire building process (objects, works and resources) will be codified, described and named in a uniform manner, through collection schemes of shared and transparent information for all individuals operating in the sector, in order to optimize for the whole "building system", also through actions of feed back (from the construction site to the final user), the energy performance of the"building product" and subsequently the entire "building process" in general.*

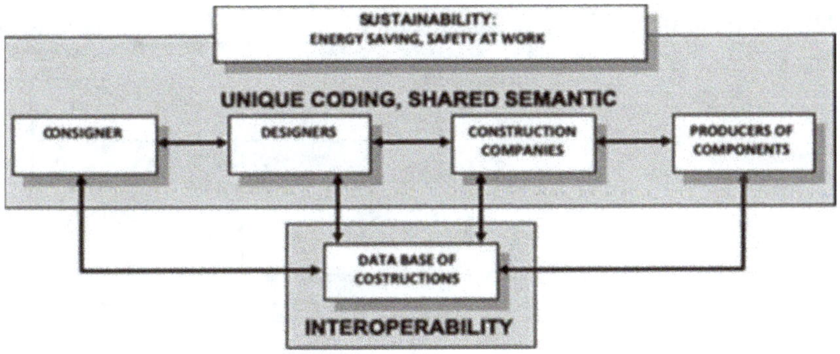

Fig. 1 Unique codified

> *The system of codification, nomenclature and filing of the products and processes will be object of technical national regulation within the UNI and UNI-CTI (the first on a European level) and will contribute to the overall reorganization, in semantic terms, of the technical information (already occurring on a European level). The interoperability platform and usage of the BIM technology (Building Information Model) will ensure a considerable competitive advantage for Italian companies abroad who will operate on superior consolidated qualitative standards compared to the competition…* (see Fig. 1)

Right from the start, the partnership has focused on interoperability as a basic requirement in such a fragmented and varied supply chain.

> *…A BIM3D archive is a virtual model of the building in a digital format, from which to extract the necessary documentation during the planning, executive and management phases.*
>
> *Different process operators can access the data contained in a BIM with different applications, and share and use constantly updated information, without data loss and without the need to continuously check their consistency and, in extreme cases, to reconstruct from nothing the database when the software used is completely incompatible.*
>
> *The concept of software interoperability is fundamental in order to create an operating environment like the one described above…*
>
> *…The main format for software interoperability in construction is the IFC standard - Industry Foundation Classes - developed by the BuildingSMART Alliance, also known as the IAI - International Alliance for Interoperability* www.iai-international.org *-. The IFC standard is based on the ISO STEP 10303 standard, and is itself an ISO/PAS 16739 protocol…*[1] (see Fig. 2)

Since its proposal (2008), the INNOVance project has, therefore, placed itself at the forefront for the period (the PAS 1192 arrived only 5 years later, the BIMTolkit after 7, there were still none of the CDE on the market today, BIM360, BIMx, BIM plus, etc.), focusing on collaboration and open languages as its first goals.

The project started formally in 2011 and ended as established in 2014, with the intervention of more than 300 collaborators belonging to the 14 partners:

[1]The original parts taken from the technical proposal of the project INNOVance (2008) are in italics.

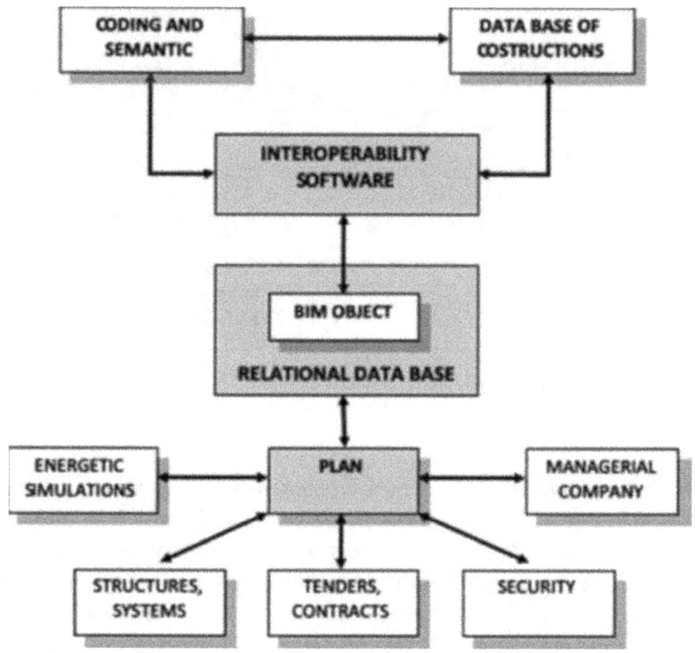

Fig. 2 INNOVance platform structure

- First proposer: Consorzio ANCEenergia;
- Universities and research institutes: Politecnico di Milano and the Politecnico di Torino, Università Federico II di Napoli, ITC-CNR, ISTEDIL, Istituto Sperimentale per l'Edilizia, S.p.A., Consorzio TRE;
- Software houses: SAP Italia S.p.A., ONE TEAM S.r.l., DERGA S.r.l.
- Component manufacturers: Edilstampa S.r.l.—(ANCE), Aedilmedia S.r.l.—(UCSAAL), Laterservice S.r.l.—(ANDIL), Concreto S.r.l—(ATECAP), Federlegno Arredo S.r.l.—(Federlegno).

3 The Prototype

INNOVance is an open platform on which any other software (design, calculation, management, etc.) can freely operate extracting and redeploying data and, unlike other potential competitors, is at the same time: BIM library, catalog of products and building systems; BIM server (CDE), work environment for collaboration, and information sharing between subjects; BIM platform, an environment for storing and structured data and information management (see Fig. 3).

INNOVance is a collaborative platform that collects in a structured way and relates all kinds of information serving the user and their software which, through open protocols, can operate on it extracting and redeploying raw and/or aggregated data, improving the process thanks to the continuous increase of the contained knowledge, whatever it may be.

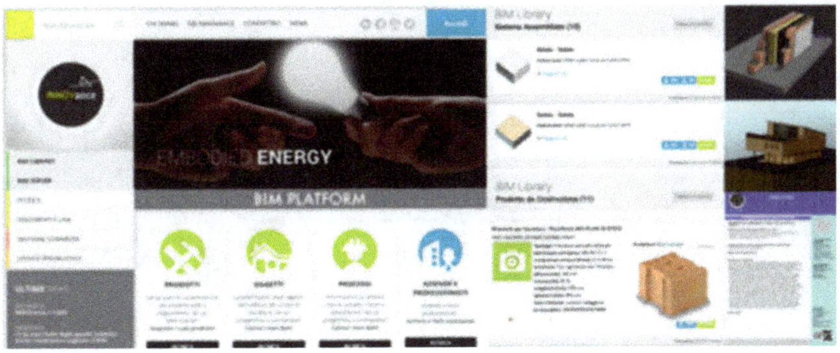

Fig. 3 INNOVance web portal

INNOVance manages data and information on:

- Products (from sand, mortar or brick, to radiators like the boiler, the glass, and the window);
- Subsystems (from the layer of plaster to that of masonry or insulation, water tightness, etc.);
- Systems (closing wall or partition packages, partition or roof slab packages, etc.);
- Works (houses, industrial complexes, green areas, infrastructures, etc.);
- Spaces (from residential or tertiary volumes to hot/cold/fire zones, rooms, premises, etc.)
- Work (laying, construction, installation, maintenance, etc.);
- Human resources (professionals, workers, specialists, etc.);
- Vehicles and equipment (cranes, excavators, concrete mixers, grinders, drills, etc.).

In INNOVance, it is possible to find the digital object "BIM" but also the video of the installation, the dimensional tolerances, the type of control activity or the preparation work of the necessary support, the nature and type of a typical team, the kind of license required by the operator for that machine such as which safety devices must be worn, the type of packaging and the quantity of waste it generates and the disposal operations to be carried out, the installation times such as scheduled maintenance periods, the location of the production plant, incompatible products, the regulatory dimensions of a room according to the standards of reference, the costs of a product as a process up to the accounting of a work, its volume, the floor area, the charges of urbanization, the load and outreach of a crane, the power of an excavator and the type of accessories that can be installed: bucket, compacting machine, caliper, etc., and all their characteristics (see Fig. 4).

Compared to common libraries of objects, the core of INNOVance is not the component products but the resulting product of the supply chain: the work (building or infrastructure). The work is defined in the platform through the model to which the following chapters of general attributes are connected:

Work result (building or infrastructure):

1. Name + INNOVance code
2. CPV identification code

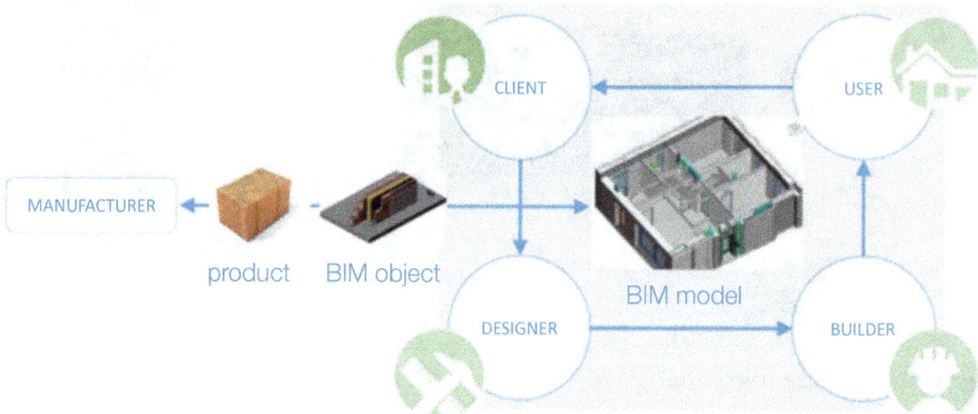

Fig. 4 INNOVance information flow

3. Intended use
4. Functional and spatial division of the work
5. Project parameters
6. Environmental well-being
7. Plant system
8. Energy efficiency
9. Fire extinguishing system
10. Distribution systems
11. As Built

And nine related specialist files:

A. Descriptive dossiers:

   1. Registry Dossier
   2. Territorial dossier framework

B. Operational Dossiers:

   1. Procedural Dossier
   2. Economic framework Dossier
   3. Timeline dossiers
   4. Construction site Dossier

C. Specialized dossiers:

   1. Geological and geotechnical dossier
   2. Structural dossier
   3. Maintenance Dossier

In INNOVance, it is possible to search and download a current industrial production object and insert it into the model with all its attributes. In the same way you can also virtually build a wall (system) by first defining the layers, selecting materials directly from the portal using the products datasheets and then sending all the information to the BIM authoring software (for the direct test project carried out with Revit) for it to create (itself) the geometrical BIM object or system. Or one can model a wall in the BIM authoring software and send it to INNOVance for it to build (itself) the necessary information sheets to fill out the attributes. Attributes that will remain inextricably linked to the BIM objects but will not weigh down the graphic file (except for those that directly serve the authoring software management) (see Fig. 5).

In INNOVance, every component object (product, system or subsystem) is identified by its geometric information (managed by the BIM authoring software and viewable from the portal) such as

1. Identification information of the manufacturer
2. Product identification information
    - 2.1. Denomination (UNI 11337: TS1)
    - 2.2. Identification code (UNI 11337: TS1)
    - 2.3. Trading name
    - 2.4. CPV code
    - 2.5. Other internal codes assigned by the manufacturer
    - 2.6. Intended use (obtained from the harmonized technical specification)
    - 2.7. Harmonized technical specification (hEN-EAD): name, classification, definition, code, number and standard year
    - 2.8. Description from specifications
    - 2.9. Description from the price list
    - 2.10. Synonyms
    - 2.11. Keywords
3. Technical information
    - 3.1. Morphological-descriptive characteristics
        - 3.1.1. Geometry and shape
        - 3.1.2. Visual and constructive aspect
        - 3.1.3. Dimensions
        - 3.1.4. Physical–chemical: qualitative, quantitative
        - 3.1.5. Main components of the product
    - 3.2. Declared performance characteristics
        - 3.2.1. Essential features
        - 3.2.2. Voluntary characteristics

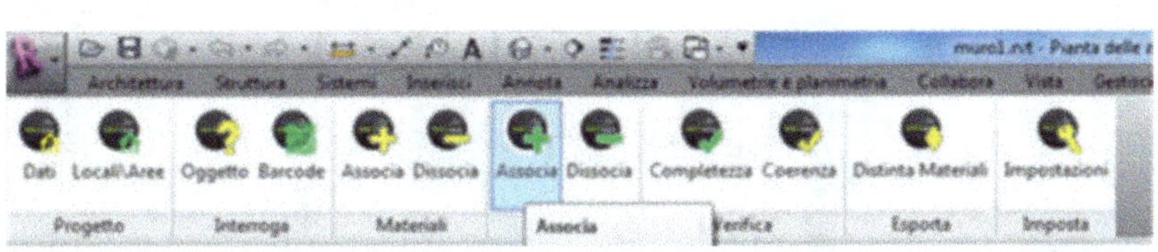

Fig. 5 Revit new command directly work on INNOVance platform

    3.3. Information on sustainability
    3.4. Safety information

4. Information on packaging, handling, storage at the factory and transport
5. Business information
6. Additional technical information
7. Attachments
8. Information on data reliability

    8.1. Date of realization of the technical sheet
    8.2. Compiler ID
    8.3. Date of revision of the technical sheet
    8.4. Auditor ID (see Fig. 6).

The attributes of the work results and of the products or systems are interrelated to each other through the execution information that allows the control and use of the BIM models not only in design or planning but also on the construction site:

Processing attribute chapters:

- General information
- Analysis of the resources necessary for processing
- Accounting for resources
- Operational constraints
- Operating procedures and controls in progress (preliminary, in progress, final)
- Environmental management of the site
- Safety management and risk analysis
- General provisions
- Additional information
- Summary and annexes
- Accounting for resources.

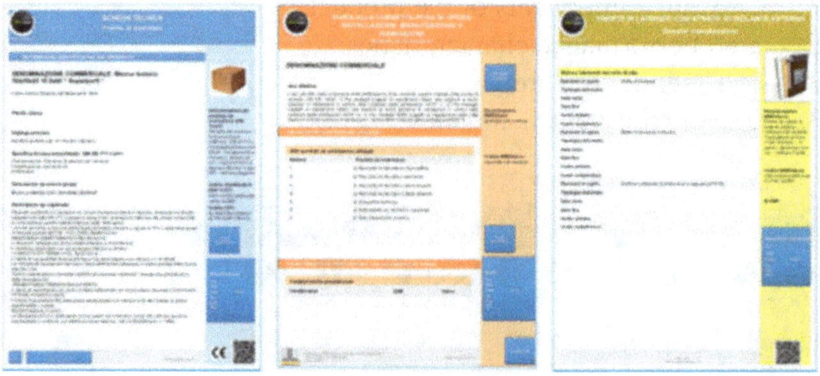

Fig. 6 INNOVance products sheet

Means and equipment attribute chapters:

- Use and labor
- Emissions
- General security measures
- Instructions to reduce (or eliminate) risks
- Accessories
- Documentation present on site during the time of use of the machine
- Additional information
- Photographs, videos, drawings, graphic details.

4 Conclusion

The INNOVance project represented the testing field for BIM in Italy, anticipating many of the common intended or consolidated topics today (first real CDE in the world) as in others that are still beginning to appear on the international scene like the European digital platform for the construction sector 2019–2023 (call DT-ICT-13-2019, DigiPLACE Project, Politecnico di Milano). INNOVance is also at the heart of the new UNI 11337: 2017 regulatory body and its works have, therefore, contributed to the writing of the ISO 19650: 2018 standard. The same can be said for the European regulatory work on the CEN still in existence. The debate that has arisen at every level around INNOVance has favored the introduction of BIM in public procurement in Italy DM 50/2016 and DM and its diffusion among the private sector. It is interesting to note that the future developments of BIM have been partly anticipated in INNOVance and still find meaning today by providing valid inspiration (see Figs. 7–9).

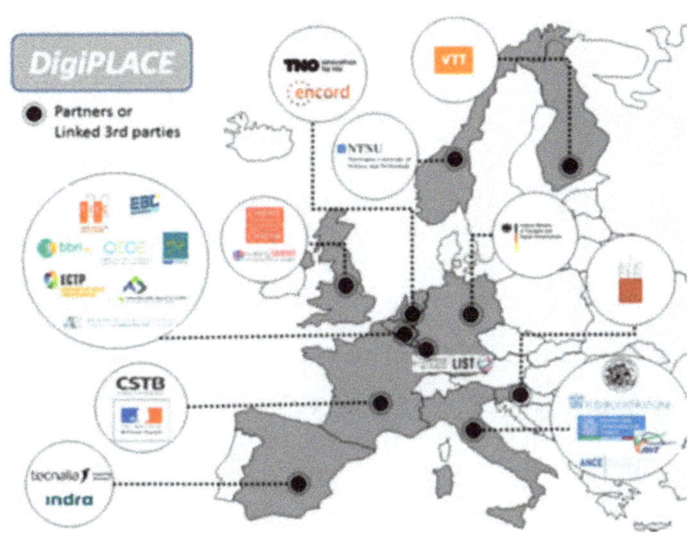

Fig. 7 Partner of European digital BIM construction platform

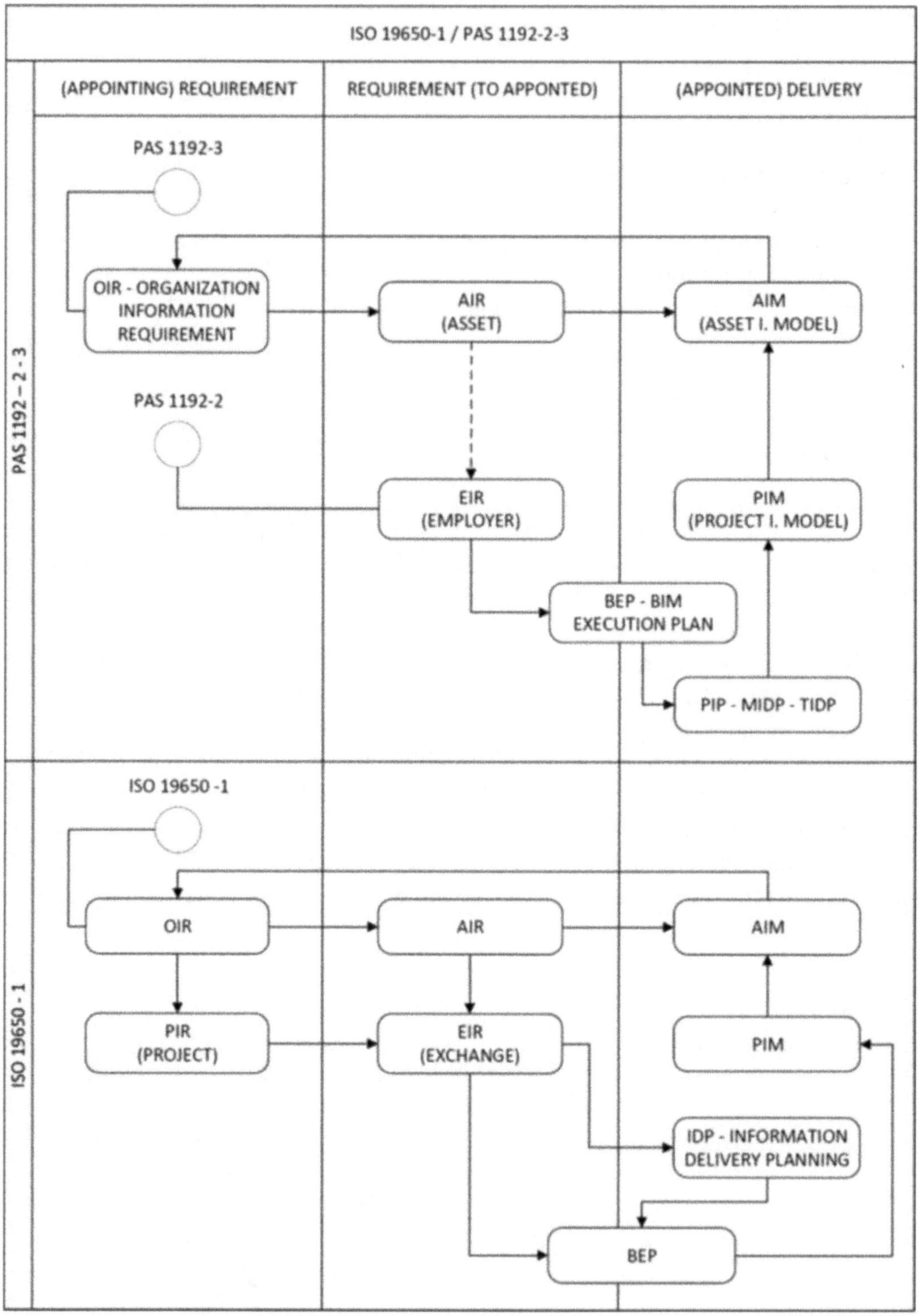

Fig. 8 ISO 19650-1/PAS 1192-2-3

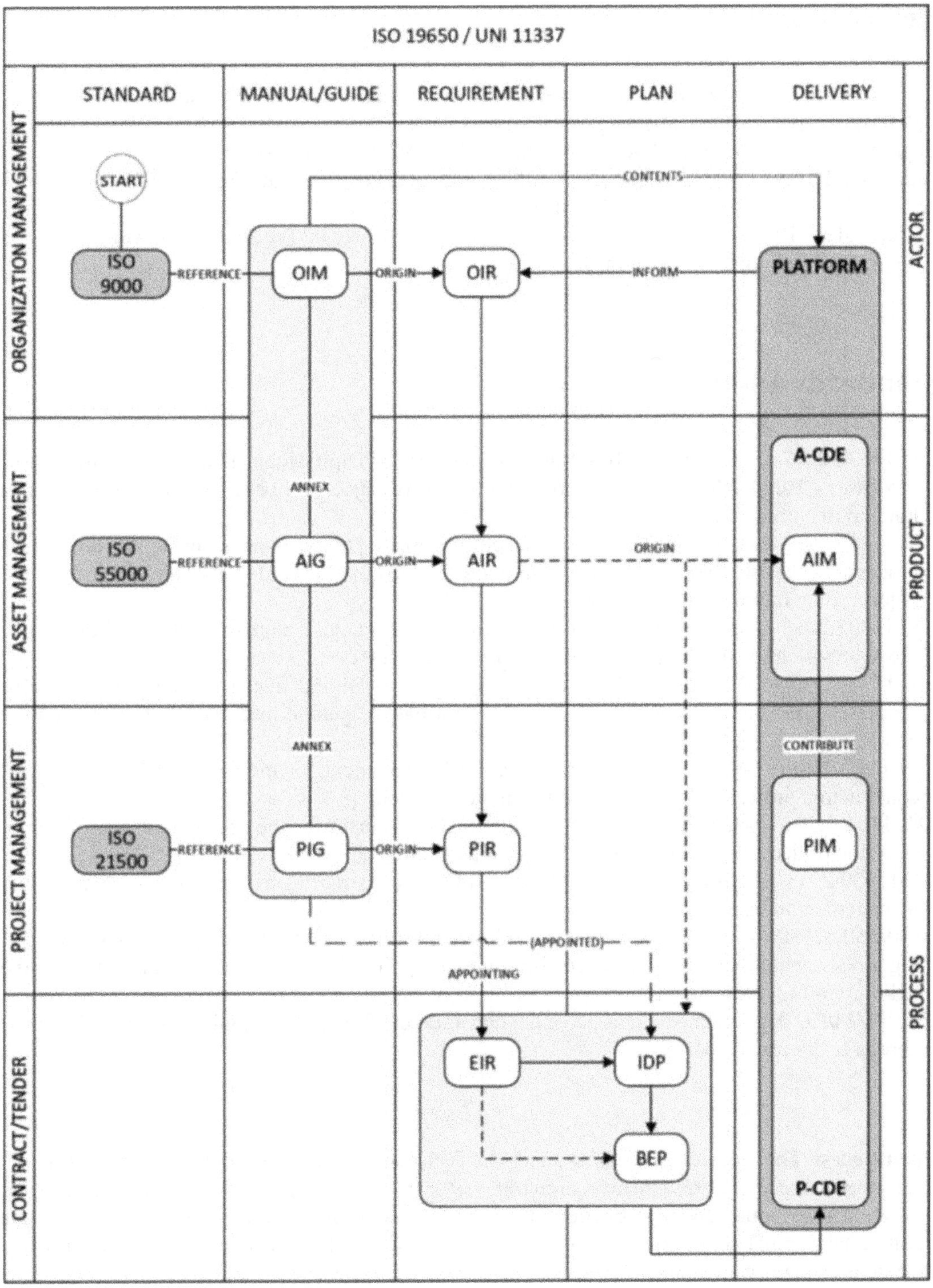

Fig. 9 ISO 19650/UNI11337

References

Eastman C, Fisher D, Lafue G, Lividini J, Stoker D, Yessios, C. (1974) An outline of the building description system. Carnegie-Mellon University, pp 1–23

Negroponte N (1970) The architecture machine. ISBN 9780262140089

Sutherland IE (1964) Sketchpad a man-machine graphical communication system. SIMULATION 2, R-3-R-20. https://doi.org/10.1177/003754976400200514

Sutherland IE (1968) A head mounted three dimensional display In: Proceedings of AFIPS fall joint computer conference. Thompson Books, Washington D.C.

Standards and Laws

UNI 11337:2017-1—Building and civil engineering works—Digital management of the informative processes—Part 1: Models, documents and informative objects for products and processes. Ente Italiano di Normazione

UNI 11337:2017-4—Building and civil engineering works—Digital management of the informative processes—Part 4: Evolution and development of information within models, document and objects. Ente Italiano di Normazione

UNI 11337:2017-5—Building and civil engineering works—Digital management of the informative processes—Part 5: Informative flows in the digital processes

UNI 11337:2017-6—Building and civil engineering works—Digital management of the informative processes -Part 6: Guidance to redaction the informative specific information. Ente Italiano di Normazione

PAS 1192-2:2013—Specification for Information Management for the capital/delivery phase of construction projects using Building Information Modelling

PAS 1192-3:2013—Specification for Information Management for the operational phase of construction projects using Building Information Modelling

ISO 16739:2013—Industry Foundation Classes (IFC) for data sharing in the construction and facility management industries

ISO19650-1:2018—Organization and digitization of information about buildings and civil engineering works, including building information modelling (BIM)—Information management using building information modelling

D. 2014/24/EU del Parlamento europeo e del consiglio del 27/02/2014 sugli appalti pubblici e che abroga la direttiva 2004/18/CE

From Cloud to BIM Model of the Built Environment: The Digitized Process for Competitive Tender, Project, Construction and Management

Franco Guzzetti, Karen Lara Ngozi Anyabolu, Lara D'Ambrosio and Giulia Marchetti Guerrini

Abstract The new methodologies of surveying and modelling allow for important advantages in the whole life cycle of a building. Specifically, the point cloud, with its richness in detail, can be the basis to set an interesting and useful work of creating a BIM model. Inside, all the objects with a well-structured informative part allow for the interoperability between the different figures involved in the process. One of the main aims can be the use of a BIM model to manage the competitive tender in the public works in order to control time and costs. The case study is about urban open space: analysis of digital and integrated management of a built environment.

Keywords Built environment · BIM · Urban open space · Competitive tender · Facility management

1 Digitization Process of Building Construction

The digitization of construction projects is now standard within the process that leads from concept to construction and then management over time.

In the past years, the use of BIM software has allowed users to simplify project development in its entirety, especially for new buildings. This great change in building construction is characterized by the opportunity to manage the whole process with a single tool in a clear and transparent manner. The data can be modified in real-time by anyone who is licensed, appointing party and appointed party.

As is known, most of BIM software such as ©Autodesk Revit, is designed for projecting new constructions; plans, sections and components are studied in order to have an accurate drawing and standard elements. In a way, BIM can also be used for existing buildings; specifically, many issues regarding historical buildings arise but there are signs of progress in modelling irregular elements.

F. Guzzetti (✉) · K. L. N. Anyabolu · L. D'Ambrosio · G. Marchetti Guerrini
Architecture, Built Environment and Construction Engineering—ABC Department, Politecnico di Milano, Milan, Italy
e-mail: franco.guzzetti@polimi.it

B. Daniotti et al. (eds.), *Digital Transformation of the Design, Construction and Management Processes of the Built Environment*, Research for Development,
https://doi.org/10.1007/978-3-030-33570-0_2

This paper illustrates another part of the construction that is not currently considered for digitization: the open urban space.

Modelling the open space, as a square, a garden or simply a portion of a street, is actually an interesting challenge because the bases are still to be invented and several experiences remain to be analysed.

While for buildings there is a form of legislation issued by the International Organization for Standardization and national governments, regarding open spaces the current information derives, in our experience, from a topographic database. Working on the open space there is a new purpose to increase the database and improve the management of built constructs in general.

The study attempts to analyse in detail the creation of a part of a square and crossroad, in order to understand limits and the potential of BIM modelling, considering a hypothetical competitive tender or maintenance intervention in an urban open space. The process and an idea of object organization is described in this paper.

2 Reference Regulations

At the beginning of the year, in Italy, parts 1 and 2 of UNI EN ISO 19650 have been received and published. The legislation deals with new dynamics related to the digitization and introduction of BIM in the building construction process.

According to European legislation, UNI EN ISO 19650-1 and 19650-2 "Organization and digitization of information about buildings and civil engineering works, including building information modelling (BIM)" describe the principles and the data flow during projects. The figures involved in the process exchange data and information by using the same digital environment in which documents can be shared, verified and updated.

The UNI defines a model of work organization in teams and between the different parties, while also setting guidelines for the information flow.

In the document, classifications and standardization refer only to building projects. As in the software (for instance, ©Autodesk Revit), all the components and the organization of the elements are thought of in the same way as the construction process. This concerns structures, columns and beams, walls with relative stratification, finishes. All things considered, nowadays, it is actually a significant step forward in the management and advancement of using BIM but considering the open spaces, there still are several gaps, especially related to terminology and categorization of the elements.

The use of standards approved by all the organizations involved is important to reach the main aim of planning in BIM: the interoperability can be fast and clear, the errors can be avoided during the project and construction phases, the correctness of data is certain, and huge savings can be made by checking all the phases in advance (Ciribini et al. 2016).

Furthermore, it is interesting to have a single tool in order to obtain a wide range of cases to identify in an easier way causes of issues and solutions.

3 Level of Development: Geometry and Information

The legislation defines, in all respects, the internal organization of a project team, the relationship among the figures, particularly appointed and appointing parties, the review and delivery process (Caffi et al. 2014). Furthermore, the description of roles is presented including that of the Task Information Manager who covers an essential position relating to following standards throughout the entire process; they direct the production of information based on the agreed procedures. In fact, by legislation, there are certain standards, however, the level of development is defined in the initial client request; the detail is chosen based on the needs.

The LOD, level of development, is a significant topic that allows to uniform BIM modelling on an international level.

Thanks to this classification, it is possible to immediately recognize the level of detail, both geometrical and informative. To a LOD 100 the geometry is simple, and the correlated data is less (indispensable for the definition of geometry); instead, a LOD 400 object has a more detailed geometry and a greater amount of data, up to the producer's name of the item.

The LOI or level of information (as defined in UNI 11337) is often a priority over geometry. The possibility of managing the entire life cycle of a building is due to the presence of data and metadata that allows users to extrapolate statistics and evaluations. With a BIM model, it is possible to obtain information about costs and time, which are the most critical parameters related to a competitive tender or maintenance intervention, as well as the quality of the projects.

Based on the LOD reached, the model can be questioned in relation to different specifications and topics, and all the data can be considered within a time frame equal to the life of the building: the added value of a BIM model is time.

As an example, a competitive tender which concerns the project of a new school in Melzo is reported. The participating companies could update their documentation through a predefined platform and, in real time, the public administration would be able to check quotations and schedules and compare different offers.

One of the interesting points concerns the difference between the initial cost (essentially the same for all companies) and the cost of the building at 50 years, which creates a curious gap among the participants (Fig. 1). This gap refers basically to maintenance; therefore the importance of counting the management costs in the overall costs of work is highlighted.

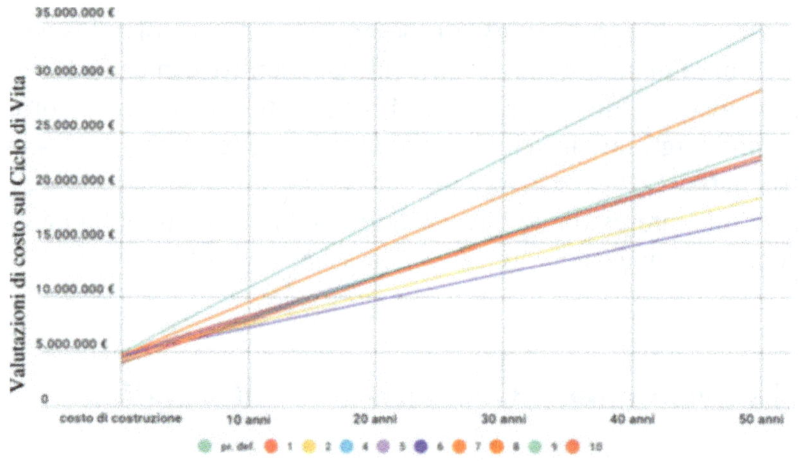

Fig. 1 Life cycle cost of a new school in Melzo: offer of different participating companies at the competitive tender (Di Giuda G.M., Villa V., Paleari F., Schievano M.)

With a tool like a BIM model, a fundamental aspect is that the data is not only predictable but also correct and consequent to the design choices made.

4 Case Study: From Point Cloud to BIM Model of a Square in Milan

Back to the analysis of open space, our experience concerns surveying and modelling; in particular, the creation of a BIM model of the external environment (Fig. 2) including all the relative elements visible and invisible (underground utilities) (Guzzetti et al. 2018).

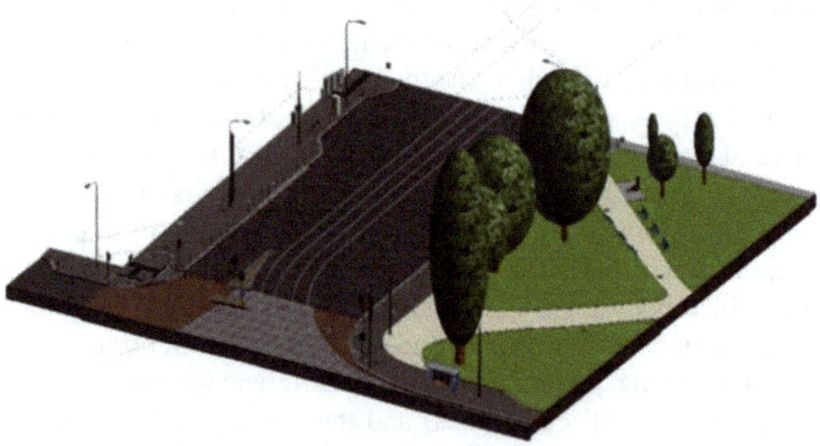

Fig. 2 View of the final BIM model with all the elements designed

The question is: what does the process of digitization involve in open urban areas?

How is it possible to manage a project of an external space using BIM in order to have advantages for a competitive tender, maintenance and renewal project? Which is the difference in using GIS?

The first phase of the work has been addressed through a laser scanner survey.

This part was particularly challenging since the urban area is vast and full of objects that make on-site scans difficult; both for objects that are necessary to survey and for objects that contribute noise to the cloud (such as vehicles or people).

The positioning of the targets was difficult next to the street because of its width, approximately 20 m; this is more than the minimum distance that must be between the instruments and their targets. Despite this problem, it is necessary to focus on the targets in order to join the scans in the elaboration phase.

After the elaboration of the 26 scans, a point cloud is obtained, which is usable as a foundation for the modelling phase. The transition from point cloud to the 3D model is a delicate stage in which the level of correspondence is defined between the two elements.

It is noticeable that the level of accuracy of this type of survey is elevated (Barazzetti et al. 2015). The instrument used, a Faro Focus 3D HDR CAM2 laser scanner, has a high precision and the point cloud ensures a 2 mm degree of accuracy. This value is very small whereas the main aim is to create a BIM model of an open urban space referred to as the hypothesized topic; there is no interest in obtaining a high-precision modelling of a sidewalk or a streetlight, so the detail is redundant compared to the needs.

This concept is entirely different compared to the level of detail.

The precision of adherence among the points of the cloud and the solid in BIM is defined at the beginning, based on the specific needs of the project.

The LOD, instead, is not connected to the point cloud but is referred only to the model. It could take, for instance, the creation of a portion of the sidewalk.

The real surface, which is derived from the point cloud is studded with noise (presence of people, moving vehicles and other disturbing elements); the modelling part is made by following 3D polylines created as guidelines on ©Autodesk AutoCAD. These paths are positioned in the middle of hundreds of points considering an adequate scale.

This type of precision does not interfere in any way with the possibility of creating a LOD 100 element rather than 400.

The definition of detail lies in the amount of data entered and, in addition, through the enrichment of geometry it is possible to improve the quantity of information.

The sidewalk can be modelled in different ways:

- geometry is extremely simplified and is created with a unique linear slope
- geometry is more complex with different slopes in order to follow the cloud more faithfully considering all the discrepancies
- the geometry is simplified but the stone ashlars are defined as single and countable elements with the same section.

From the methods listed here, it is clear that modelling can take place in different ways.

The goal is to find the right balance between the precision of representation and the elements that must be represented (Guzzetti et al. 2019).

The amount of data depends on the model's final purpose: in this case, the intent is to generate a BIM for a competitive tender or a maintenance program (shared among different figures).

The simplification of geometries is desirable, maintaining the fundamental data of elements and surfaces in order to have a more performing and specific model (Fig. 3).

The inclusion of the data necessary to be able to interact with the model and draw conclusions regarding the quantity and costs of the materials is essential.

Another example is referred to as the family of a traffic light, created specifically by our team.

The advantage of having a BIM model is that in case of maintenance the number of elements could be known immediately, the total economic amount and the condition of the product as well (only as an inserted parameter); thanks to this, it is possible to avoid failure by replacing the element over time (Osello et al. 2018).

In certain circumstances, it could be conceived that working in a GIS environment there could be the same results: points, lines, surfaces can contain the same data as a BIM model.

The differences essentially concern three topics as listed below.

- GIS is undoubtedly an adequate tool to manage data on a territorial scale. However, the cartographic base, DBT, presents a tolerance of about 40 cm due to a 1:1000 drawing scale. For instance, the correct position of manholes and underground networks is important, and BIM precision is the best solution.
- The better precision of the BIM model allows for more accurate quantification of square meters of materials. In the same way, the number of certain elements can be known, which are not represented in GIS: for example, the single stones of a sidewalk curb.

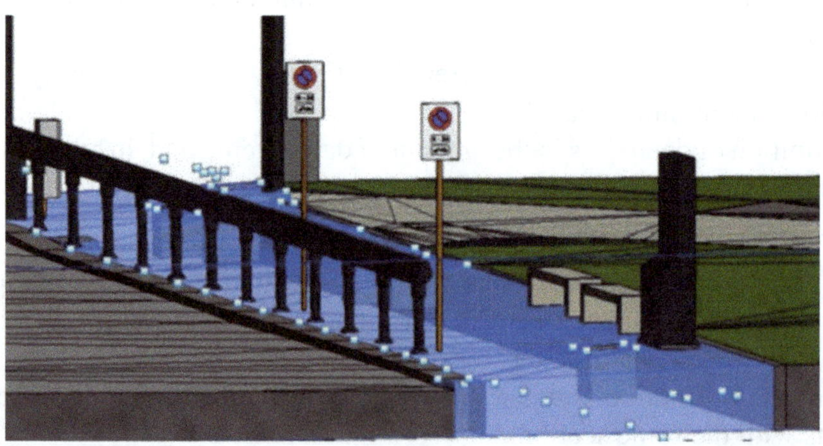

Fig. 3 Flooring parts: curb as single elements and sidewalk with a linear slope

- The visualization in a BIM model allows users to view three-dimensional objects with their materials and it is extremely useful to verify the landscape impact of a renewal, or of an installation of different urban furniture or of the organization of temporary events.

5 Categorization of BIM Elements

Following these considerations, hoping for the use of BIM in the future, the next step for the representation of external spaces is the need to create an adequate official system for cataloging objects as is the case today for topographic databases with .shp files. At the regional level, technical specifications are issued to regulate the representation of cartography in GIS.

In these specifications, in addition to the general information relating to cartographic operations, there are subdivisions to categorize the various elements represented. The data is ordered according to layers, themes and classes; the defined scheme allows users to collect all the elements present in the territory.

In a BIM environment, there is a lack in the organization of elements of open urban space (Malinverni et al. 2019). Therefore, the primary aim will be creating an informative model in order to manage the external spaces with a criterion that can be repeatable and standard.

Analysing the research carried out, the attempt is to generalize and define a logical scheme following the categories present in DBT: all the existing elements are considered.

The first macro layer defines the typology of terrain, which should cover 100% of the surface in the model: pedestrian and vehicular circulation areas, green areas, cycle lanes and objects are included in the categories.

The underground utilities are another category that includes networks, terminals and connections relating to installations on public terrain (Fig. 4).

These elements are further divided according to their position under the ground and above the ground. The emerging parts and the underground networks are defined using MEP modelling with ©Autodesk Revit MEP.

The vegetation is another section that partly refers to the terrain described before (green areas) and contains within it everything that can be called urban green; in addition to the different types of areas, the planting and sports fields are listed. In the subsoil, the plant roots of an area could be inserted in order to verify possible interferences with underground utilities (Cattaneo 2010).

This could be the basis for the management of open urban spaces in a BIM environment (Fig. 5). Obviously, the categories and the subdivisions have to be implemented, with subsequent studies, considering every type of material and product.

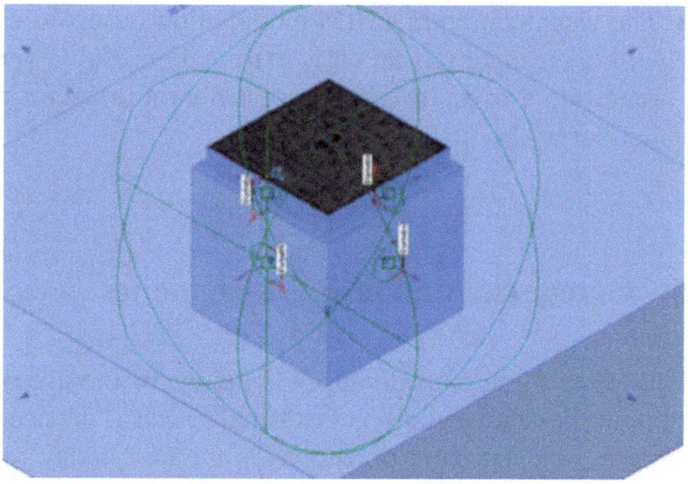

Fig. 4 The family of a manhole, the part under the ground it is connected to piping on MEP (©Autodesk Revit)

6 Conclusion

Facility management is everything that concerns maintenance and organization within a company. It is every service that facilitates business activities; it discusses safety, personnel, telecommunications.

The sector also concerns buildings, structures, facilities and maintenance.

The "management of structures" (physical or virtual) is a fundamental issue for the proper functioning of a company. It is clear how the BIM, in this case, allows for optimal management of all the information concerning every single element of the building.

For example, if all the lighting points of electrical systems with related lighting objects are inserted in a BIM model, it is possible to obtain a variety of information. The number of street lamps, but also the total electricity consumption; even more interesting can be the analysis of the elements over time, and therefore it is possible to understand how long it will take to intervene with a replacement of the bulbs.

Regarding open spaces, it is possible to apply the same methodology, furthermore, to analyse the whole areas with a single tool, a BIM model, guaranteeing a global view that is usually difficult to have, especially in urban areas.

The interoperability is an interesting feature for the management of all the services in a city by the public administration. The maintenance over the years could be analysed immediately in order to define a priority scale and to manage the economic balance sheet.

Furthermore, it would be key to involve suppliers who could share their information in order to provide accurate data and collaborate with other users.

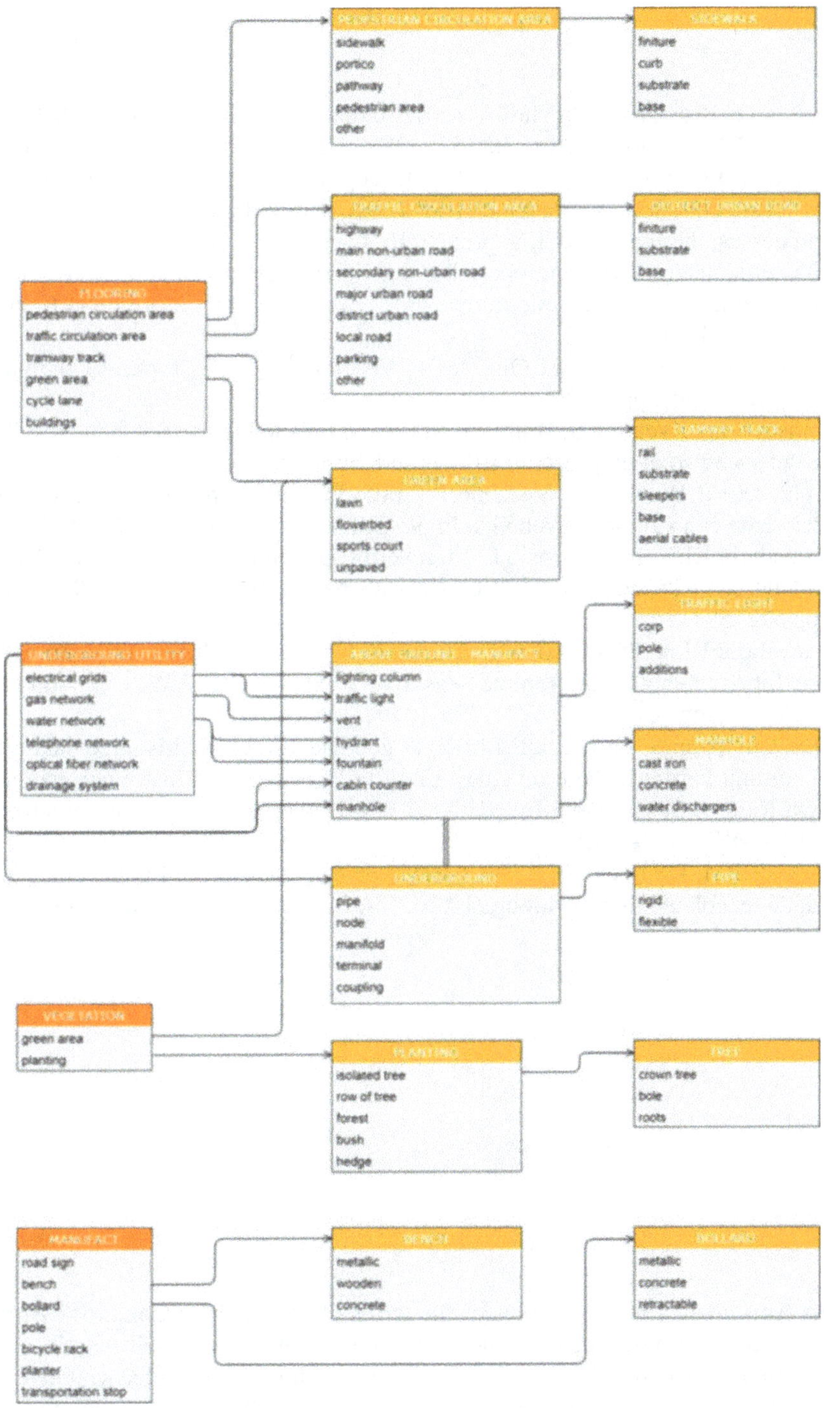

Fig. 5 The logical scheme of categorization of elements in a BIM environment

References

Barazzetti L, Banfi F, Brumana R, Previtali M (2015) Creation of parametric BIM objects from point cloud using NURBS. Photogram Rec 30(152):339–362. https://doi.org/10.1111/phor.12122

Caffi V, Daniotti B, Lo Turco M, Madeddu D, Muscogiuri M, Novello P, Pavan A, Pignataro M (2014) Il processo edilizio supportato dal BIMM: l'approccio InnovANCE/BIMM enabled construction processes: the InnovANCE approach. In: Conference: convegno ISTeA 2014 Energia, sostenibilita' e dematerializzazione operativa. La ricerca scientifica sulla Produzione Edilizia nell'era della digitalizzazione e delle nuove sfide ambientali nel Settore delle Costruzioni At: Bari, Politecnico di Bari

Cattaneo N, Di Maria F, Privitera A, Guzzetti F, Viskanic P (2010) Terreno Comune. ACER n 6/2010. Verde Editoriale, pp 51, 53

Ciribini A, Mastrolembo Ventura S, Bolpagni M (2016) La validazione del contenuto informativo è la chiave del successo di un processo BIM-based. https://doi.org/10.14609/Ti_2_15_1i

Di Maria F, Guzzetti F, Privitera A, Viskanic P (2005) Progetto verde Milano: il censimento e la gestione del verde con strumenti Web Gis. In: 9° Conferenza Nazionale ASITA, Catania

Guzzetti F, Anyabolu KLN, D'Ambrosio L, Marchetti Guerrini G, Sarrecchia S (2018) Dal rilievo al modello BIM di una piazza. In: XXII Conferenza Nazionale ASITA, 27–29 novembre 2018, Bolzano, pp 577–584

Guzzetti F, Anyabolu KLN, D'Ambrosio L, Marchetti G (2019) Built environment: modelling the urban space. Int Arch Photogram Remote Sens Spatial Inf Sci XLII-2/W11:595–600. https://doi.org/10.5194/isprs-archives-XLII-2-W11-595-2019

Malinverni ES, Chiappini S, Pierdicca R (2019) A geodatabase for multisource data management applied to cultural heritage: the case study of Villa Buonaccorsi's historical garden. Int Arch Photogramm Remote Sens Spatial Inf Sci XLII-2/W11:771–776. https://doi.org/10.5194/isprs-archives-XLII-2-W11-771-2019

Osello A, Ugliotti FM, De Luca D (2018) Il BIM verso il Catasto del Futuro potenziato tramite l'utilizzo della tecnologia. https://doi.org/10.26375/disegno.2.2018.15, pp 135–146

The Construction Contract Execution Through the Integration of Blockchain Technology

Giuseppe Martino Di Giuda, Giulia Pattini, Elena Seghezzi, Marco Schievano and Francesco Paleari

Abstract This paper aims to analyze the Blockchain level of implementation, focusing on the AEC sector that has always suffered from lack of trust, incomplete sharing, and transparency of information flow throughout the process execution. In this context, the progressive introduction of BIM based on the Blockchain technology can provide a trustworthy infrastructure for information management during the design, tender, and construction phases.

Keywords Blockchain · Smart contracts · Public tender · BIM · Information trust · Information immutability

1 Introduction

Blockchain can currently be considered as the main technology characterizing the digital transition observed within the most advanced world economies. Blockchain is a methodology capable of managing contracts and transactions through which assets are organized and guarded, social actions are governed and relations between nations, institutions, and individuals are guided (Shen and Pena-Mora 2018). Thanks to its nature, many public administrations are taking interest in the technology implementation, supporting Blockchain initiatives, setting goals and pursuing approaches that keep up with the technology (Carson et al. 2018; Kshetri 2018). Although various sectors are now exploring and testing Blockchain applications in their processes, its investigation in the construction sector is still meager and at a conceptual level (Mason 2017).

In an expanding construction sector, the maintenance of trust among stakeholders is difficult, the links are too complex hindering the information sharing with consequent waste of time and process costs (Institution of Civil Engineers, Blockchain

G. M. Di Giuda (✉) · G. Pattini · E. Seghezzi · M. Schievano · F. Paleari
Architecture, Built Environment and Construction Engineering—ABC Department, Politecnico di Milano, Milan, Italy
e-mail: giuseppe.digiuda@polimi.it

B. Daniotti et al. (eds.), *Digital Transformation of the Design, Construction and Management Processes of the Built Environment*, Research for Development,
https://doi.org/10.1007/978-3-030-33570-0_3

technology in the construction industry 2018); traditional information-sharing methods can influence negatively trust among participants, supply chain control, and asset management. In this context, the actual digital transition modifies and evolves the communication methods between project contractors. The actual BIM approach to the project guarantees the digital information exchange, offering a single database containing all the data created and shared by the operators during all phases of the construction process (Hsiao 2016). Building Information Modeling is in fact much more than an asset digital model because it defines the modalities of the model integration in the general construction system and the procedures through which information is added or extracted from the same and the creation, use, and management criteria of the informative model (Di Giuda and Villa 2016).

The progressive implementation of BIM models has fostered the project information exchange and collaboration among parties; however, the reliability and the transparency of each transition are not always guaranteed (Turk and Klinc 2017). Issues such as model property, right of modification and error responsibility, make it legitimate to assume an integration between BIM and Blockchain in order to overcome those limitations. In this dynamic context, Blockchain stands as a solution ensuring a transparent and precise information distribution among the participants by diverting the control of information by a single subject.

2 Blockchain Technology

Blockchain as a Distributed Ledger Technology (DLT) is a distributed data logging and maintenance system, that depends on and is ensured by the *consensus mechanism* implemented by the agents. The autonomy and updating of the information contained in the blocks are in fact subject to verification and authorization by all participants (Garzik and Donnelly 2018).

Consisting of a chain of blocks that develops within a distributed database, Blockchain prevents the structural alteration and the content violation, because the same information entered and validated in the chain are distributed, then stored in all nodes that compose it. Blockchain can be defined as a system that allows the data acquisition in a computer format, making it true and unchangeable thanks to the verification, validation, and control process carried out by the entire network through the *consensus mechanism*, and not by a third party (Kshetri 2017). The recent fast spread of the technology is mainly due to the benefits offered by its use, such as

- Intermediaries elimination;
- Information inalterability;
- Information traceability (Fig. 1).

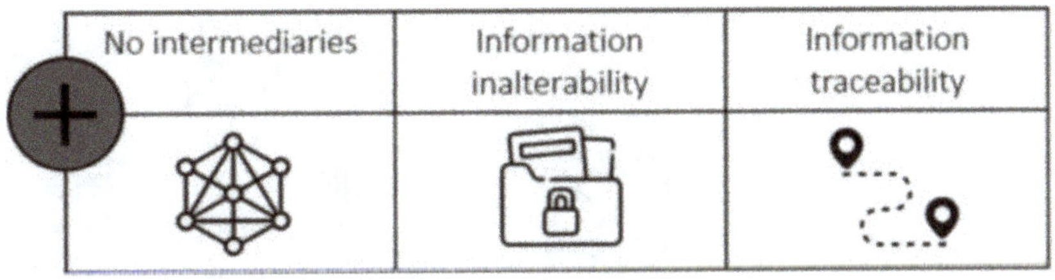

Fig. 1 Blockchain opportunities

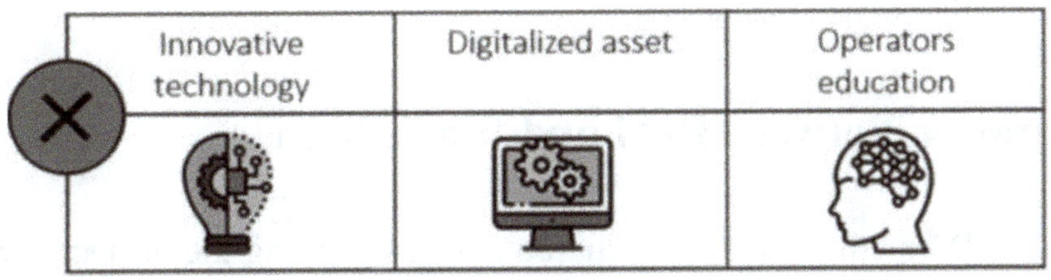

Fig. 2 Blockchain limits

Despite the advantages that Blockchain use offers, its youth creates challenges and obstacles that involve different areas and that can be identified with

- Innovative technology;
- Digitalized asset;
- Operators education (Fig. 2).

While Blockchain could solve some of the issues that arise with the use of BIM, one of its protocols has the potential to revolutionize the relationship between construction projects and to establish longstanding contractual procedures. Thanks to its programmable nature, the Blockchain allows the proper use of Smart Contract, i.e., contracts written in code capable of executing the clauses established and shared by the contracting parties automatically and independently. The implementation of a Smart Contract promotes the representation of clauses in the form of structured data executable by means of computer protocols with a high degree of accuracy compared to those defined by traditional language (Giancaspro 2017) (Fig. 3).

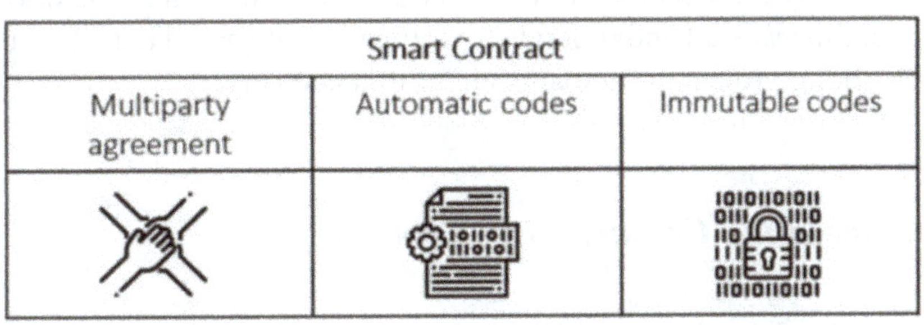

Fig. 3 Smart contract based on Blockchain main features

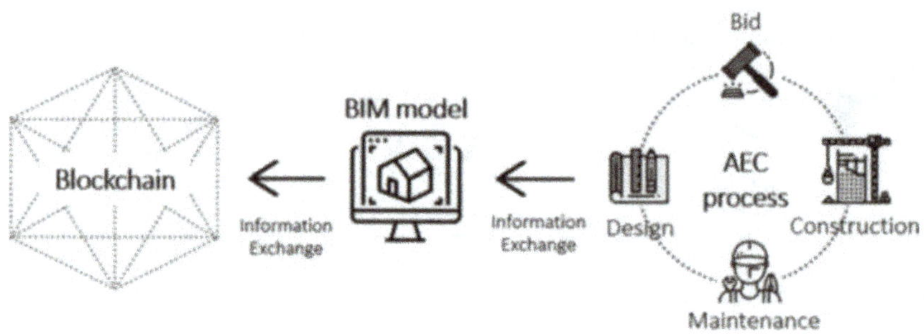

Fig. 4 BIM and Blockchain coordination

3 Integration Between BIM and Blockchain

Considering BIM methodology as a shared database among the project participants aimed at exchanging information on each process phase, it is fair to assume the activities of sharing, managing, and recording data by means of an information model supported by the Blockchain (Turk and Klinc 2017). The sharing information digital register offered by the technology ensures data reliability, integrity, and transparency by fostering loyal cooperation and trust among operators.

Since both the BIM approach and the Blockchain technology are based on the creation and management of a single source of information related to the process, it is legitimate to assume and investigate how to integrate them during the execution of the construction process. Due to the comparison and coordination of multiple disciplines, construction projects are often characterized by a large amount of data that can be stored in the BIM model, and ensured from the point of view of reliability and transparency if stored in the shared register offered by Blockchain. The combination of the information model and the distributed digital database makes it possible to create a single effective shared source of information relating to the project. This source can, therefore, be considered as the only source of truth that guarantees the data reliability, the congruence of the sources of information and the identity of the subjects responsible for the activities.

In this innovative context, the project BIM model is the only source of information, accessible and consultable by all participants, composed of reliable and unchangeable data. Since any information stored in the Blockchain database is traceable and unchangeable, time wasted and redundant verification of shared data due to frequent lack of trust among project participants are eliminated (Fig. 4).

3.1 Benefits in the Construction Process

The potential offered by the use of Blockchain allows removing the limitations that have recently discouraged the creation of the BIM model, such as reliability, traceability, disintermediation, recording of changes and data ownership. It is, therefore,

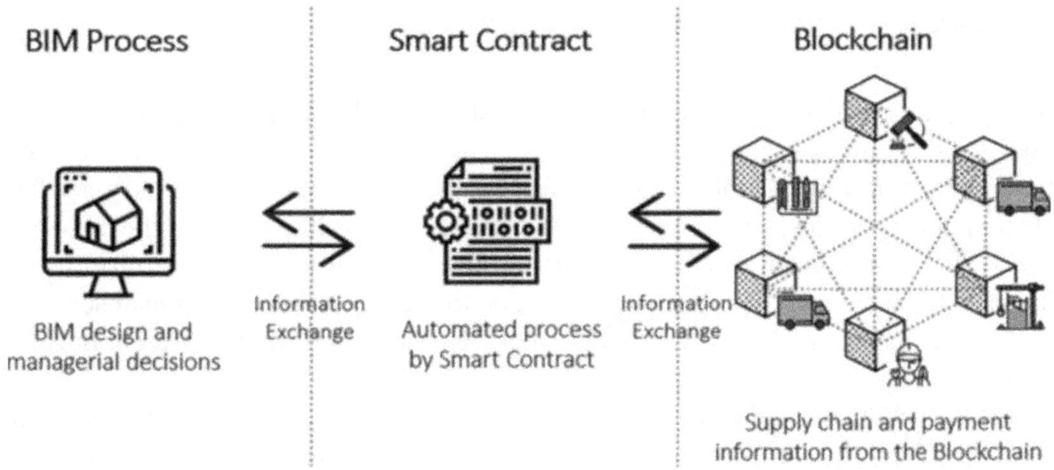

Fig. 5 BIM enables Blockchain technology

possible to highlight three main benefits that can be obtained through the joint use of BIM and Blockchain. First, the coordination between the BIM model and the distributed database containing all the process information ensures the creation of a single and reliable register, establishes a collaborative environment among all participants and defines transparently the responsibilities and duties of each, reducing or eliminating the emergence of any misunderstandings and subsequent conflicts between the parties. Second, to support the creation of a collaborative environment, the distributed database allows to store and trace the information intellectual property contained in it and entered by each party of the process. Third and last, if the two benefits illustrated above are mainly attributable to the design phase, the implementation of Smart Contract associated with the evolution of the BIM model is also relevant during the construction phase. The simultaneous progress between the BIM model and the execution of a Smart Contract makes it possible to automate all the delivery phases (Fig. 5).

The relationship among the parties involved in the whole process is often characterized by the presence of asymmetrical information during the design, tender, construction, and management phase, generating conflicts and mistrust with consequent impediments to the contract execution. For these reasons, the integration between the BIM model and the Blockchain illustrated so far is useful in making the activities carried out during the process explicit and visible, highlighting the honesty of those who act. Considering the benefits, various public administrations, including the Italian one, are committed to understanding the potential uses of the Blockchain in order to eliminate and manage in an optimal way the traditional criticalities.

3.2 Implementation in Contract Execution

The BIM model development based on a distributed digital register, updated and modified during contract execution, allows us to archive all the transitions made by drawing up a non-modifiable chronology of all the construction process stages. The Blockchain properties, guaranteeing the unchangeability of the data and the presence of a widespread control of every single step of the procedure, are, therefore, well disposed to face the waste of time and cost due to the lack of trust and the absence of a transparent sharing of information between the participants in the different process stages.

The possible uses and therefore the advantages that can be obtained through the BIM model development to support all the contract execution phases—design, commissioning, construction, and management of the asset—based on the distributed database Blockchain are highlighted in Fig. 6.

3.2.1 Design

The design levels preparation based on BIM methods allows us to share and exchange all the project information through a single digital platform. The integration of these digital methods with the shared Blockchain database allows to create a truly collaborative process, that limits the gap between digital information modeling and the management of project information—administrative, financial, insurance—typically expressed in sheets. The possibility of associating automatic payments to the data in the model through the use of Smart Contract guarantees an exhaustive implementation of the engineering and architecture service contracts and therefore the respect of the requests expressed by each person involved. In this regard, the usual BIM model is no longer considered as a central data-sharing environment based on a cloud platform governed by a third party, but rather as a peer-to-peer data exchange environment, in which each participant can define and control the ownership of information in an immutable way.

Design phase	Bid phase	Construction phase	Life cycle phase
Sharing and exchanging project information in a collaborative environment	Immutable and transparent tender documents	Updating and monitoring supply chain and construction progress	Monitoring maintenance and circular economy principles satisfaction

Fig. 6 BIM and Blockchain in contract execution phases

3.2.2 Bid

Due to the frequent ambiguities that arise during tender procedures, the presence of a distributed ledger in which all information is stored in a transparent, permanent, and accessible manner helps to contain any misunderstanding. The digital tools implementation provided by the distributed platform, typical of the Blockchain technology, would make it possible to store in an immutable way all the main tender documents: that is, the information models presented both by the client and by the bidders.

The procedure described above guarantees the immutability and transparency of the tender documents published by the client. The procedures and criteria for evaluating the bids are accessible to all participants, making the reasons for the contract award explicit and eliminating any possible operational ambiguity. In the same way, the tenderers bids are also deposited in an unchangeable way on the distributed register, the client can access them only at the end of the reception phase, thus ensuring effective competition and giving all participants the opportunity to observe all the bids and to analyze them compared to the predefined award criteria.

3.2.3 Construction

The 4D and 5D dimensions model development allows to update and monitor the construction phase progress. The BIM model allows us to understand the actual construction works state, and the continuous information storage allows to understand the causes and trace those responsible in case of delays or budget excess. In this way, the client can control the actual activities progress and the sharing of information relating to the construction site is immediate and transparent.

The Blockchain used during the construction activity is also useful in controlling the supply chain: materials that arrive at the site are traced along the entire route and therefore in case of defects or delays you can consult each stage of processing that was recorded on the Blockchain database and then connected to it internally to the BIM model (Kouhizadeh and Sarkis 2018).

Finally, the record of both supplies and work status information allows the Smart Contract to be performed correctly. The connection between the information model, i.e., the activities progress, and the computational contract allows automatic payments to be issued whenever the milestone set by the work program is reached.

3.2.4 Life Cycle

Once the endeavor is completed, it is possible to create the as-built model of the building, containing all the information relating to each component actually built. Through the entire asset life cycle, the BIM model implementation based on Blockchain facilitates the satisfaction of the circular economy principles (Marzouk et al. 2018). Some building component data, collected during the construction phase and stored in the distributed registry, can support future maintenance, replacement, and dismission activities during the operational phase. The constant components conservation and maintenance state updating make it possible to use the BIM model as a database of materials and therefore to facilitate a considerable reduction in waste.

In addition, in the presence of system terminals equipped with smart interfaces, it is possible to envisage the execution of Smart Contract for maintenance work on plant devices in the building. Once the intervention has been carried out, in fact, the maintenance technician can insert the activity carried out into the machine, which confirms the fulfillment of Smart Contract clauses, automatically releasing the payment.

Considering this preliminary analysis of the integration of the BIM approach and Blockchain technology in the development of the procurement process, it is visible that one of the most encouraging advantages is the reduction of information asymmetry between all contracting parties. The data definition, updating, and validation by the participants allow them to access and consult, throughout the entire process, complete and truthful information, thus increasing mutual trust. Such a technological enhancement would ensure that operators have a transparent understanding of the procedures for the award of contracts, and that the client has a reliable check of the data relating to competitors, their offers and commitment.

4 Conclusion

The topics discussed in this paper demonstrate how the BIM model and Blockchain technology integration can offer benefits and add value to the main stages of contract execution. The BIM model can either incorporate information from Blockchain distributed ledger, or send to Blockchain information about model changes that need to be updated and used later in the execution of Smart Contract, for the automatic release of payments or the definition of new supply orders (Carson et al. 2018). The construction sector has always been characterized by aspects that, due to ambiguity or incompleteness, often generate conflicts during the contract execution. The potentials offered by the Blockchain and BIM model development allow to address positively the aforementioned problems thanks to applications that allow to streamline and improve the contract management, ensuring transparency in information sharing and traceability.

Despite the advantages, the main Blockchain obstacles are not only of a technical nature but involve other elements, including the man who decides to adopt the technology, economy, and institutions that must support and regulate the digital transition. In addition, because of the impossibility of predicting certain construction process variations, efficient Smart Contracts execution is rather difficult.

For these reasons, unlike some industries, the construction sector has not yet developed an effective digital ecosystem and investments in new technologies. The integration between Blockchain technology and BIM is a key step in the sector progress: it creates a single source of true information about all aspects of the process, giving the digital project model the only reliable tool to support the development and management of all phases of the construction process—from the design phase to the operation along the whole lifecycle.

The Blockchain technology can, in fact, guide a lean construction process, by reducing the industry fragmentation and complexity, making it a single trusted entity. Blockchain technology is here and although it is at an early stage of development with many challenges it presents a relevant opportunity for all companies in the construction industry to emerge as more effective, transparent, and sustainable entities (Institution of Civil Engineers, Blockchain technology in the construction industry 2018).

References

Carson B, Romanelli G, Walsh P, Zhumaev A (2018) Blockchain beyond the hype: what is the strategic business value? McKinsey Co. pp 1–19

Di Giuda G, Villa V (2016) Il BIM. Guida completa al building information modeling per committenti, architetti, ingegneri, gestori immobiliari e imprese

Garzik J, Donnelly JC (2018) Blockchain 101: an introduction to the future. In: Handbook of Blockchain, Digital Finance and Inclusion, vol 2, pp 179–186. https://doi.org/10.1016/b978-0-12-812282-2.00008-5

Giancaspro M (2017) Is a 'smart contract' really a smart idea? Insights from a legal perspective. Comput Law Secur Rev. https://doi.org/10.1016/j.clsr.2017.05.007

Hsiao JI-H (2016) Smart contract on the blockchain—paradigm shift for contract law? https://doi.org/10.17265/1548-6605/2017.10.002

Institution of Civil Engineers, Blockchain technology in the construction industry (2018)

Kshetri N (2017) Blockchain's roles in strengthening cybersecurity and protecting privacy. Telecomm Policy 1027–1038. https://doi.org/10.1016/j.telpol.2017.09.003

Kshetri N (2018) Blockchain's roles in meeting key supply chain management objectives. Int J Inf Manag 39:80–89. https://doi.org/10.1016/j.ijinfomgt.2017.12.005

Kouhizadeh M, Sarkis J (2018) Blockchain practices, potentials, and perspectives in greening supply chains. https://doi.org/10.3390/su10103652

Marzouk M, Azab S, Metawie M (2018) BIM-based approach for optimizing life cycle costs of sustainable buildings. J Clean Prod 188:217–226. https://doi.org/10.1016/j.jclepro.2018.03.280

Mason J (2017) Intelligent contract and the construction industry. J Leg Aff Disput Resolut Eng Constr 9

Shen C, Pena-Mora F (2018) Blockchain for cities—a systematic literature review. IEEE Access 1–33. https://doi.org/10.1109/access.2018.2880744

Turk Ž, Klinc R (2017) Potentials of blockchain technology for construction management. Proc Eng 196:638–645. https://doi.org/10.1016/j.proeng.2017.08.052

BIMReL: The Interoperable BIM Library for Construction Products Data Sharing

Sonia Lupica Spagnolo, Gustavo Amosso, Alberto Pavan and Bruno Daniotti

Abstract BIMReL is an online platform for construction industry operators (primarily manufacturers, but also designers, construction companies, facility managers, trade associations and owners) because it allows using and sharing all the technical and commercial data of construction products. Moreover, it helps collecting and exchanging data about performances, durability and environmental indicators in a simple and user-friendly way. The creation of such a tool became necessary because, despite many interesting developments in Information and Communication Technologies (ICT), the data exchange along the building process is hardly efficient and still mainly based on a transmission of paper documents, on a plurality of classification systems, as well as on a use of different criteria and practices. The lack of a common structure for data storage involves different inefficiencies and it also may affect such essential aspects for users as comfort, safety, health, energy saving and environmental sustainability. Inefficiency means a waste of time (and therefore a waste of money) because it often may cause complicated variations during the construction stage or misunderstandings, which may lead to legal disputes. Therefore, BIMReL has been developed in order to provide a new digital tool for sharing product data in a BIM environment and such as to enable the user to choose the most proper product but also to easily retrieve all useful information during its use, its maintenance or its disposal.

Keywords BIM library · Product data template · CE marking · Interoperability

1 The Development of an Interoperable Platform

Digital solutions are often adopted for managing information in an ordered structure. Several efforts have been devoted for developing solutions and optimising methods for sharing information about buildings. Referring to the level of products, two

S. Lupica Spagnolo (✉) · G. Amosso · A. Pavan · B. Daniotti
Architecture Built Environment and Construction Engineering—ABC Department, Politecnico di Milano, Milan, Italy
e-mail: sonia.lupica@polimi.it

B. Daniotti et al. (eds.), *Digital Transformation of the Design, Construction and Management Processes of the Built Environment*, Research for Development,
https://doi.org/10.1007/978-3-030-33570-0_4

improvements in the management of information are represented by the development of BIM libraries (Duddy et al. 2013; Kim et al. 2015; Pasini et al. 2017) and the definition of Product Data Templates (PDTs) (Gudnason and Pauwels 2016). BIM libraries are tools published on the web for hosting objects describing products; PDTs are structures for defining standardised data content, listing attributes required for describing objects from a geometrical and alpha-numerical point of view (Lucky 2019). The BIMReL platform is freely accessible by a website and it has been implemented to maximise and simplify its use by different stakeholders in the construction industry who can upload and exchange all the technical and commercial product data in a user-friendly way (Lupica Spagnolo et al. 2017). With the proposed system, it is in fact possible to check easily all the parameters that are necessary to guarantee the seven basic requirements of constructions works (i.e. mechanical resistance and stability, safety in case of fire, hygiene, health and the environment, safety and accessibility in use, protection against noise, energy economy and heat retention and sustainable use of natural resources) in a completely new, effective and efficient way.

As a result, end users such as property owners and clients (individuals or public administrations), as well as designers and facility managers, for whom it is now extremely difficult to find a lot of data to help in choosing products, will have access to information in an easy, transparent and cost-free way, to be able to choose and define the technical solutions that best fit their needs and requirements. Exploiting BIM and from a perspective of its active and effective promotion, also in view of the application of the Italian Procurement Code (Legislative Decree No. 50/2016), the platform associates this information to the BIM objects, thereby allowing their efficient management.

The information system created is profoundly different from the current libraries of BIM objects that can usually be found online, because it creates the first life cycle management of all BIM-ready information, namely to ensure the exploitation of the efficiency of calculation, simulation and modelling automation. This system uses techniques and tools capable of simplifying the interaction between physical objects (the building elements) and the digital world in which these objects are represented. QR-codes, integrations with smartphones and tablets and advanced research systems can support all the actions of the supply chain operators in their daily operations. The user can easily and intuitively compare materials and technologies on the basis of different validated and available information in a transparent and, therefore, manageable way, along the life cycle (from production to disposal/recovery) of the constructed facility where they are placed. An interoperable library is not based on proprietary formats for exchanging information, but it has to undertake the development of solutions for connecting information through interoperable web-services so to collect data in a neutral format (Pasini et al. 2017). In such a way, the solution is not software-dependent, therefore, also the efforts and the difficulty in maintaining consistency between the representations for the different software over time are reduced (Duddy et al. 2013).

BIMReL, therefore, ranks as an innovative and integrated supply chain tool of business processes and models from an export perspective, thanks to the systematisation aimed at the digitalisation of the entire construction sector in the wake of

Industry 4.0, as well as a smart tool for simplified and transparent management for all of the choices regarding products, materials and systems made by the different actors involved in the building process.

2 Research Project Objectives

The main objective of the project is the reduction of environmental impacts at every stage of the product life cycle ("from cradle to grave") as BIMReL first of all allows choosing construction products by comparing them not only on the basis of their technical-descriptive features, but also in the light of its environmental impact along the entire life cycle (Signorini et al. 2019): a special section is dedicated to the collection of environmental sustainability indicators in accordance with the UNI EN 15804:2014 (UNI, 2014) and ISO 21930:2017 (ISO, 2017), finally allowing the user to be able to make an informed choice in terms of sustainability. Nationally and internationally, there are databases with data for life cycle assessment data (LCA), but they are very poor of data, not well known, non-interoperable and difficult to use for non-experts, so BIMReL stands as the first and only full, interoperable and user-friendly international platform for sustainability for individual citizens, and of course for all other actors of the construction sector who require such information.

From the perspective of the smart supply chain, an integrated process chain is pursued by means of the total rationalisation of information flows from manufacturers to designers, to construction companies down to individual end users. That is so to streamline all phases of the construction process, as well as to ensure to the end-users, public or private clients full transparency concerning the safety, well-being, health and eco-sustainability requirements of products, entire technological packages or installation elements. A tracking of the changes made over time by the data owner, first of all, the manufacturer who will need to update technical specifications, trade or installation instructions, will ensure data reliability for the user. The platform also has the primary objective of promoting the manufacturing companies at national and international level, using an innovative and comprehensive communication channel and latest generation marketing tools that are already consolidated on other international web portals by the partner TraceParts.

A further pivotal purpose of the project is to push strongly toward virtual design systems so detailed as to provide the user with a realistic view. With BIMReL the approach to Building Information Modelling is amplified not only in the design phase but also in the delivery and operational phase of constructed works over time.

3 Compliance with the Reference Framework

In BIMReL, the aim is to exploit ICT's potential to digitise the construction sector with the main intention of making it effectively sustainable, in the broadest sense of the term, economically, environmentally and socially.

Aware of this need, different players in the construction industry have recently developed a consensual standard, UNI/TS 11337-3-2015 (UNI, 2015), which provides key guidelines for developing a product data template. However, the tools that are in accordance with these rules are missing. The BIMReL project is an innovative tool that is compliant with the UNI standards, as well as already in line with the latest international standards on interoperability (developed by ISO TC 59 SC13 and CEN TC 442), durability (developed from the ISO standard 15686) and sustainability (developed by the ISO TC 59 SC17 and CEN TC 350 standards).

The platform, built on a regulatory basis and using the format of the product technical datasheet indicated in UNI/TS 11300-3:2015, allows the unambiguous identification of the products (Category, Type and Standard) and the essential and voluntary characteristics obtained from the reference product standards, enriched by all the information deemed necessary by the market. On one hand it allows manufacturers to use an advanced matrix of data collection, and on the other hand it allows users to be able to effectively compare the features of different products. If you then consider that data are structured like database fields, it is also clear that the efficiency of applying all the research filters, as well as automatic retrieving or calculating using software tools offers significant benefits in terms of result reliability and processing speed.

In Fig. 1, the logical process of unambiguous object classification is shown: it follows a precise workflow that starts from the individualization of the family, moving then to the selection of its macro-category (if any), its category, its typology and the reference standard (if any).

In this way, it is, therefore, possible to ensure maximum speed of circulation for new digital work methodologies, first of all, the one provided by the introduction of BIM, but also and especially to guarantee the possibility of checking by end-users.

4 The Development of an Advanced BIM Library for the Construction Sector

The platform comes as a library of BIM objects, but with an innovative component. In fact, BIMReL three-dimensional digital objects not only possess the graphic geometries, but they also contain every descriptive attribute of the product so as to be useful to the user himself. Moreover, specific tools allow a fast comparison of requirements and performances of products. These objects are then associated to a standardised data sheet already compliant with UNI/TS 11337-3:2015 and using a unique code.

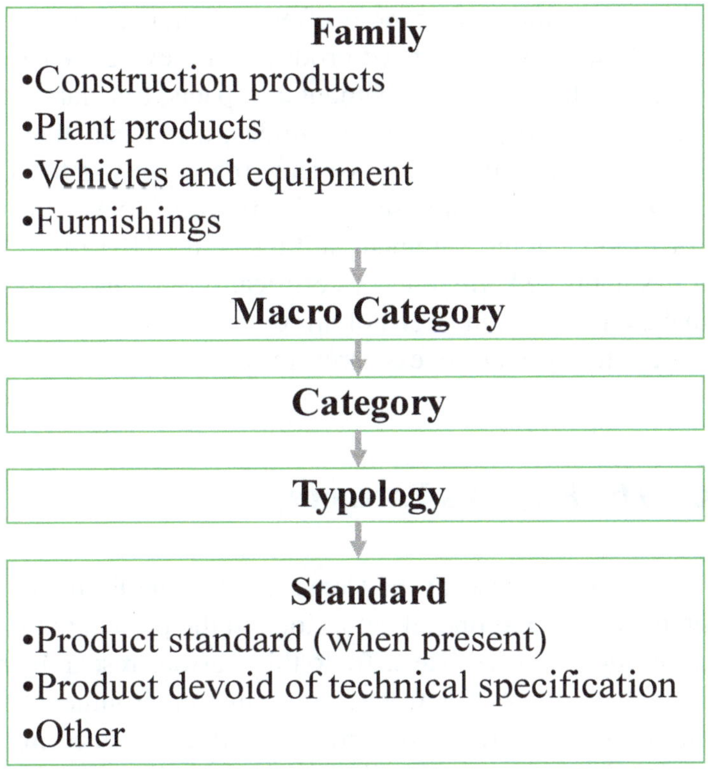

Fig. 1 Logical process of unambiguous object classification

Thanks to the developed implementation of the BIM, a part of the platform enables direct dialogue between users/designers and manufacturers in order to customise certain products in the catalogue. This is in order to create increasingly defined 3D prints of products, whether they are building or plant structures, as well as scale models of entire buildings or parts thereof, in order to give the user an accurate and detailed preview of the construction asset. Conversely, from the BIM breakdown model it is also possible to prepare all the information for additive manufacturing, thus allowing new and pioneering manufacturing companies to fit smoothly into the building market, and at the same time for the users themselves to finally exploit the potential offered by this production approach.

5 The Role of the Citizen in the Pursuit of Sustainability

With the BIMReL tool, even citizens can exert a new active role in a building process, thanks to an innovative system of information management that allows them to make aware choices that are not limited to the mere expression of aesthetic tastes or the pursuit of energy savings, but also contemplate the environmental implications.

Currently, in fact, the user who wants to make some environmentally friendly choices cannot help but rely on generic advertising claims such as 'environmentally friendly', 'sustainable', 'zero emission' or fathom out the real meaning of the various sustainability labels (the so-called eco-labels) that are gradually emerging.

This information, however, is often misleading because it is generally related to just one phase of the advertised product's life cycle. What we need is to allow a global vision of the actual environmental impacts, translating the generic adjectives above mentioned into numerical magnitudes, which can, therefore, be mutually compared. This is possible if we can associate the objects to the values of the individual indicators of environmental sustainability, which have already been identified with their corresponding measurement unit inside the UNI EN 15804:2014 standard. Thus, as it has happened with energy certification, the most discerning users will be able to make choices on the basis of an 'environmental qualification' while still being able to effectively educate less careful users.

6 The Project's Partners

The leading partner of the Consortium is the Politecnico di Milano, which worked on the research proposal definition, on the project management and on the scientific development of BIMReL. To do this, various research groups were involved, mainly traceable to the ABC (Architecture, Built environment and Construction engineering) and DEIB (Electronic, Information and Bioengineering) departments.

One Team S.r.l. is an Italian company dedicated to consulting services and turnkey supplies in the design field (CAD, GIS, EDM/PDM/PLM, FM). It has been the main Italian Autodesk partner for more than 30 years, holding for it the Autodesk Platinum Partner and Autodesk Developer Network qualifications.

Inside the BIMReL project, One team has worked in particular on the experimental development, supporting the Politecnico in analysis and research tasks; it has implemented the software toolkit and also developed the technological interfaces interchange with the BIM&Co portal.

Eventually, TraceParts is a leading global provider of 3D digital content for engineering. The company provides Internet-based solutions for creating and managing libraries of components, catalogues and product configurators. It also provides digital marketing services to help manufacturers promoting their products and services better.

7 The Web Portal

7.1 User Management

To have full access to the features of the web portal, registration is required. The new user could be a 'User' or a 'Manufacturer'. The latter figure is different from the former in that it will also insert a first corporate profile and will be enabled to upload content.

Special features include the possibility of inserting the manufacturing plants, essential to the assessment of the environmental impacts according to the distance to be covered, and the company certifications of the registered office or of any production facility.

7.2 *New Product*

The insertion of a new product immediately shows the user ten points (see Table 1). The goal of BIMReL is to sensitise the user to compile all the points in their entirety. With reference to the amount of information provided in the BIM portals on the market, often limited to a small amount of data, it was considered appropriate not to make compiling all the points binding, with the exception of the clear definition of the product in point 1 and entry of the 'Commercial name'.

Currently, BIMReL contains more than 1 thousand of different classes of product, as graphically highlighted in Fig. 2.

7.3 *Edit and Review*

The 'Edit' button, as previously mentioned, allows to complete or change the inserted information.

While a company may continue to place products with the same properties, the declaration of performance number and lot number can change over time. Following the previous consideration, the possibility of generating a DoP history was introduced through the 'Review' action.

7.4 *XML (eXtensible Markup Language) and Digital CE*

All the information submitted is also made available in XML format, as this format is readable on any type of computer and will ensure a high level of interoperability.

In addition, what the BIMReL calls the 'Digital CE', i.e. the ability to extract the version in XML format with all of the DoP data, has been implemented.

7.5 *BIMReL IFC*

In the product section, besides having the original IFC file available, the user will have the option to download the "BIMREL IFC' containing information from the portal filtered according to the chosen LOD. An example is shown in Fig. 3.

Table 1 Description sections of the product entry wizard

Step	Description
1	It is a section that identifies the product within the BIMReL classification, based on the identification of the same class (so if it is a construction or installation product, a furniture, a vehicle or an equipment), then the macro-category (present, for example, for the furniture, to be able to be grouped, for example, into office furniture), the category (such as, for example, plaster mortar, thermal insulation, etc.), the product type and the reference standard (if present). The possible combinations already implemented (Family, Macro-category, Category, Type and Standard) enable the user to uniquely classify each product and, depending on that, allow the database to propose specific data in the following points
2	In this step, there is the opportunity to define the product commercially, to attribute keywords, synonyms and encodings according to: Uniformat, OmniClass (Table 23), MasterFormat, Uniclass and ETIM, to select the corresponding production plant and insert the images that best represent the element
3	CE Marking and DoP (Declaration of performance). If you are inserting a construction product subject to a harmonised standard, at this point, the fields of Regulation (EU) No 305/2011 (the European Parliament and the Council, 2011) will be shown, along with the words 'optional field' for those sections that are not mandatory based on delegated Regulation (EU) No. 574/2014 (the European Commission, 2014), with the addition of the essential basic requirements described in Annex ZA of the same harmonised standard (Pavan et al. 2019). It will then be possible to generate the DoP of the product
4	The point shows the fields of the technical data sheet as defined in the UNI/TS 11337-3:2015 standard. The only exception is the section on sustainability, more specifically described through the insertion of all environmental sustainability indicators in accordance with the UNI EN 15804:2014 standard, in order to grant the end user the ability to quantify the ecological impact at every stage the product life cycle ('from cradle to grave')
5	Not selectable for all types of products, it contains the guide dossier, a document that aims to contain all the information concerning the transport, handling and storage, works/installation, use and maintenance, disposal and prevention and security.
6	In this section, there is the option to insert 2D and 3D models in both the proprietary and the open-source formats by indicating the LOD (Level Of Development)
7	Document management aims at ensuring that the manufacturer/dealer attaches any additional certifications and attestations, providing a further opportunity to transparently leverage its product. Doc, pdf, docx, xls, xlsx, png and jpg files types can be inserted
8	In this step, the partner portals with which technological interchange interfaces have been developed are proposed; it will, therefore, be possible to move the product to portals such as the ones belonging to TraceParts, project partners boasting 15 million users, thus exponentially increasing visibility on the market
9	This section aims to sensitise the compiler, showing the percentages of completeness achieved in the previous points
10	The last point allows to create a version of the technical data sheet and guide dossier, if any, anonymous. This activity wants to meet the requirements under public contracts, i.e. providing these documents free from any information that would allow the specific product or manufacturer to be traced

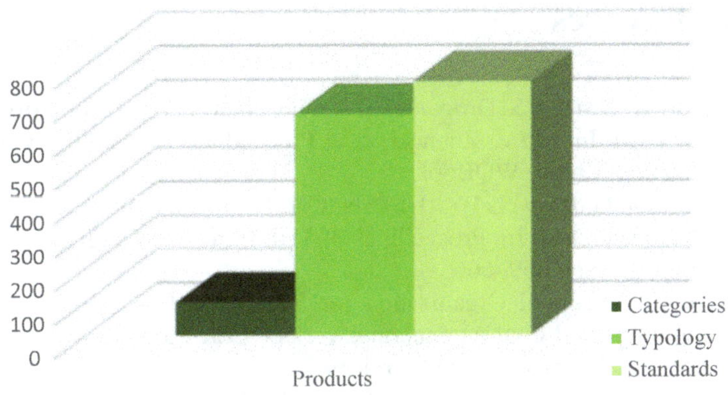

Fig. 2 Graphical distribution of the product classification

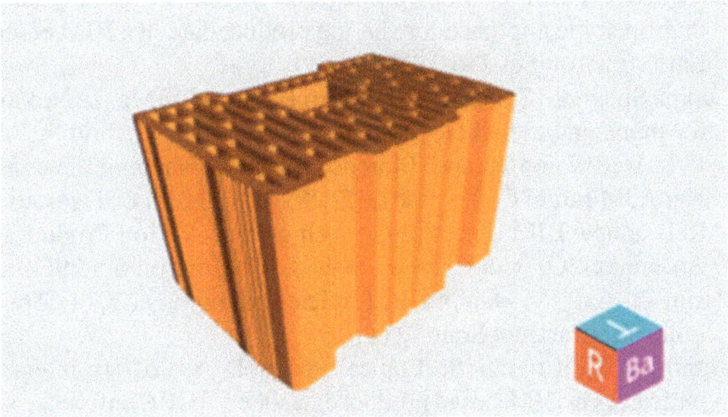

Fig. 3 Example of a masonry brick element using the portal's IFC viewer

8 Concluding Remarks

The impacts of the use of this portal are numerous and improve the whole chain of the construction industry. The platform is positioned as the top product portal for the construction industry, meeting the specifications of the new UNI/TS 11337-3:2015 standard that lays down the criteria for coding construction works and products, activities and resources. Thanks to BIMReL, component manufacturers for the construction industry have the opportunity to publish their catalogues not only on the Italian market but also internationally. In addition, the simple and immediate access to the BIMReL platform for citizens allows the possibility to easily check and compare each characteristics and performance of materials and products, creating concrete awareness of their environmental impacts. In conclusion, the interoperable platform for the management of the BIM library in the Lombardy Region will, therefore, not be a simple 'objects container' but, thanks to the information breakdown, adherence to the new standards and the systematisation of the latest ICT tools system, it will improve the entire supply chain in the construction industry, providing citizens with a smart channel of data access and use.

References

Duddy K, Beazley S, Drogemuller R, Kiegeland J (2013) A platform—independent product library for BIM. In: Ma Z, Zhang J, et al Proceedings of the 30th CIB W78 international conference (Beijing, China) WQBook

Gudnason G, Pauwels P (2016) SemCat: Publishing and Accessing Building Product Information as Linked Data. In: Proceedings of 11th European Conference on product & process modelling (ECPPM), pp 659–666

ISO—International Organization for Standardization (2017) ISO 21930:2017 Sustainability in buildings and civil engineering works—Core rules for environmental product declarations of construction products and services

Kim JB, Jeong W, Clayton MJ, Haberl JS, Yan W (2015) Developing a physical BIM library for Building thermal energy simulation. Automat Construct 50:16–28

Lucky MN, Pasini D, Lupica Spagnolo S (2019) Product data management for sustainability: an interoperable approach for sharing product data in a BIM environment. In: IOP conference series, Earth Environ Sci 296 012053

Lupica Spagnolo S, Pasini D, Daniotti B, Pavan A, Mazza N, Valra A (2017) A BIM-based platform for managing the whole building life cycle. In: Ciribini A, Alaimo G, et al (eds) Proceedings of ISTeA 2017 conference (Florence, Italy) Re-shaping the construction industry, pp. 112–121

Pavan A, Mirarchi C, Amosso G, Lucky MN, Pasini D, Daniotti B, Lupica Spagnolo S (2019) BIM-ReL: a new BIM object library using Construction Product Regulation attributes (CPR 350/11; ZA annex) IOP Conference Series, Earth Environ Sci 296 012052

Pasini D, Caffi V, Daniotti B, Lupica Spagnolo S (2017) The innovance BIM library approach. Innovat Infrastruct Solut 2(1):p15

Signorini M, Frigeni S, Lupica Spagnolo S (2019) Integrating environmental sustainability indicators in BIM-based product datasheets. IOP Conference series, Earth Environ Sci 296 012028

The European Parliament and the Council (2011) Regulation (EU) No. 305/2011 of 9 March 2011 laying down harmonised conditions for the marketing of construction products and repealing Council Directive 89/106/EEC

The European Commission (2014) Commission Delegated Regulation (EU) No 574/2014 amending Annex III to Regulation (EU) No 305/2011 of the European Parliament and of the Council on the model to be used for drawing up a declaration of performance on construction products

UNI—Ente Italiano di Normazione (2014) UNI EN 15804:2014 Sustainability of construction works—Environmental product declarations—Core rules for the product category of construction products

UNI—Ente Italiano di Normazione (2015) UNI/TS 11337–3:2015 Building and civil engineering works—Codification criteria for construction products and works, activities and resources—Part 3: Models of collecting, organizing and recording the technical information for construction products

Life Cycle BIM-Oriented Data Collection: A Framework for Supporting Practitioners

Anna Dalla Valle, Andrea Campioli and Monica Lavagna

Abstract For updating the construction sector in line with current trends, Life Cycle Thinking (LCT) has to be integrated into the building process from the beginning. In this perspective, the digitalization increasingly assists practitioners in the task, taking advantage in particular of the now widespread Building Information Modeling (BIM). To face construction challenges, the research suggests conceiving BIM as a life cycle database that evolves over time in conjunction with building process development. In this way, to support the players involved in the shift both in thinking and in process, the result of the research is a data collection framework that activates over the process a rigorous life cycle oriented information flow to build up the expected life cycle project-based BIM database. Indeed, since to guarantee the effectiveness of the digitalization process collaboration environments, each stakeholder has to know the requirements of the information content, it points out, based on current practice, the set of life cycle information, the actors in charge and the derived life cycle information flow demanded during the process. The aim is to establish, in one single BIM record, a project-based and well-framed set of data of the facility during the whole life cycle.

Keywords Life Cycle Thinking · Building Information Modeling · Life cycle database · Information flow

---

A. Dalla Valle (✉) · A. Campioli · M. Lavagna
Architecture, Built Environment and Construction Engineering—ABC Department, Politecnico di Milano, Milan, Italy
e-mail: anna.dalla@polimi.it

B. Daniotti et al. (eds.), *Digital Transformation of the Design, Construction and Management Processes of the Built Environment*, Research for Development,
https://doi.org/10.1007/978-3-030-33570-0_5

1 Change of Paradigm of BIM Toward LCT

To handle buildings as complex systems, the construction sector is increasingly confronted with a BIM approach. However, to face sustainable goals straight away, it should be complemented with LCT and related methodologies, which are now recognized as pivotal in improving resource efficiency with environmental, social and economic benefits (UN environment 2018). There is the need to shift the shared mental model of buildings from stuff (i.e., products and technologies) to purposeful systems- and life cycle-thinking (Boecker et al. 2009; Rusu and Popescu 2018).

In this perspective, the advancement of technology, computation, and digitalization consistently boosts the transition of the building sector in that direction (Jiao et al. 2013). A wide range of methodologies and tools is now available to help practitioners in the understanding of buildings as systems and as parts of a larger system of their context, extending thus their influence beyond the site and toward the whole environment (Rezgui et al. 2011; Riese 2012; Chirjiv and Ben 2017). Moreover, as anticipated, BIM is adopted even more in design and construction practice to address the challenging tasks which characterize the building sector (Becerik and Kensek 2010; Dupuis et al. 2017). Despite the technological, cultural, and legal barriers, the great potential offered by BIM is that it is conceived as a database that embeds, displays, and calculates graphical/tangible and non-graphical/intangible information, ensuring the connection of information and data to the related objects. These functionalities appear pivotal in providing a reliable basis for decisions during the whole building life cycle, especially because, serving as a shared resource for information, it encourages communication, collaboration, and cooperation across the broad spectrum of disciplines and stakeholders involved in the process (Succar and Kassem 2015). Nevertheless, since BIM is tailored to fit a multitude of practices and projects, if, on one hand, it allows maximum flexibility to practitioners, on the other, it requires considerable effort to customize and arrange all the data in an efficient and effective way. For this reason, industry players agree that they are only just beginning to explore the full capabilities of BIM, claiming to achieve far more than they currently do (McGraw Hill Construction 2009), regarding, for instance, the amount of data and information included in the building model.

In this context, to support building practitioners toward the requested system- and life cycle-thinking, the paper shows a data collection framework aimed at creating a life cycle project-based BIM database (Dalla Valle et al. 2018). It is the result of a 3-year research project, based on the in-depth analysis of current practice taking, in line with construction trends, a life cycle perspective. Indeed, to face the forthcoming construction challenges, it emerges the need for a change of paradigm in the conception of BIM towards LCT. To take full advantage of the potential of digitalization, BIM has to be advanced from a database that evolves over the building life cycle in terms of quality and quantity to a life cycle database that properly progresses over time from its inception onward. The transition of BIM into a building life cycle database consists of the integration of the life cycle information content into the model showing, as normally happens as the phases are perfected, different degrees of depth (detail), stability (development), and reliability (approval). The improvement during the process of the set of intangible information attached to the tangible BIM objects and model with the related life cycle information and data has a dual objective. First, from the outset it provides additional specifications on the selected technological elements, not limiting them to the construction phase but involving their entire life cycle. Second, it reveals and traces the set of criteria adopted during the decision-making of the whole building process, expanding the typical performance and aesthetic parameters with the connected life cycle information to make aware decisions, avoid shifting problems and gain a long-term perspective.

The paper therefore presents the data collection framework that, by organizing the required life cycle information and actors in charge and envisioning the resulting information flow during the process, supports in practice construction players in the establishment and development of their life cycle BIM database, creating over time a valuable source of information for all the stakeholders involved in the building process.

2 Joining Life Cycle Information, Building Elements and Actors

The data collection framework is based on the list of life cycle information required for performing building Life Cycle Assessment (LCA) studies (EN 15978: 2011; EPD PCR UN CPC 531: 2014). In this way, LCT, which represents a general mindset, is explained taking as a reference frame LCA, providing an added value since it depicts an international standardized methodology. However, it is important to stress that the paper focuses only on the data collection of "foreground systems" (EC-JRC 2012), involving the set of quantitative information since directly demanded by building practitioners and therefore to bear in mind during the process. The environmental information of the "background system" are thus actually omitted, since they are not tied to practice but rather attributed to literature, external database (e.g. Ecoinvent) or primary data (e.g. EPD).

Once re-elaborated from the standards, the life cycle information are revealed by the data collection framework, referring them to the main technological elements in question at the building level. In particular, for construction materials they concern: structure, cladding, envelope, walls, floors, finishes, equipment, and furnishing; while for building systems they concern: HVAC, plumbing, electrical, and renewables systems. Each life cycle information is thus further detailed in terms of the technological elements involved, in some cases encompassing them in their entirety (e.g. transport) while in others partially, since the life cycle topic affects only certain building elements (e.g. maintenance).

Besides the systematic arrangement of the life cycle information breakdown in the connected technological elements, the data collection framework promotes their progressive implementation into the expected life cycle BIM database from the beginning of the process: the design phase. Indeed, it figures out the life cycle information for each design subphase according to the process development of current practice, explored through an ethographic approach (Pink et al. 2013), joining an internationally affirmed architectural and engineering firm and analyzing a sample of representative case studies. In this way, the data collection framework specifies to practitioners the information to be collected in each phase, distinguishing the ones already considered in practice, based on the decision-making of the reference projects, from the ones to be implemented to turn into a life cycle oriented practice. Concerning the latter, the targeting of the life cycle information is established depending on the information and requirements now available and outlining the associated life cycle topics to be included in the different process phases.

Based on the progressive information setup, narrowed—due to the partnership agreement—on the design process, the data collection framework entrusts the design team with integrating into BIM right from the beginning the advised life cycle information and data according to the specific phase in progress. In this way, the aim is to orient the whole decision-making in line with life cycle perspective, soliciting not only a gradual implementation but also a growing level of detail and accuracy of the life cycle information in conjunction with the process development. In fact, to boost the optimization of the process, the whole set of life cycle data, inputted following the data collection framework recommendations, turn out to be the thresholds not to be exceeded in the subsequent process phases.

In addition, due to the broad spectrum of information required toward LCT and the wide range of competencies involved in the building design process, the data collection framework is structured in such a way that the roles of gathering the recommended life cycle information are shared with the main design competences. The actors in charge are explained in accordance with a recognized classification (Omniclass 2012), joining them with the highlighted technological elements based on current practice. In particular, they involve: architecture for cladding, envelope, walls and floor; interior design for furnishing; health/laboratory design for equipment; structural engineering for structure; mechanical engineering for HVAC systems; plumbing engineering for plumbing systems; electrical engineering for electrical systems; building energy design for renewables systems and building operational energy use; and environmental design/sustainability for building operational water use. In this way, the joint combination of all individual efforts allows to obtain an overall and systemic vision of the designed building, resulting in the life cycle BIM database with the associated opportunities offered (Fig. 1).

Indeed, all these competencies are assigned to design the related technological elements including life cycle information as further criteria of the decision-making, as prescribed by the data collection framework, contributing to and extracting information from BIM. In this perspective, to encourage the different team of actors in adopting a systemic approach towards life cycle design, personal worksheets are inferred for each competence from the data collection framework, setting out the information required at each phase of the process for each assigned technological element. The worksheets thus enable each individual actor to become aware of the

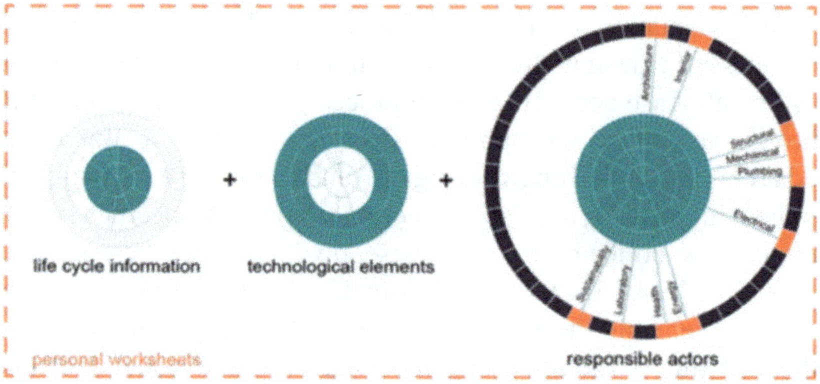

Fig. 1 Relationships of information, elements, and actors toward life cycle data collection

life cycle information and data under their responsibility and of their step-by-step collection to create the life cycle BIM database. In this manner, they simultaneously affect the decision-making process, calling upon the practitioners involved to find design solutions by accounting for the listed life cycle information and data.

3 Life Cycle Information Flow During the Process

Appointing the specific design competencies as responsible for the defined life cycle information, the proposed data collection framework activates, during the process, a rigorous life cycle-oriented information flow to build up a project-based life cycle database within a BIM working environment. To this end, it is important to underline a key strength compared to current design practice. Indeed, even today very little life cycle information are actually ever considered in practice by the design team, although they are exchanged in one-to-one relationships and cited fragmentarily in various supporting documents. By contrast, following the data collection framework, all the responsible actors enter the entire set of life cycle information and data into the life cycle BIM database of the facility. In this way, all life cycle information are collected in one single record, enabling all practitioners not only to fill in their assigned information but also easily to find all the available data for the widest range of purposes (Fig. 2).

To provide a synopsis, the following paragraphs illustrate the information flow demanded by the data collection framework to build up the life cycle BIM database, pointing out for each phase of the design process the overall set of actors involved with the committed life cycle information. In this connection, note that the assigned life cycle information are tailored to the requirements and level of detail achieved in the specific phase in question (DVA 2010) and that the engaged actors represent the main competencies that have to be supported, foremost to integrate LCT into the construction sector starting from the early phases of the process.

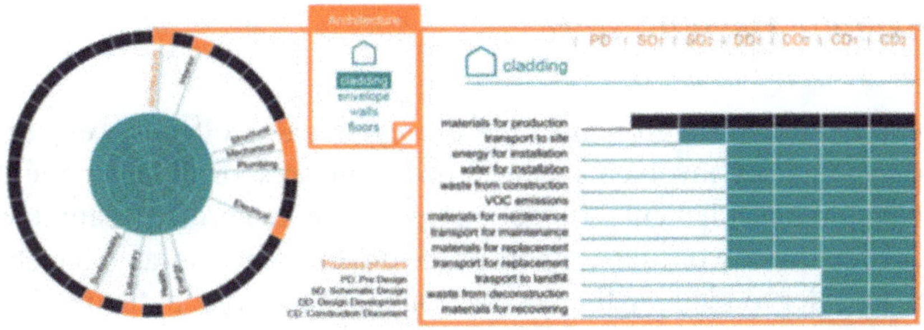

Fig. 2 Life cycle information of cladding assigned to architecture during the process

3.1 Pre-design Phase

The pre-design phase represents a crucial step, especially for the definition of the project sustainability goals to be pursued and reached throughout the whole process. In fact, besides heavily affecting the decision-making process, sustainability goals also have an impact on the building life cycle data collection because, being established in conjunction with clients as targets of the project, they compel the design team to consistently check with them the life cycle information and data entered into BIM. In this way, if the data does not comply with the targets set, the team has to call into question the previous design choices, changing at the end the proposed design solutions. For this reason, it is practical to express from the start the sustainability goals in accordance with the list of life cycle information defined by the data collection framework, in order to encourage their monitoring and verification during the project development. Sustainability goals should thus combine the targets commonly set by clients and requirements for the operational phase (e.g. energy use, water use, and greenhouse gas emission reduction), with the ones relevant to the other life cycle phases (e.g. construction waste and recycling materials), taking as reference the thresholds set by the various Green Building Rating Systems criteria and Green Public Procurement requirements.

Concerning the required life cycle information flow, due to the very preliminary state of the project, it involves few competences and a limited set of life cycle information. Indeed, in this phase the design team identifies alternative volumetric solutions, allowing clients to select one of the proposed concepts, outlining the general massing and site location of the facility for further development. In this context, only architecture and building energy design competencies are encouraged by the data collection framework for considering during the decision-making, in the first case, the envelope surface and, in the second, the main building energy consumption for heating, cooling, and lighting systems. However, it is worth mentioning that, since the data collection is conceived within a BIM working environment, these data (although not only) are automatically generated by the building model or related plug-in, avoiding rework and time wasting and improving the coherence of the information provided.

3.2 *Schematic Design Phase*

During the Schematic Design phase, the life cycle data collection engages all the principal competencies involved in the process and responsible for the design of the main technological elements. In fact, in this phase, the project team develops the approved concept plan, working as appropriate on different spatial and technological alternatives in order to identify the best options for the project at hand. In this way, during design, the actors in charge are committed to an evaluation of different options for the assigned technological elements, involving the selection of a wide range of criteria, including also the life cycle ones.

In particular, the implementation of life cycle information is stressed from the outset for engineering competences, requesting mechanical and electrical engineering as well as building energy design to provide specifications on the maintenance and replacement processes of the proposed solutions, respectively for HVAC, electrical and renewables systems. Instead, since architectural competencies are mainly focused on aesthetic and functional building decisions, only transport distances are demanded as life cycle information for the preliminary cladding solutions. The same happens for structural engineering, called for considering transport distances during the material selection of the proposed structural systems. Moreover, combining the set of early architectural and engineering solutions, building energy design has to estimate the energy consumption of heating, air conditioning, lighting and ventilation systems, as well as the energy generated by renewables systems, showing also the associated emissions. Meanwhile, environmental design/sustainability has to estimate the water consumption of sanitation and irrigation systems. In addition, as a design process within a BIM working environment, all technological solutions are progressively quantified as appropriate in terms of area, volume, item, or linear development, systematically arranging the production information for both construction material and building systems.

3.3 *Design Development Phase*

The design development phase is intended to add an increased level of detail for all aspects of the project to further define the design and the attribute of interior and exterior spaces. In this context, the data collection framework plays a key role in supporting the design team in selecting the set of material solutions most appropriate for the project also from a life cycle perspective, improving and enhancing the accuracy of the collected life cycle information and data.

Indeed, the design team effort is focused on the definition of all technological components, soliciting the team of actors responsible to explore the main life cycle topics and to collect the connected life cycle information and data at least of the typical items. Concerning the project construction stage, architecture has to consider transport distances and construction waste quantity for envelope and cladding and structural engineering for building structure. In addition, architecture is in charge of the estimate in overall terms of the energy used for the installation of products and for the water used for on-site production. About the use stage, architecture and interior competencies have to be included for cladding and finish solutions, respectively, the related VOC emissions as well as the materials and transport required for maintenance and replacement operations. The latter information have to be updated also for HVAC, electrical, and renewables systems by the appropriate competencies, represented by mechanical, electrical engineering, and building energy design. Moreover, building energy design has to recalculate, based on the latest design solutions, the operational energy consumption by end-user, considering in detail: heating, air conditioning, hot water supply, ventilation, lighting systems, pumps and auxiliaries and miscellaneous equipment, then compared with the energy generated both by renewable and nonrenewable resources. Likewise, the environmental design/sustainability has to rerun the operational water consumption breakdown in the usage for sanitation, domestic hot water, and irrigation systems. In this way, the resulting life cycle BIM database turns out to be enriched with a broad range of life cycle information, relating to the production, construction, and use stages of the principal technological elements.

3.4 Construction Document Phase

During the construction document phase, the project is given further detail by the design team in terms of specific product solutions to provide the necessary documents for bidding. At this point, following the data collection framework, the team is entrusted with completing all the required life cycle information, later converted into the set of specifications to be clearly communicated to the contractor.

To this end, all the previously inputted life cycle information and data are checked, adjusted, and updated both in relation to the design development and to the data sources, opting preferably for primary data of specific products rather than database or literature data. Furthermore, additional life cycle information are required, enlarging the set of technological elements to be considered and the life cycle topics in question. In this way, transport distances are demanded not only for structure, cladding, and envelope components but also for walls, floors, finishes, equipment, furnishing, and for all the main building systems, involving all the competences of reference. The same happens for the amount of construction waste, in which the technological elements included are extended to walls, floors, and finishes, involving in particular architecture and interior design. The latter competences, with the support of health/laboratory design, also have to insert the VOC emissions of cladding, finishes, equipment, and furnishing. Materials and transport distances for maintenance and replacement processes have to be finalized for both the architectural and engineering elements, engaging the different competences according to their respective areas of expertise. With regard to building use consumption, as in the previous phase, the expected operational energy split by end-user has to be reprocessed by building energy design in collaboration with mechanical engineering. Instead, the operational water use has to be further specified as follows: by environmental design/sustainability for water sanitation and irrigation systems, by plumbing engineering for domestic hot water and building auxiliaries systems, by mechanical engineering for HVAC systems and by health/laboratory design for the main building appliances. Finally, the end of life stage information has to be implemented involving, on one hand, architecture and interior design for all construction materials and, on the other, mechanical, plumbing, and electrical engineering for all systems materials. These competencies are thus called upon to assess, with reference to the assigned technological elements, the connected transport to landfill, waste derived from the deconstruction process and the potential materials for reuse, recycling, and energy recovery.

4 Outlook of the Resulting Life Cycle BIM Database

Established under the proposed data collection framework, the development during the process of the life cycle BIM database has a twofold aim for the construction sector. From the life cycle perspective, it allows a project-based and well-framed set of data of the facility during the whole life cycle to be tracked in one single record, conceived in a BIM working environment. In addition, it appears to be of added value for clients, stakeholders as well as for design firms, enabling a continuous build-up of know-how, not only in terms of the specific life cycle project-based BIM database but also for that of future projects, and also represents the input data especially for Life Cycle Assessment and Life Cycle Costing analysis. By contrast, from the design process perspective, it allows the overall monitoring of the process and the verification of its optimization, since the declared life cycle information and data stand for the thresholds not to be exceeded in the subsequent process phases. Moreover, the data collection framework helps to build sector practitioners to comply with sustainability goals and towards LCT, calling for a life cycle information flow through a sharing of roles and responsibility. However, if on one hand, this organizational setup represents a point of strength, not overburdening design competences in their assignments, on the other, it constitutes a point of weakness, since it involves a wide range of competencies that must be trained and successfully managed to account for life cycle information in practice.

Indeed, it is important to underline that the integration of LCT into the design and building process represents an upcoming and challenging task for the involved actors, demanding a shift both in thinking and in process. In this context, the data collection framework should be considered as the springboard for a life cycle-oriented practice, to be advanced by follow-up investigations. These concerns, for instance, the technical implementation of the BIM information content, the identification of the possible specific data sources and the exploration of the now available LCT tools. Furthermore, it is worth mentioning that, based on current practice, the data collection framework supports the establishment of the most virtuous life cycle BIM database currently feasible. As a result, it can be overcome in the near future in relation to potential new life cycle information and data and it can be simplified in relation to building project complexity.

Consistent with the data collection framework, the progressive implementation of the life cycle information resulting in the life cycle BIM database, can be perceived by building practitioners in two different ways: first, as minimum submission requirements demanded at each design phase for internal purposes, leading to a voluntary process put into effect by the design team to support the development of life cycle oriented practice; and second, as minimum submission requirements for external purposes, followed by a mandatory process in the event that it is assimilated directly by project requirements. In both cases, from a BIM perspective, the life cycle information set by the data collection framework becomes part of the model information evolution, being perceived as a sort of Level Of Development of the information content related to the digital building objects. To this end, those primarily affected are the design team and the manufacturers, who have the chance to achieve directly the BIM objects of their products with the relative life cycle information, thus enhancing data transparency. In this context, defining what to get out of the project information model at the end and at each phase, it is expected that the so-called Employer's Information Requirements extend the typical responsibility matrix and detail the information requirements with the personal worksheets inferred from the data collection framework to build up, over time, the life cycle project-based BIM database.

References

Becerik-Gerber B, Kensek K (2010) Building information modeling in architecture, engineering, and construction: Emerging research directions and trends. J Profession Issues Eng Educat Practice 136(3):139–147

Boecker et al (2009) The integrative design guide to green building. Wiley, New Jersey

Chirjiv KA, Ben A (2017) Recent developments, future challenges and new research directions in LCA of buildings: a critical review. Renew Sustain Energy Rev 67:408–416

Dalla Valle A, Lavagna M, Campioli A (2018) Matching Life Cycle Thinking and design process in a BIM-oriented working environment. In Paper presented at XII Italian LCA network conference, Messina, 11–12 June 2018

Department of Veterans Affair (2010) A/E submission requirements for VA medical center major new facilities, additions & renovations

Dupuis M, April A, Lesage P, Forgues D (2017) Method to enable LCA analysis through each level of development of a BIM model. Proc Eng 196:857–863

EC-JRC (2012) ILCD Handbook. Towards more sustainable production and consumption for a resource efficient Europe. European Commission—Joint Research Centre, Luxembourg

EPD: UN CPC 531:2014 Product Category Rules: Buildings

Jiao et al (2013) A cloud approach to unified lifecycle data management in architecture, engineering, construction & facilities management. Adv Eng Inform 27:173–188

McGraw Hill Construction (2009) The business value of BIM. Getting Building Information Modeling to the Bottom Line. http://bim.construction.com/research/SmartMarket_Report.pdf

Omniclass (2012) A strategy for classifying the built environment. Table 33 – Disciplines

Pink S, Tutt D, Dainty A (2013) Ethnographic research in the construction industry. Routledge, London and New York

Renz et al (2016) Shaping the future of construction. A Breakthrough in Mindset and Technology, World Economic Forum

Rezgui et al (2011) Past, present and future of information and knowledge sharing in the construction industry. Comput Aided Des 43:502–515

Riese M (2012) Technology-augmented changes in the design and delivery of the built environment. Commun Comput Informat Sci 242:49–69

Rusu D, Popescu S (2018) Decision-making for enhancing building sustainability through life cycle. Appl Math Mech Eng 61:191–202

Succar B, Kassem M (2015) Macro-BIM adoption: conceptual structures. Automat Construct 57:64–79

UN environment (2018) Life cycle initiative. http://www.lifecycleinitiative.org

UNI EN 15978:2011 Sostenibilità delle costruzioni. Valutazione della prestazione ambientale degli edifici. Metodo di calcolo

Decision-Making BIM Platform for Chemical Building Products

Gabriele Gazzaniga, Luigi Coppola, Bruno Daniotti, Claudio Mirachi, Alberto Pavan and Valeria Savoia

Abstract This paper presents the innovative solution based on BIM technology proposed to BASF Construction Chemicals Italia for the digitization of chemical products for buildings. It deals with the implementation of a system able to support stakeholders during the selection phase and with the digitization of product information without geometrical features such as concrete admixtures or repair products. The most significant problems linked to the construction sector are deep fragmentation and the characterizing inefficiency of the building process; these phenomena are often due to uniqueness, unevenness, and inconsistency and they result in quality decay, lengthening of working time and claims. The construction sector needs properties like dynamicity and a structured and easily consultable database in order to reorganize information. BIM could represent the solution to these requirements since it is able to manage knowledge and information among the involved stakeholders and manufacturers during the building process and the useful life of a building itself. Thanks to the digitization of information made possible by the BIM process, it is possible to define the products' attributes that are key to informing the relevant stakeholders and to import them into a digital model in order to reduce inaccuracies during the constructive phases and to draw a guideline for a potential restoration.

Keywords Building information modeling · Chemical products for construction · Products without geometrical features

G. Gazzaniga (✉) · L. Coppola
Department of Engineering and Applied Sciences, University of Bergamo, Bergamo, Italy
e-mail: gabriele.gazzaniga@unibg.it

B. Daniotti · C. Mirachi · A. Pavan
Architecture, Built Environment and Construction Engineering—ABC Department, Politecnico di Milano, Milan, Italy

V. Savoia
BASF Construction Chemicals Italia Spa, Treviso, Italy

B. Daniotti et al. (eds.), *Digital Transformation of the Design, Construction and Management Processes of the Built Environment*, Research for Development,
https://doi.org/10.1007/978-3-030-33570-0_6

1 Introduction

Despite the fact that during the last decade a lot of innovative proposals concerning constructive techniques, materials and project management have been presented, the construction sector has not always been able to incorporate these innovations (Taxén and Lilliesköld 2008). Differently to other industrial processes, the building process is strongly characterized by a historical component made of old traditions and techniques that inexorably conflict with the available innovations that have been usually seen as negative only because they are not aligned with those always used. Another problem related to the construction sector is project management (Clough et al. 2008): many different figures have to cooperate; this lack is getting bigger due to the uniqueness and the unevenness that are typical of the building process (Kifokeri and Xenidis 2017).

A new instrument capable of managing all the different information and of establishing a positive communication system between the different stakeholders is just what the construction sector needs. Property of dynamism and immediacy are typical of Building Information Modeling. Initially BIM has been seen as a new Computer-Aided Design (CAD) able to represent in a better way the different project in terms of 3D (Succar 2009); nowadays, BIM is understood as an overarching variety of activities in an object-oriented dimension which supports the representation of building elements in terms of their 3D geometric and non-geometric attributes and relationship (Tookey 2017). BIM embodies a series of new technologies and solutions with the aim of an interorganizational collaboration among the different stakeholders to improve design, construction, and maintenance (Miettinen and Paavola 2014). According to the National Institute of Building Science, BIM is defined as "a digital representation of physical and functional characteristics of a facility and a shared knowledge resource for information about a facility forming a reliable basis for a decision during the project's life-cycle" (NIBS 2018). The main functionality of BIM is to create a database of information using all the inserted data which, concurrently with its 3D representation, could be able to improve design construction, prefabrication and the operations of a facility (Charehzehi et al. 2017). The innovative potential of BIM allows stakeholders to exploit different project levels (UNI 11337 2018): by connecting 3D geometric models to the schedule data it is possible to obtain a 4D aspect that is able to provide several efficient advantages in order to facilitate site planning and management. Another level of BIM technology is 5D which allows for the monitoring of costs during the constructive process; the 6D aspect describes the building in relation to the management of the building itself in terms of usage, administration, maintenance, and future disposal. Finally, the interest in sustainability and environmental impact is contained in the 7D aspect.

Although BIM has yet to be fully adopted for the reason mentioned above, a lot of studies have described the potential of this instrument (Tookey et al. 2018; Miettinen and Paavola 2014; NIBS 2018; Bryde et al. 2013). Following real examples of constructions manufactured with BIM technology, have analyzed the benefits of using BIM in relation to some important criteria such as cost, time, communication, coordination, quality, risk mitigation, organization, and relation to software and found out that all of these criteria have been positively mentioned especially in terms of cost reduction and time-saving.

With the introduction of BIM as a new powerful instrument for integrated design, a lot of different companies engaged within the construction sector have tried to introduce BIM applications in their business in order to allow designers and builders to operate with a BIM approach. Although this effort attempts to move in an innovative direction, a lot of inaccuracies and mistakes regarding data management are noticeable (Fu and Zhang 2014); the main deficiencies concern the possibility to modify or update certain data actively among the different stakeholders, the incorrect allocation of data about materials' composition which usually causes an unclear generalization and discrepancy between the data collected in the technical datasheets and the ones stored in BIM objects. These inaccuracies lead to confusing BIM libraries made up of BIM object with lacking information, moreover, if data is inserted using BIM authoring software without a logical system of organization, BIM objects cannot be used efficiently. BIM could become more effective only if the BIM object is connected with an external structured database to reduce the inaccuracies about information exchange, to integrate the BIM approach with the company's actual information management system and make it easier to upgrade the data.

2 Current Italian Situation Regarding BIM Utilization

The evolution over the past 3 years of the number of procedures that prevent the use of BIM is growing rapidly (OICE 2018): in 2015 the BIM calls were only 4, in 2016 they rose to 26, in 2017 their number increased to 83 with an increase of 70%. In 2018, the calls for applications for the BIM methodology were a staggering 268 with a leap of 229%. The particularly positive trend recorded finds its reasons both in a weak economic recovery but above all in the changes contained in the code of public contracts (D. LGS 50/2016) which came into force in April 2016 and provided for the obligation to entrust the work on the basis of the executive project, bringing to the market shares of services that were included in integrated procurement. The growing trend in the calls for applications for the BIM methodology should be reinforced in the immediate future with the entry into force of the Ministry of Infrastructure and Transport decree governing the obligation to use BIM in the design of public works.

Another important parameter for a better understanding of the BIM dissemination during 2018 is the analysis of the distribution of calls in which the BIM methodology is required for the number of activities entrusted. The "design activity" item alone represents about half of the total types of assignments, followed by the object's "seismic and structural safety assessment". A peculiarity that emerges from these analyses and that is linked to the object's "construction supervision", is, in fact, that in terms of numbers it covers a percentage equal to 1.5% while in terms of the economic value it reaches up to 15%. This discrepancy between the two parameters can be interpreted as an effort by the client to entrust the BIM methodology with great importance in terms of operations management, in order to underline the great potential of this instrument.

The last interesting aspect concerns the destination of the type of contracting authority; in light of what has been previously stated, it is easy to expect that most of the contracts to be carried out using the BIM methodology are under the administration of the state which alone covers more than 50% of the allocations. Other institutional forms, such as municipalities and provinces, give a fair relevance to the methodology while for private tenders the figure settles around a value equal to 3.7%.

3 The Information Management Process for Chemical Products in a BIM-Based Approach

The powerful aspect of projects implemented by BIM technology lies in an accurate representation of different construction elements and in a rich database accompanying the element itself; BASF Construction Chemicals Italia, international leader in the production of different types of chemical products, wants to implement a new process able to digitize all the information linked to construction materials without precise geometrical features for which until today it has been difficult to find a precise allocation within the BIM panorama for difficulties relating to an effective representation (Fig. 1).

The first step of this study concerns the choice of the most appropriate product and its attributes to digitize; starting from an accurate analysis of the company's entire portfolio, it was necessary to expand on two consequential points. First, tree diagrams have been defined to follow a Fault Tree Analysis in order to reproduce the logical process that leads to the manifestation of a need for the choice of product. Fault Tree

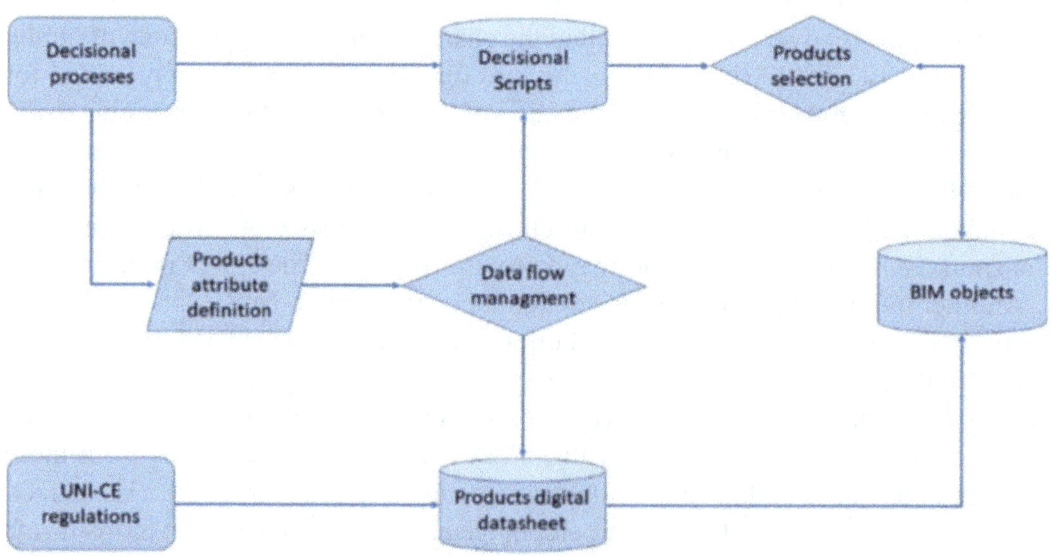

Fig. 1 Research method

Analysis is a simple widespread model in the field of reliability and it can be analyzed by implementing the Binary Decision Diagram; BDD converts the fault tree model into a binary decision diagram which encodes an if-then-else structure that could be referred to a Boolean equation (Sinnamon and Andrews 1997). Second, going through the structured logical process and tracing back through the tree diagram, it is now possible to define unanimously the specific product or the particular construction technique thanks to the use of iterative decisional scripts; these are nothing more than small applications that allow the user, by answering simple questions, to retrace their way through the Fault Tree diagram instantaneously up to the choice of the final product in order to digitize the information via a BIM-based approach (Fig. 2).

The first group of construction materials addressed is the concrete admixtures; these chemical products are one of the components of concrete mix design with cement, aggregates, water, and possible additions. UNI EN 934-2 (UNI EN 934-2 2012) brings together every type of admixture according to their main function. Nowadays, there are a lot of available concrete admixtures capable of ensuring the user with the expected performance related to mix design optimization in order to reduce water demand and consequently increase compressive strength or workability over time, or capable of managing particular operating needs since it could be necessary to intervene on the rheology side of particular cases and climatic and environmental conditions by giving concrete specific properties to prevent malicious events linked to chemically aggressive substances or external temperature. Once all the relationships have been established, a decisional datasheet based on a multi-criteria analysis is implemented in order to produce the definition of technical specifications automatically for a new mix design starting from the input provided by the users (Fig. 3).

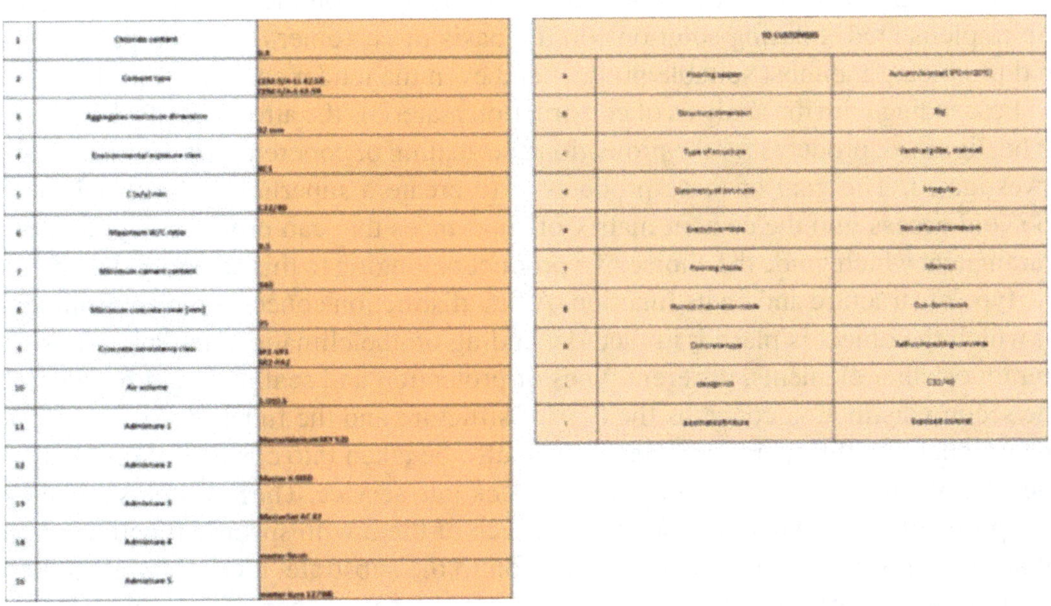

Fig. 2 Example of the decisional script

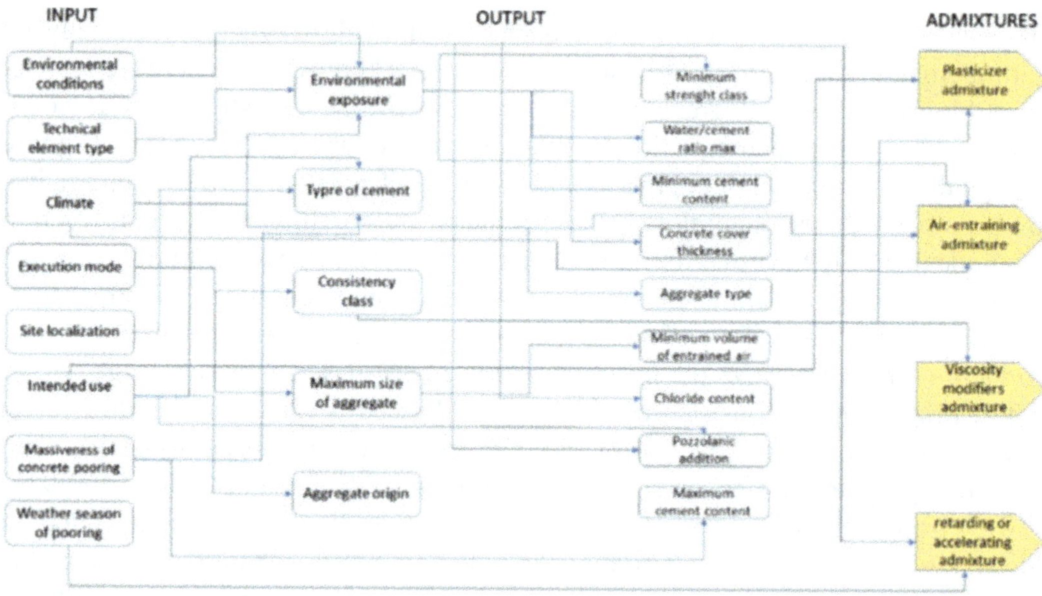

Fig. 3 Relationship requirements/boundary specifications for concrete attributes selecting

The second class of analyzed materials concerns restoration, maintenance, and reinforcement products (UNI EN 1504 2006). Differently from concrete admixtures, in addition to environmental and climatic condition, restoration and maintenance interventions depend on the failure mode and on its deterioration triggers. The decisional path to lead customers to a specific product and its correct placement starts from the potential causes and it is definitively set by combing fault tree knowledge results with other considerations about operating aspects and particular constructive techniques. Subsequently, decisional applications based on the decisional diagrams are implemented, filtering solutions on the basis of customer-compiled parameters and producing the most suitable product and even the tender specification (Fig. 4).

Proceeding with the analysis of construction materials the attributes of which have to be digitized, products for the protection and sealing of concrete surfaces have been investigated. The goal of these products is to create a superficial film that is able to avoid access into the cement matrix of substances that can promote degradation. Parameters which guide the choice of a product belonging to this category depend on the type of structure and on its function. A first distinction concerns the environment in which the object is placed; in fact, depending on the climatic situation, exposure to rain or other elements, different types of protection and sealing could be chosen; the second distinction concerns the type of structure and the function envisaged for the product; finally, operational needs must also be taken into consideration, such as the time of covering a product or putting it back into service. These systems are often composed of more than one product type, each of them with specific functions: first of all the primer guarantees a better adhesion to the substrate, it can vary according to the type of substrate and the conditions are the same; then there is the sealing membrane, which can be applied by spray or brush in a number of variable hands

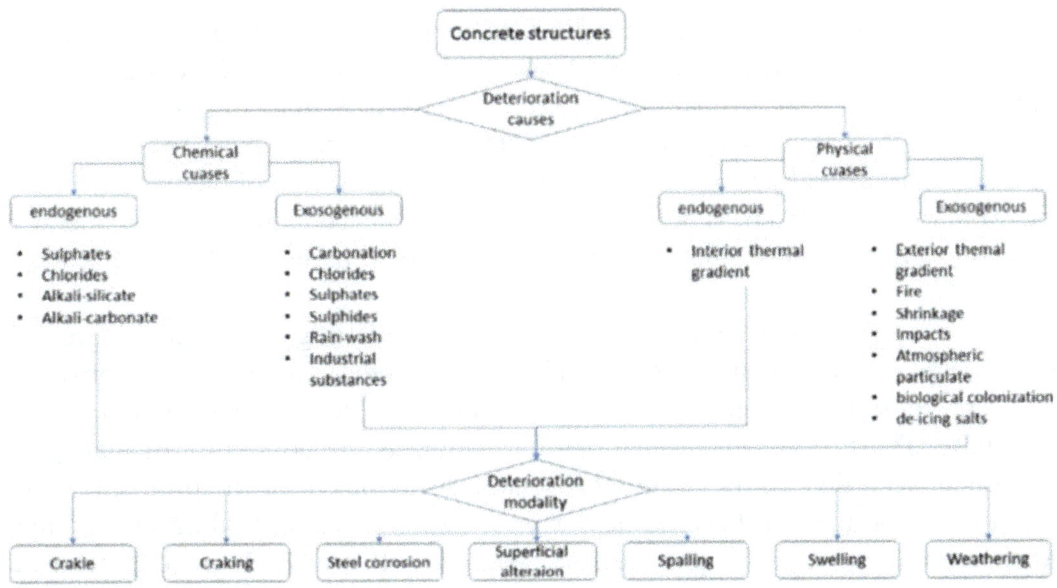

Fig. 4 Causes and types of failures in reinforced concrete structures

according to the aforementioned exposures, and finally the finish ensures a pleasant color rendering and an additional protection against UV rays.

The last category of analyzed products is the flooring finishes; environmental conditions, the position in the building, loads to be carried but also other specifications such as hygiene, health, and safety requirements determine the selection of the proper flooring solution. Precisely because of the vastness of the range of products available, it has been essential in the decision-making process to create a close relationship between the technical performance of the product, the customer's requests and the conditions of use; multicriteria analyses, that allow the stakeholders to choose the proper product for each case, start from the definition of importance values for each performance based on a relationship with decisional parameters, go through a normalization of measurement to obtain the ideal values and end with the calculation of the sequence to select the final product with a combination of importance values and normalized indicators (Cavalcante 2014). Similarly to what has been previously described for the protective and sealing products, the floor finishes also require the creation of a system composed of different elements: the primer creates a monolithic bond with the support, a coating made of reactive resin determines flooring mechanical properties, the filler spread on the fresh coating helps to improve the coating power on specific floor and finishing layers create a scratch-resistant surface and provide an aesthetic value.

Once all the decisional processes have been defined and digitized, the customer now has at their disposal a precise mix design or a series of products with the specific technical specifications according to their needs. At this point, all the information relating to the products that emerged from the previous step has to be translated into a BIM approach: to be able to do this, templates were created for each type of product. Templates are the BIM models to be used to start the design process and, by containing the data fields provided for the different product categories, can, therefore, receive BIM objects, virtualizations of geometric and non-geometric attributes of finite entities, physical or spatial, referred to a project, or to a complex of projects and to their processes. The digitization of the data of each product resulting from the decisional script is performed by a plug-in which takes care of the transfer of the above data to the related BIM objects present in the project creating a correspondence with an external database and the same BIM objects through a codification system (Fig. 5).

This new process for the digitization of objects without geometrical features, as well as allowing a punctual transfer of all information related to the product, provides the user with other potential uses in order to optimize both the design and executive phase. The first advantage that can be favored is the possibility of managing the project according to the different levels of development of the project (LOD); each type of profile will have particular information needs according to the functions carried out within the process: the designer will want to know the technical performance characteristics of the products, the company will also need information regarding the execution methods and the preconditions for installation, the manager will need instructions for maintenance. However, the building process, despite the multiplicity of actors and specializations involved, must lead to the creation of a unique product, and the uniqueness and the consistency of the choices during the various phases is an essential prerogative for quality in the final result. It was, therefore, decided to structure the information in such a way that the most frequently consulted data is directly allocated to the BIM model and, on the basis of this, the LODs have been determined in order to define, for each product category, the representative configurations of the BIM objects and the informative contents made immediately accessible on the model.

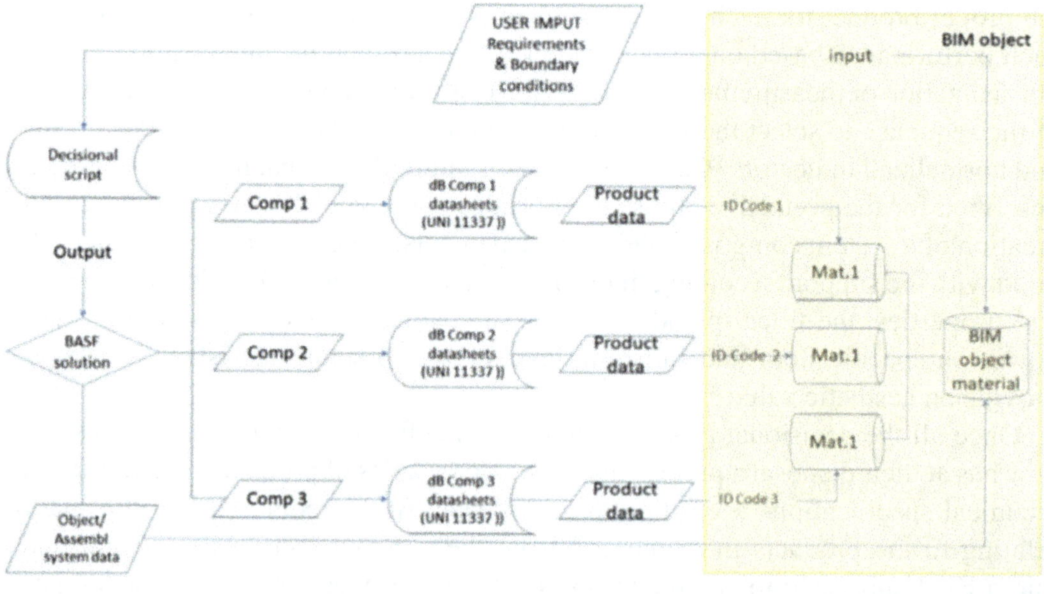

Fig. 5 General information flow for data digitization

The second advantage is still linked to the optimization of the production process in relation to the different figures involved and this time concerns the prefabrication of the concrete. The correct realization of a mix design goes through a careful design process, a careful choice of admixture, a correct mixing, and finally an implementation carried out in a workmanlike manner; designer has to establish the class of exposure and the performance characteristics of the concrete mixture, the concrete producer has to choose the precise definition of the components making up the mixture, the dosages and the granulometric curve, the manufacturer of additives has to propose the right admixtures for the required performances and the executor is responsible for implementing the final product. This BIM approach allows users to put into communication the different figures involved in the prefabrication thanks to the drafting of a specific board, where there are specific items for each figure and for each step of the concrete packaging; at each step it is possible to make changes to this sheet and consequently update the database linked to the concrete in production and obtain from time to time the BIM objects that re congruent to the changes made.

The last advantage provided by this new digitization process is connected to ordinary and extraordinary maintenance works. Indeed, by offering the chance to create a complete model in which the replay operations made with regards to layers without any geometric properties can also be digitized, it is now possible to leave an unequivocal trace of every action performed on the structure; this new possibility granted by the digitization of these elements allows an in-depth knowledge of the product in the design phase which translates into a careful design that takes into account any previous interventions and that regulates accordingly for the choice of materials and restoration procedures.

4 Conclusion

This paper proposes a new process of digitization of data concerning chemical products without geometric consistency in a BIM approach. This process manages to go beyond the inaccuracies found in the previous methodologies thanks to the use of an external database capable of sharing data with BIM objects and thus creating a continuously updated synergy. The proposed process is configured as a tool that, in addition to managing data and providing it throughout the process during the phases of interest, acts as a support for the user for the purpose of making choices, guiding him in identifying the products that are suitable for their needs through the simulation of a decision-making process. The process also makes it possible to integrate information relating to the entire life cycle of the products, by setting up a maintenance card integrated with the user's BIM model, which traces its maintenance history by archiving the recovery products used over the years for functionality maintenance.

References

Bryde David, Broquetas Martí, Volm Jürgen Marc (2013) The project benefits of Building Information Modelling (BIM). Int J Project Manage 31:971–980

Cavalcante C (2014) Multicriteria decision model to support building maintenance planning. Advanc Mater Res 1468–1477

Charehzehi Aref, Chai ChangSaar, Yusof Aminah Md, Chong Heap-Yih, Loo Siaw Chuing (2017) Building information modeling in construction conflict management. Int J Eng Business Manag 9:1–18

Clough RH, Sears GA, Sears SK (2008) Construction Project Management: A Practical Guide to Field Construction Management

Fu R, Zhang J (2014) Social Involvement to Empower a Better BIM Content Library. Computing in Civil and Building Engineering, pp. 9–16

Kifokeris D, Xenidis EY (2017) Constructability: outline of past, present, and future research. Jd construction engineering and management, vol. 143, n. 8

Miettinen R, Paavola S (2014) Beyond the BIM utopia: approaches to the development and implementation of building information modeling. Autom Constr 43:84–91

National Institute of building sciences (NIBS). United States National Building Information model standard (2018)

OICE, Annual report on the 2018 BIM competition for public works

Sinnamon RM, Andrews JD (1997) Improved accuracy in quantitative fault tree analysis. Qual Reliab Eng Int 13:285–292

Succar B (2009) Building information modelling framework: a research and delivery foundation for industry stakeholders. Automat Construct 18(3):357–375

Taxén L, Lillieskӧld J (2008) Images as action instruments in complex projects. Int J Project Manage 26(5):527–536

Tookey J et al (2017) Building Information Modelling (BIM) uptake: clear benefits, understanding its implementation, risks and challenges. Renew Sustain Energy Rev 75:1046–1053

UNI 11337 (2018). Building and civili engineering works- Digital management of the informative. UNI, Ente Nazionale di Normazione

UNI EN 1504 (2006) Products and systems for the protection and repair of concrete structures. Definition, requirements, quality control and evaluation of conformity

UNI EN 934–2 (2012) Admixtures for concrete, mortar and grout. Part 2: concrete admixtures- Definition, requirements, conformity, marking and labelling

BIM Electric Objects Plug-in for Industry 4.0

Alberto Pavan, Andrea Cunico, Claudio Mirarchi, Dario Mocellin, Elisa Sattanino and Valentina Napoleone

Abstract The use of digital MEP objects in BIM models is a reality in the building world. This approach allows for the extrapolation of information relating to digital models, the simulation of their operation and the calculations of the sizing of plant circuits. The final goal of this research project is the creation of BIM object libraries for the 4.0 Industry, which includes the modelling of BIM objects for manufacturing connected with a company system. The Revit EasyBIM plug-in, developed for Vimar SpA and object of this research paper, allows for the creation of BIM objects, connected with a company system for the control of customer satisfaction throughout the production chain, design of use, installation, use in life cycle and disposal.

Keywords Building Information Modeling (BIM) · MEP electrical objects · Plug-in · Vimar SpA · Autodesk Revit

1 Introduction

The widespread presence of the BIM work methodology has involved several sectors of the world of constructions, one of which concerns products. Many manufacturers have invested, and still investing, on the digitization of their objects, in response to the increasingly widespread demand of designers to access their BIM libraries directly.

The digital object (very often referred to simply as "BIM Object") contains the technical characteristics of the product (geometry, performances, areas of use), but

A. Pavan (✉) · C. Mirarchi · E. Sattanino
Architecture, Built Environment and Construction Engineering—ABC Department, Politecnico di Milano, Milan, Italy
e-mail: alberto.pavan@polimi.it

A. Cunico · D. Mocellin
Vimar S.p.A., Marostica, Italy

V. Napoleone
Milan, Italy

B. Daniotti et al. (eds.), *Digital Transformation of the Design, Construction and Management Processes of the Built Environment*, Research for Development,
https://doi.org/10.1007/978-3-030-33570-0_7

one of the main advantages is the fact that it is not only a mere container of information. Every BIM object, in order to be considered really as such, should have very specific requirements,[1] such as the size in terms of kB, the coherence of the informative attributes[2] with the three-dimensional model, the level of detail differentiated according to the object LOD[3] scale, etc. (UNI11337).

Many of the BIM objects currently present and downloadable from the main BIM libraries are mostly objects that contain dimensional and informational data of the country in which they were modelled.

2 The Project

The research project "*Study of modelling flows inherent in the production of BIM objects for MEP plant engineering*", carried out in collaboration with Vimar s.p.a., has the goal of developing digital objects (MEP objects) according to the Italian market.

The need of designers (in particular plant engineers and architects) was analysed and, in the initial phase, it was decided to digitize the objects of the Vimar catalogue in order to create some models available for the users. After the first hypothesis of modelling the product, some criticalities emerged. The proposed strategy was to model the components in single objects: boxes, plates, and electrical modules (fruits) and then put them in a "container" family. The main disadvantage of this solution was the high weight of the "container" object (several MB), because in order to cover the whole range of possible combinations, it was necessary to nest all the fruit categories (sockets, control devices, data devices, etc.) to allow the end user to personalize the composition.

In addition, the digital object was difficult to use because the user had to manage parameters in order to choose both the 3D geometries and the 2D annotations to be displayed in the plant. To overcome this problem, it was decided to invest in the development of a Revit plug-in, with the aim of providing a real tool for the designer and not a simple preassembled library of digital objects.

In fact, the plug-in allows the user to instantly create the composition,[4] choosing objects directly from the Vimar commercial catalogue.

[1]These requirements are illustrated in "BIM Object Standard NBS".

[2]*Informational attribute* means non-geometric information relative to a digital object (UNI 11337-1-2017).

[3]*LOD* means the level of development (or definition) of a digital BIM object (UNI 11337-1-2017).

[4]A composition means an object formed by a box, a plate, and one or more electrical modules.

3 Method

Before reaching the decision to develop the plug-in, to understand the real needs of the reference market (designers, installers, companies, etc.) preliminary researches were conducted, through surveys and interviews with the main customers Vimar. From these first contacts, it emerged that the main software used by professionals was Autodesk Revit. Moreover, one of the needs that remained unresolved was to have "light" objects (file size of a few KB) that have 2D indications in a clear and well-readable plan in the different drawing scales.

Most of the currently modelled electrical objects have very detailed 3D geometry and a set of two-dimensional annotations nested within the object itself, which are managed through object parameters. To understand how to structure the workflow we analysed the catalogue of Vimar objects (divided into civil series, domotics, and video entry), their breakdown in categories according to their function and correlation with the respective 2D electric symbolism.

The categories identified are, for example, single-pole switch, bipolar switch, single-pole diverter, etc. This breakdown of the products made it possible to identify which informational attributes to link to the object and the geometry modeling strategy. The goal was to quantify the number of three-dimensional objects characterized by their own geometry, which differed according to the commercial series (Eikon, Plana, Arkè, Idea, and other series), to the number of modules (one, two, three, etc.) and to the specific function of the object (pushbutton, switch, electrical outlet, data socket, etc.). In parallel, the information attributes for each LOD scale were also analysed for the categories of electrical objects to be scanned.

The selection of the informative attributes involved the different scales of LOD, that is, American (BIMForum 2013), English (Pass1192-2 2013) and Italian (Norma UNI 11337).

Subsequently, two 3-dimensional prototypes were created with their informative attributes, evaluating the respective strengths and weaknesses of the different solutions.

The first object prototype, named "All in One", was modeled according to the modularity of the composition: then distinct objects were modeled based on the maximum number of devices you could insert. The object provided the possibility of manual choice of the 3D electrical modules to be displayed inside the box and the association of the corresponding 2D symbol displayed in the label. This label was modeled as a generic nested annotation within the "All in One" object (Figs. 1 and 2).

As you can see from the previous image, the user can select the electrical device to be displayed in "Position 1" and the corresponding 2D symbol to be displayed in the drawing label in "Position 1 2D". Each single moulded device contained the respective electrical connector MEP.

The second object hypothesis (Fig. 3), called "Drag and Drop" provides the possibility of dragging the electrical symbols onto the label, allowing for more flexibility in the design.

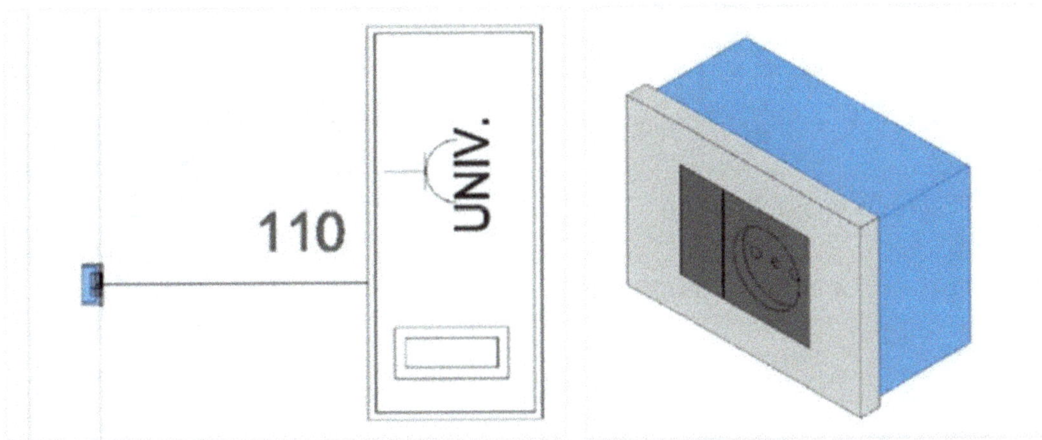

Fig. 1 Example of a plan view and a 3D view of an object, "All in One"

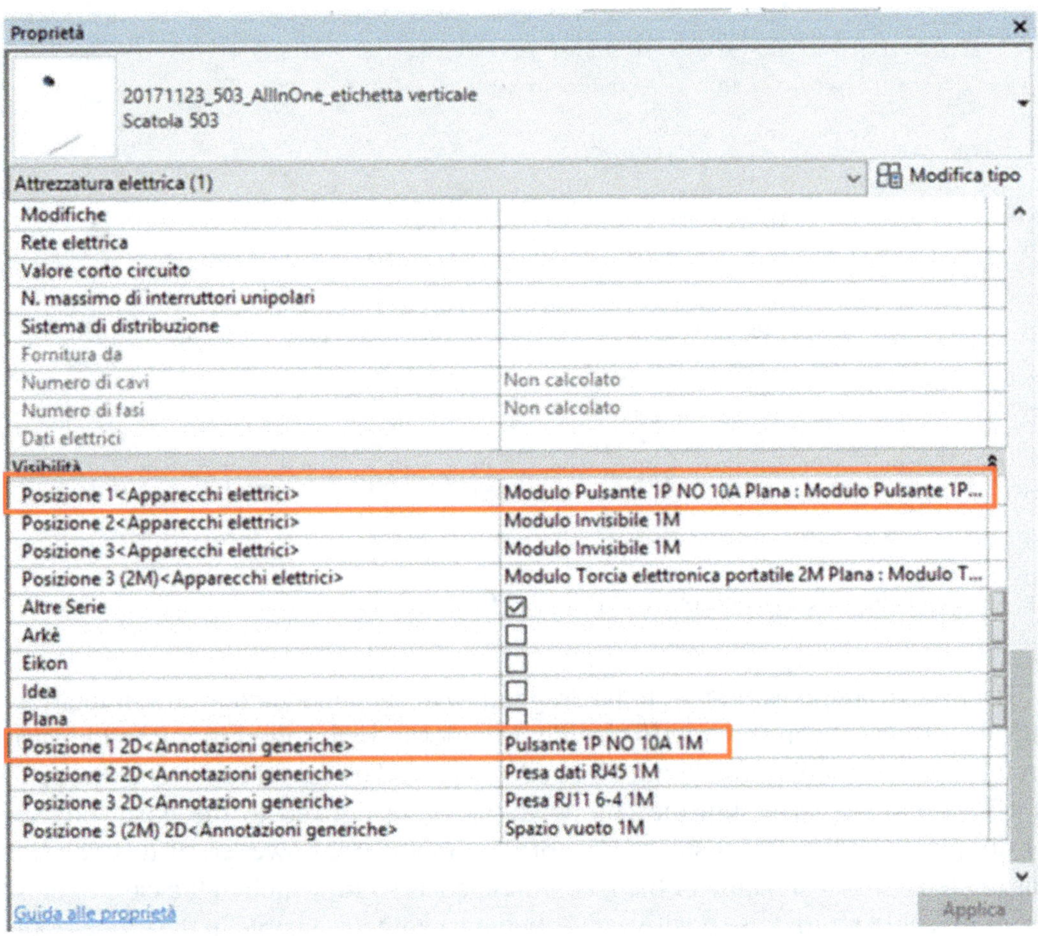

Fig. 2 Object properties window "All in One"

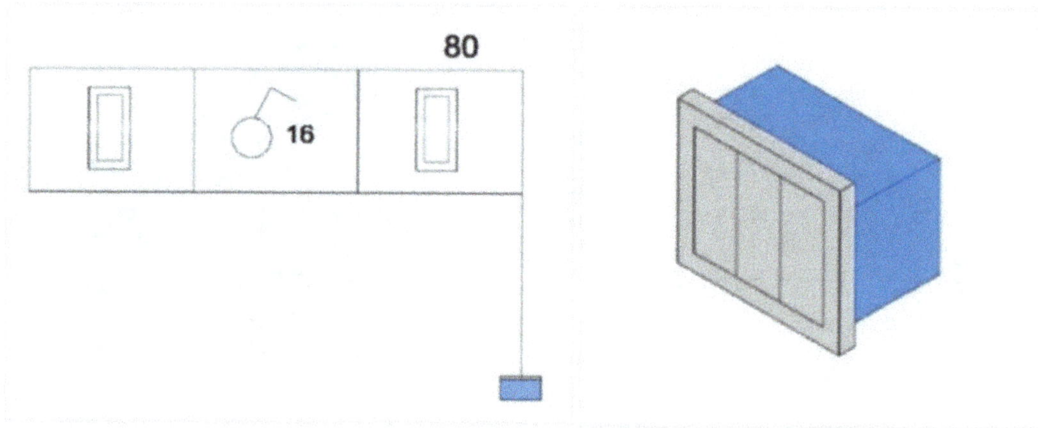

Fig. 3 Example of a plan view and 3D object, "Drag and Drop"

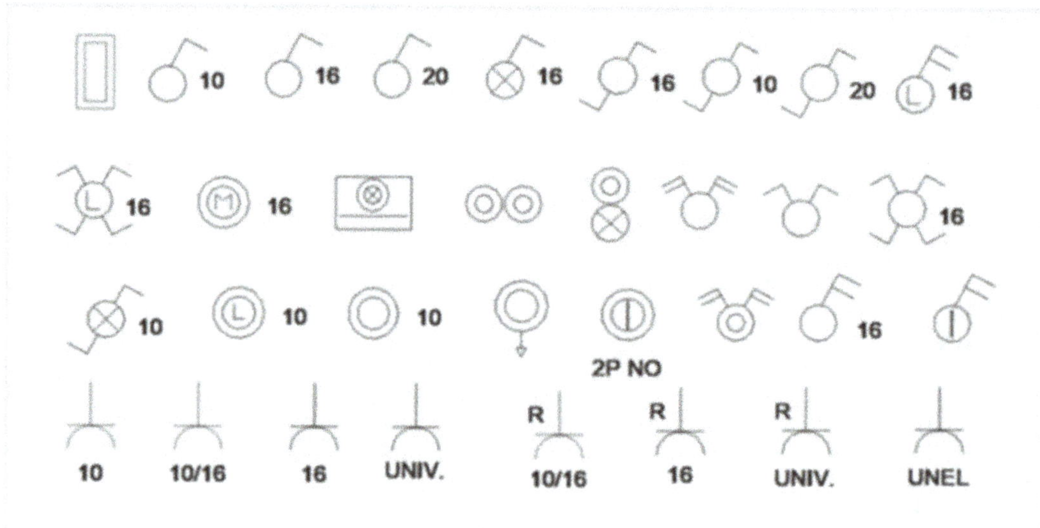

Fig. 4 List of electrical symbols object, "Drag and Drop"

The electric symbols (Fig. 4) to be dragged have been modeled as 2D annotations with the electrical connector MEP inside, in order to be inserted in a Revit electric circuit. These symbols, after being dragged and placed on the label surface (modeled as a 3D solid nested within the composition), were hooked to the host and then solidly placed in the object shifts. One of the drawbacks, however, of this solution was related to the real nature of the label modeled as solid, which in the case of clash detection could return false positives,[5] as being a geometry visible in the space-could interfere with other objects in the model (e.g. Label/wall interference).

The label modeled in the two three-dimensional objects hypothesis shows the installation quota from the finished floor expressed in centimeters, automatically

[5]They mean positive results in the control of geometric interferences, but they do not mean a real problem of the object.

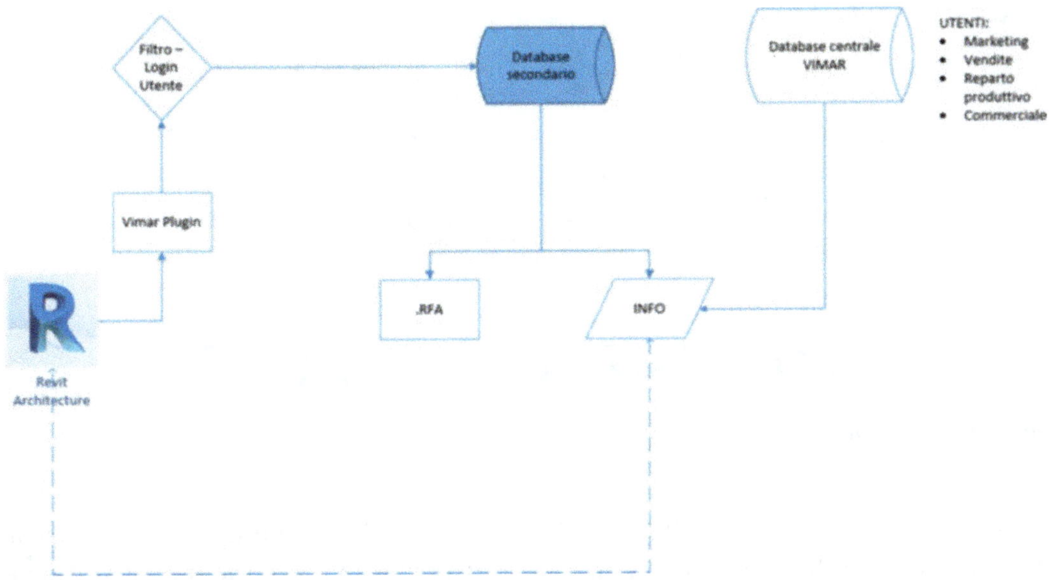

Fig. 5 Hypothesis workflow, Revit Plug-in

detected by the parameter set in the object. In addition, the shape of the label varies according to the choice of the commercial series (for example, the rectangular shape indicates the Eikon series).

Both of the proposed solutions have a serious disadvantage, which is the thousands of possible combinations (based on modularity, commercial series, and type of electrical modules). To realize a single object that can cover the entire case means making a very heavy object and therefore difficult to use.

To overcome this problem, a workflow was hypothesized to use a Revit plug-into manage both the composition of the three-dimensional electrical objects and the information relating to them (Fig. 5).

4 Results

The end result obtained from the research project was the creation of a Vimar plug-in (Fig. 6) able to automatically compose the three-dimensional object (consisting of a box, plate, and electric modules) according to the user's needs. It has been arranged into a graphical interface that is identified with the plug-in EasyBIM.[6]

Within the EasyBIM menu, there are several specific commands, grouped under "Compositions", "Logical Links", and "Separate".

[6]In order to download the plug-in from the Vimar website, it is necessary to register on the company's website and, once installed locally on the user's platform, it appears in the Revit taskbar under the EasyBIM tab.

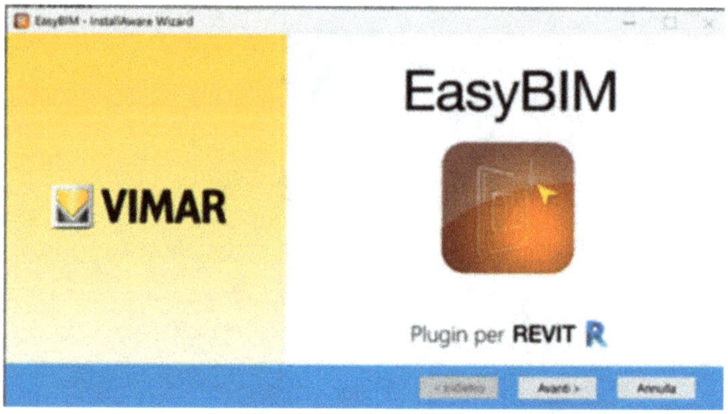

Fig. 6 EasyBIM plug-in for Revit

The main function of the Revit add-in is to be able to configure the electrical objects, choosing them from the Vimar price list or from a generic one. It was considered appropriate to give users the possibility to create objects that are "generici" (generic), or without make or model, to be used for public procurement or in preliminary/definitive projects, in which the commercial model of the product has not yet been chosen (Fig. 7).

In the graphical interface of the Configurator, there are some fields automatically managed by the plug-in (for example, N. Box) While the others are at the discretion of the user, such as "abbreviation" and "description". The "Elevation" field allows you to set the installation quota of the composition from the floor of the finished floor. In the dropdown menus below you can choose the price list (generic or Vimar), the commercial series (in the case of Vimar products the commercial series currently

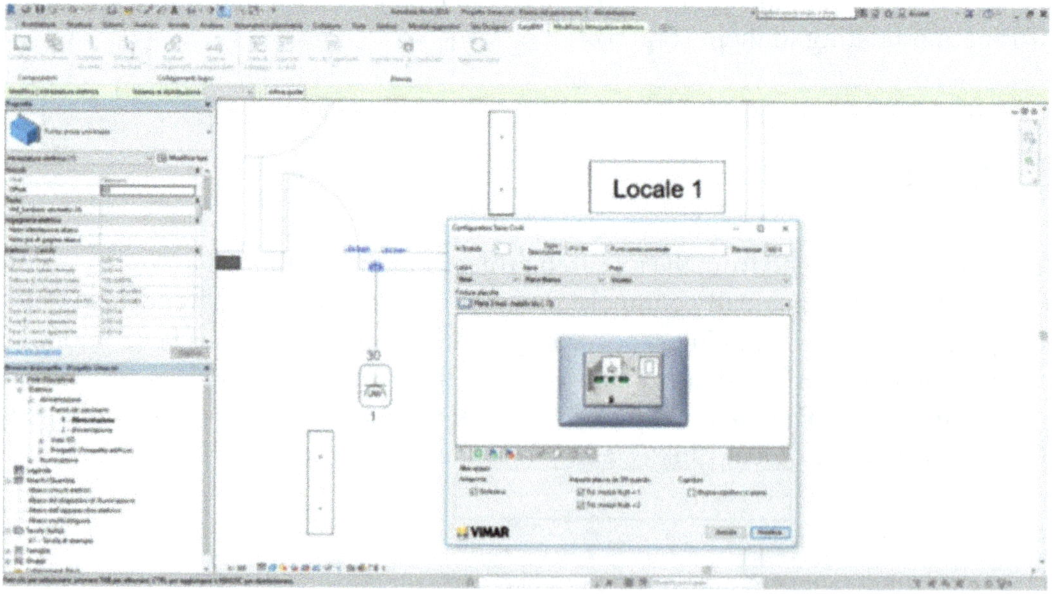

Fig. 7 Examples of graphical interfaces in the case of a Vimar object configuration

available are Eikon, Arkè and Plana) and finally the type of laying (built in traditional walls, incasso in light walls and wall installation with IP protection degree). In the Configurator interface there is also a two-dimensional preview box that allows you to visualize the object you are composing and to modify the electrical modules through the commands in the bar. Through the use of these commands you can:add a new device; automatically add the hole covers in the modularity left free; remove all the hole covers added to the composition; duplicate an existing device; replace the device with another; replace the device with a similar one for function; delete selected devices; assign an abbreviation (tag) to the selected device; change the position of the device within the composition.

Through the use of the plug-in it is possible to create logic circuits assigning tags to the objects to be connected (composition, switchboard and utilities) (Fig. 8).

It was decided to implement this function in the plug-in because, during the research and development phase, there were some bugs using the electric connections of Revit MEP. For example, it is not possible to connect a user to multiple command devices (such as a light point controlled by two or more different points). For this reason, it was decided to insert this function of the plug-in into the standard Revit command until Autodesk will implement the Electrical MEP section of Revit.

Once the logical connections have been made, the plug-in allows users to extrapolate the "wiring list", an abacus created ad hoc by plug-ins with all the parameter information that allows for the unequivocal identification of the composition and the utilities connected to it within the project. Without the help of the plug-in, it would be impossible for the user to be able to compose an abacus with the same information, as

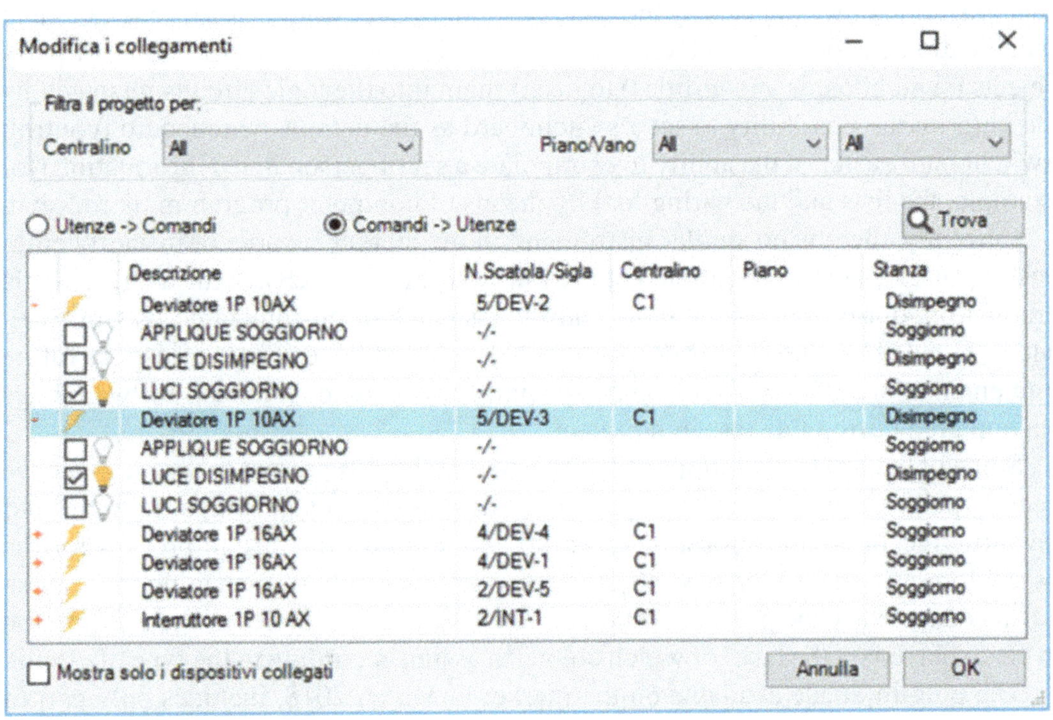

Fig. 8 EasyBIM logic links menu

they belong to multiple categories of Revit objects. Finally, in the "separate" section of the plug-in you can select different types of lists (reports) to be displayed in Revit in the form of Abachi or that are exportable in XLS format.

The typologies of selectable lists are conceived in order to be able to participate in public procurement (generating items of specifications without reference to the brand and model) either for preliminary and final projects, or to be able to make a calculation for resources (material list).

5 Conclusion

Most of The MEP electrical objects currently modeled and available through BIM object libraries satisfy a number of design requirements that are different than those of the Italian national market. The main advantages related to the use of the plug-in are related to the possibility of creating and managing the logical connections (electrical circuits) between the objects and the project users, overcoming the current limitations of the Revit Software.

Moreover, an instantaneous generation of infinite compositions is possible with contained weights (few KB) with their informative attributes; there is also the possibility to create both generic objects (without reference to the manufacturer or commercial series) and Vimar objects selectable from the catalogue; along with the automatic assignment of the labels with the respective 2D electrical symbols, with the shape of the frame that varies according to the commercial series; generation and customization of wiring lists by the user; generation of custom lists for schedules and specifications, exportable in Excel; ability to assign parameters (tags) to third-party objects (switchboards and utilities) to insert them into the logic circuits managed by the plug-in; the possibility to set a switchboard as the default switchboard (control: Switchboard current); the ability to extrapolate a set of personalized information (for example, the lists and the wiring list) by means of computer programming in Revit.

Currently, there is no similar instrument on the market capable of instantly composing an electric composition with a low weight (few KB), with a 2D graphic production, optimized for the table and the issuance of customizable lists in Revit and in Excel. The plug-in was designed and developed to be able to transpose future implementations of Revit MEP, and especially solve some Autodesk software bugs encountered during this research project.

Some of the implementations of the current commands present in the plug-in have been hypothesized. For example, you could develop a management system of your favourite/recent compositions, to speed up and simplify the design process for the user. Moreover, one of the next objectives is the possibility of managing through an advanced interface the personalization of labels such as the choice of the information to be displayed on the label or which colour to assign according to the type of circuit.

The plug-in, made available on the market in March 2018, includes only part of the Vimar catalogue. The post-launch phase envisages the modelling and informative management of the objects of the Domotic series and the video entry categories.

References

BIM: a cosa serve e chi lo userà, Marra A., Edilportale, 2015
BIM: metodi e strumenti, Pavan A., Mirarchi C., Giani M., Tecniche nuove, 2017
BIMForum Specification. Level of development, specification: The guide, Part 1, 2 (2017)
Il BIM e la digitalizzazione dei prodotti, Marra A., Edilportalc, 2018
http://www.bimidea.it/
http://www.bebim.it/2018/03/18/standardizzare-gli-oggetti-bim-la-proposta-di-nbs/
https://www.nationalbimlibrary.com/en/nbs-bim-object-standard
http://www.ibimi.it/lod-livello-di-dettaglio-per-il-bim/
https://harpaceas.it/il-bim-e-il-concetto-di-dettaglio-e-lod/
https://www.ingenio-web.it/18667-sistema-dei-lod-italiano-uni-11337-4-2017
https://www.bimandco.com/it/management-lod
https://bimforum.org/lod/
https://www.ingenio-web.it/18926-i-loi-nelle-scale-di-lod-usa-uk-ita
https://store.mepcontent.com/it/product/details/5
https://store.mepcontent.com/it/product/details/44
https://www.gewiss.com/it/it/services/support/softwares/plugin-suite/revit
https://www.vimar.com/it/it/easybim-revit-autodesk-vimar-14955162.html

Standards and Laws

Level of Development (LOD) Specification: 2019
NBS BIM Object Standard
Norma CEI 64-8:2011 Guida alla Norma CEI 64-8, soluzioni ABB per gli impianti residenziali
Norma CEI 64-8:2007 Norma CEI 64-8 per impianti elettrici utilizzatori
Pass1192-2:2013 Specification for Information Management for the capital/delivery phase of construction projects using Building Information Modelling
UNI 11337-1:2017 Edilizia e opere di ingegneria civile - Gestione digitale dei processi informativi delle costruzioni - Parte 1: Modelli, elaborati e oggetti informativi per prodotti e processi
UNI 11337-4:2017 Edilizia e opere di ingegneria civile - Gestione digitale dei processi informativi delle costruzioni- Parte 4: Evoluzione e sviluppo informativo dei modelli, elaborati e oggetti

Da.Ma.Tra: Material Traceability Database

Ilaria Oberti and Ingrid Paoletti

Abstract The chapter summarises the content and objectives of the Da.Ma.Tra (Material Traceability Database) research project, funded by the Lombardy Region within the "Smart Living" competition call. The purpose of the project is to build a web-based digital platform prototype that can handle the constructive traceability of materials for civil buildings. The project's attention to new materials has a specific focus on the use of bio-based materials and agricultural waste in construction. The economic value of this waste can be increased by encouraging reuse and recycling. This aspect places the research at the edge of the initiatives that favour the concept of sustainable architecture within the broader vision of a circular economy.

Keywords Traceability · Materials · Innovation · Reuse · Recycling

1 Introduction

The possibility of tracking products and artefacts is very common in areas that are far from architecture; for example, consider the food products (Engelseth 2013) or the raw materials of industrial components. Today, however, in virtue of a renewed interest, the development of dedicated standards and the greater public awareness make it increasingly important to also trace the materials in construction.

Thomas Rau states that it is time to imagine material rights, thus making sure not to neglect the effects of our economic activity on the planet and giving each material a sort of identity card to avoid it becoming anonymous as waste and losing track of it. To maintain this system, products must be consumed in ever-increasing amounts (Rau and Oberhuber 2019).

Major global issues are linked to our economic system organised in a linear way where we extract, use and design without thinking about raw material waste. For this

I. Oberti (✉) · I. Paoletti
Architecture, Built Environment and Construction Engineering—ABC Department, Politecnico di Milano, Milan, Italy
e-mail: ilaria.oberti@polimi.it

B. Daniotti et al. (eds.), *Digital Transformation of the Design, Construction and Management Processes of the Built Environment*, Research for Development,
https://doi.org/10.1007/978-3-030-33570-0_8

reason, knowledge and traceability of materials can become an information tool, but also one of control and awareness, which can move new economic levers.

The idea and the goal of the research project is, therefore, to develop a simple yet effective method to trace the materials present in a building to be able to know always their origin, and therefore the need for maintenance, replacement or disposal.

2 The Research Project

The Da.Ma.Tra (Material Traceability Database) project involves the construction of a web-based digital platform prototype that manages the constructive traceability of materials for buildings. In particular, the new service product, developed on the basis of Open Innovation, is directed towards the use of natural materials and agricultural waste as well as naturally based components.

The project's attention to new materials has a specific focus on the use of agricultural waste in construction. The economic value of this waste can be increased by encouraging reuse and recycling.

The project builds on a previous experience in public shipbuilding: Mo.Ma.Tra (Material Movement Traceability) which concerns a prototype software platform that manages the constructive traceability of a roadway infrastructure, developed with the logic of Open Innovation, and which has already been submitted to the first trial of an initial set of road works.

Since control instances, like safety and efficiency, are not fundamental for public construction and do not stop at the time of construction, there was a desire c model and the Mo.Ma.Tra. experimentation along the following lines:

- extension to private civil construction;
- extension to the operational time of the ordinary and extraordinary exercise of the buildings' life;
- extension to the use of unconventional materials in construction, particularly bio-based materials and straw.

Project objectives are therefore to

- prepare a new prototype in accordance with the above-mentioned extensions that specifically take account of construction methods and innovative type materials;
- carry out an experiment on a significant set of buildings;
- pre-market the new software solution in the form of a product-service.

This development takes place with the integration of ICT technologies, with the aim of increasing safety and the functional and operational efficiency of the buildings.

In particular, the development issue which was directly addressed is "Development of innovative technologies for the industrialisation of production processes in construction and for the integration of information between the various parties involved in the construction supply chain, through the development of innovative digital solutions", namely the Da.Ma.Tra. software module, which is inserted as a documentation and management element with BIM systems.

The level of the technology maturity of this project has risen from a TRL (Technology Readiness Levels) 4 value (Component and/or Breadboard Laboratory Validated) to the completion of level 7 on a scale of 9 (Heder 2017).

It is thus understandable that a system like Da.Ma.Tra. intends to respond in a timely manner to the need of providing BIM model systems dedicated to private civil construction design, and in particular to the precise traceability of each moment and material of the time of construction.

The opening at this stage to the use of innovative materials and technologies has important repercussions for the project. These relate to the ecological impact of the materials used in the building industry and, in particular, in the assembly phase within the building itself, as well as producing a significant documentary system of the use of these new materials and technologies useful to evaluate their functional-energy value from the moment they are installed.

Da.Ma.Tra. makes it possible to record building energy and the building's static characteristics from the time of construction (or restructuring), in order to provide timely information on the building's history in case of damage, disasters or earthquakes, as well as in subsequent modernisation, adaptation, renovation and restoration interventions.

Da.Ma.Tra. specifically addresses the documentation of new building construction, but it can also be applied to the documentation of interventions on buildings that are already built. Thus, it provides a significant contribution to the precise documentation concerning seismic adjustments, building renovations, and energy and/or structural upgrading.

Finally, since the software solution to be implemented has a configurable architecture such as SAAS, through a hypertext protocol secure interface, this project aims to also contribute to the intelligent management of information via the cloud, laying the basis for the implementation of intelligent analysis modules of aggregate information for all the buildings documented by Da.Ma.Tra. (Big data analysis).

However, this software solution's real forte lies in the possibility of effectively supporting the infrastructure's maintenance management once constructed. The detailed, documentation of the time and the point of origin of each material, equipment, system component, facilitates the planning of ordinary operations and promptly and efficiently intervenes in the event of a failure; this is currently not possible because the modelled property of infrastructure is not sufficiently documented, organised or available.

Given that this feature is able to respond to a need for the control and quality of any type of construction, the company is interested in extending the model to a new software platform version which would support the following improvements:

- the incorporation of the actions and materials of private buildings;
- a more stringent extension during building operations.

For the best implementation of these improvements, the project partnership consisted of companies and the university using the following criteria:

- the direct involvement of the company that developed the first software version (Mo.Ma.Tra);
- the acquisition of the latest innovations in the field of architecture and of new materials;
- the identification of a company in the construction sector has an innovative approach, capable of providing first-hand information on the construction dynamics of private buildings and for field testing the new platform.

This project stems from a liaison between companies, research centres and institutions promoted by Confartigianato Lombardia. In fact, this organisation is in this case responsible for scouting category project proposals, matching skills and the dialogue with the academic entity, in particular, the Politecnico di Milano. This project, therefore, assumes a paradigm value of a dynamic, which is to be encouraged, for the involvement of SMEs around the themes of innovation in a fruitful exchange with universities and research centres and in a perspective of a system spill-over on the entire artisan sector. Even before the specific impact on the construction sector supply chain, it is therefore intended to emphasise this project's "industry" scope in a sector such as construction, which has a low integration and innovation rate.

The following points can be highlighted as regards this research's direct impact:

- the meeting of the moment of construction and that of documentation in the act of building, in direct connection with upstream planning and downstream management;
- the availability of a BIM module for direct and immediate use;
- the preparation of a platform capable of connecting the different actors in the construction industry, around each individual construction, with the intrinsic increase in control and safety, and therefore, in the final analysis, the qualitative value of a building.

The platform becomes a synthesis and macro-assessment tool at the time in which it enables the intelligent aggregation of information of all the surveyed and documented buildings, with clear advantages in the field, in urban planning and in the area of statistics on materials and technologies.

3 Innovative Ecological Materials and the Circular Economy

In its broad spectrum, the research project aims to promote a specific focus on innovative materials and products to be used in the building sector, so that through the achievement of the traceability objective, it is possible to build an archive of constantly evolving materials.

In particular, the focus is on all those materials that include waste from the agricultural sector among their constituent raw materials. This pre-consumer waste may

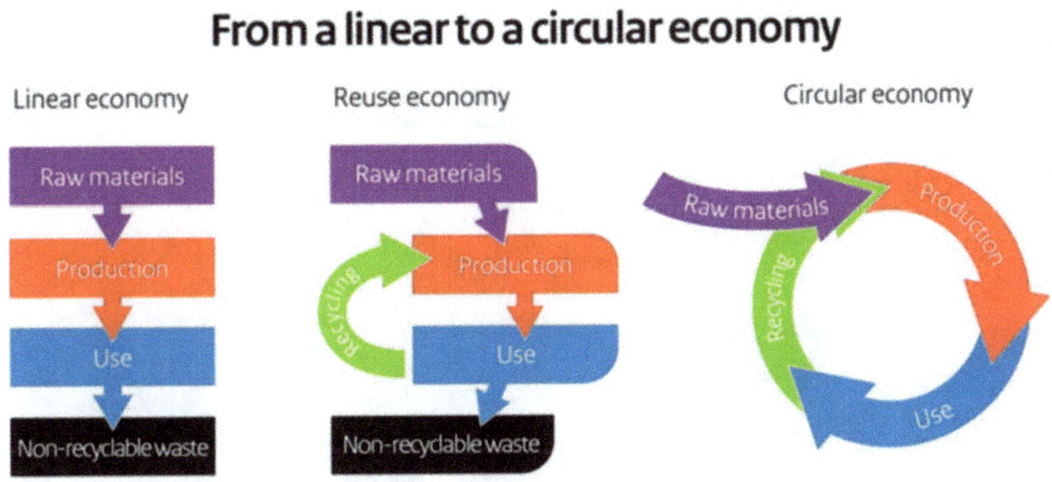

Fig. 1 Linear approach and circular economy approach. (*Source* www.government.nl/topics/circular-economy/from-a-linear-to-a-circular-economy)

constitute a valuable resource for the building industry, allowing it to produce and develop materials that are alternatives to those that are synthetic.

This aspect places the research project in the wake of the initiatives that favour the concept of sustainable architecture within the broader vision of the circular economy, as opposed to the linear economy (Fig. 1).

The first thoughts on the circular economy were produced by a Swiss architect, Walter R. Stahel, in 1976 and included in a report presented to the European Commission, entitled "The Potential for Substituting Manpower for Energy". The salient aspect of this new economic approach is the ability for self-regeneration within an industrial system through the use of renewable energies, limiting the use of chemical substances that prevent or restrict the reuse of products and the elimination of waste (McDonough and Braungart 2013). Referring to the two main types of material flows, according to the circular economy concept, biological materials must be capable of being reintegrated into the biosphere, and technical materials must be destined for upgrading, limiting interference with the biosphere (Ellen McArthur Foundation 2013, 2015) (Fig. 2).

These considerations are to be included in the design of materials, products, components and systems, so that the importance of activating circular flows in the use of resources is assessed (European Environment Agency 2016).

According to Ellen MacArthur Foundation's studies, now shared by the global scientific community, it is necessary to consider three basic principles in order to be able to guide actions in a circular economy perspective.

The first refers to "the preservation and the increase of natural capital", with the goal of keeping the availability of non-renewable resources under control and to balance the flows of renewable resources, replacing fossil fuels with renewable energy sources, or by working in such a way that the raw materials, at the end of the product life cycle, are not lost but reintegrated into the ecosystem. This means

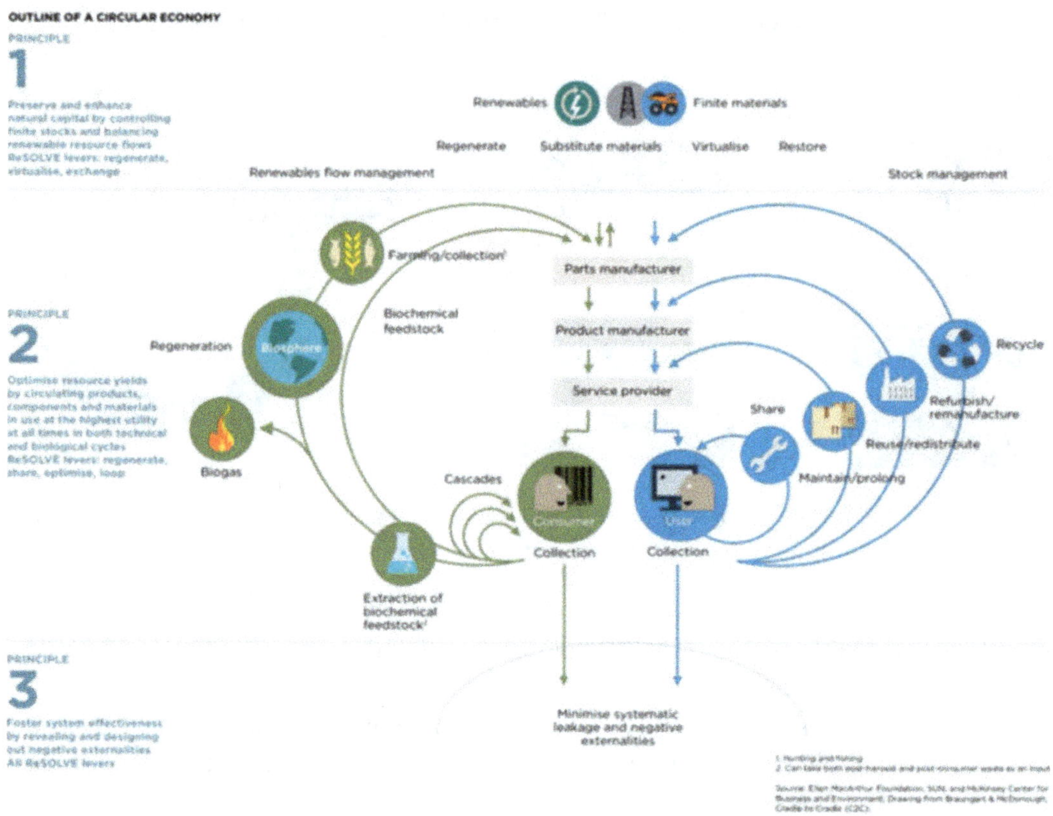

Fig. 2 Diagram of flows within the circular economy. (*Source* www.ellenmcarthurfoundation.org/circular-economy/infographic)

that products should be designed, and then manufactured, to make disassembly and reuse operations easier.

The second principle is linked to the "optimisation of resource efficiency" through the circulation of products that reach the highest level of utility in all the stages of the life cycle, both within the technological cycle and the biological cycle.

Finally, the last principle is that of "enhancing the system's effectiveness" through the identification and elimination of negative aspects related to the use of resources, such as the production of harmful substances, the pollution of the air, water and soil, and the greenhouse effect.

Certainly, the passage from the linear economic approach, practiced until recently, to the circular approach, which has emerged today, involves great efforts from everybody (Prieto-Sandoval et al. 2018) and requires the need to operate in two directions. One should be focussed on the systemic enhancement of the organisation between the different phases of the construction process and between the operators involved therein. The other, however, should push to encourage cooperation between different productive sectors external to the building sector.

This second direction opens up to the theme of pre-consumer waste management which can only find valid responses through the creation of horizontal supply chains that will allow a production sector to use as raw material that another it considers to be waste (Talamo and Migliore 2017).

Considered the continuous depletion of resources and the consequent environmental degradation caused by the increasing demand and use of virgin raw materials (UNEP 2019), the promotion of the recovery strategy and the utilisation of pre-consumer waste must be carefully evaluated. This manner of considering waste implies a revision of production processes to enable the emergence of a system logic over the linear system. In this new vision, the manufacturing process of an object, whatever it may be, is no longer analysed as a set of sequential actions, but is examined in its entirety and related to other interconnected systems.

In this logic, the waste, transitioning from this condition to a secondary raw material, can be introduced into the supply chain where it originated or into parallel supply chains, by "jumping the chain".

And this is the circumstance that the Da.Ma.Tra. project hopes to bring to light, in particular by tracking materials and building products that include, amongst others, raw materials from pre-consumer agricultural waste. The agri-food sector is potentially capable of providing material in circular economy processes that can be developed particularly for the construction sector.

Studies, already started some time ago, are specifically directed towards the use of waste from the cultivation of cereals, and particularly, rice, with an estimated global production of 501.4 million tonnes for 2018/2019 (USDA 2019).

The by-products and waste from rice processing are the husk, bran and rice straw, in particular. As regards the husk and bran, there is widespread use oriented towards the production of energy from biomass or intended for the food and farming sector. For rice straw, however, the management takes place within the same process as rice cultivation and disposal takes place through two techniques: the burning of the straw in the field and burying it in the land in order to restore to the soil some of the nutrients that were taken during the plant's growth.

The attempt to promote this waste for the production of building materials and products has been taking place since the first decades of the twentieth century, having studied the possibility of compressing the rice straw at high temperatures to obtain panels with variable density (Wei et al. 2015).

Currently, the main experiments concern the production sector for thermal insulation materials, the advantages of which can be analysed under different points of view: the consumption of other raw materials avoided for the production of alternative thermal insulation; the decrease in the consumption of primary energy required for heating buildings; the reduction of carbon dioxide into the atmosphere; the elimination of the negative effects linked to freely burning rice straw. Although the creation of materials to be destined for the production of thermal insulation seems to be the most pursued, developments can be highlighted for other innovative materials, in their composition, and with a reduced environmental impact, such as lightweight screed, mortar, and plaster with high thermal performance.

With the Da.Ma.Tra. the project, the main objective is the traceability of materials presents inside a building in order to know their origin, the need for maintenance, replacement or disposal with a view to more effective management of the same building. With the project, it is hoped that a rich archive can be created that will gather building materials and products with high innovation characteristics and reduced environmental impact, so as to respond to the principles of the circular economy.

4 Conclusions

The role of the collaboration between Politecnico di Milano and the mentioned companies has generated a transfer of information that can increase an innovative workflow.

The Politecnico di Milano has accompanied partner companies in this direction to identify the characteristic parameters of traditional materials (constant information to be identified in specific harmonised standards at European level) and in the analysis of the opportunity of finding/identifying characteristic data of those architectural and technical elements that are emerging in the panorama of modern construction and for which there are no uniform regulatory views.

Specifically encoded are the parameters of the main innovative and non-conventional insulations, as well as the finishing surfaces for bioconstruction and for sustainable architecture, which is predominantly dry assembled.

In order to allow the project tool to be able to cover a wider spectrum of interventions (including experimental ones), there has been a close integration between research and industry partners which facilitated the identification of elements for possible dynamism in the planning and implementation phases of the architectural/engineering projects that are typical in civil construction.

The systematisation of transmitted data, with those acquired by the companies, has allowed for appropriate software libraries to be created which, from the perspective of the Building Information Model and placed alongside the tracking processes, facilitate the verification of the output quantitative and qualitative parameters of the construction process. All this is to allow the testing and implementation of performance in terms of control, safety and energy efficiency at the different periods of the building's life.

In fact, and in conclusion, shared expertise was created on the tool capable of anticipating a product-service that will support the specific software on the documented use of new materials and new construction technologies.

This project's output-product was, therefore, a product-service, targeting construction companies, capable of interfacing with BIM systems on the one hand, and with real estate economic-financial management systems on the other, which is nowadays increasingly crucial.

References

Ellen McArthur Foundation (2013) Towards the circular economy. Economic and business rationale for an accelerated transition

Ellen McArthur Foundation (2015) Growth within: a circular economy vision for a competitive Europe

Engelseth P (2013) Food product traceability in value networks. Nova Science Publishers Inc, New York

European Environment Agency (2016) Circular economy in Europe. Developing the knowledge base, EEA Report, n. 2

Heder M (2017) From NASA to EU: the evolution of the TRL scale in Public Sector Innovation. Innov J: Public Sect Innov J 22(2), Article n. 3

McDonough W, Braungart M (2013) The upcycle: beyond sustainability. Designing Abundance. North Point Press, New York

Prieto-Sandoval V, Jaca C, Ormazabal M (2018) Towards a consensus on the circular economy. J Clean Prod 179:605–615

Rau T, Oberhuber S (2019) Material Matters. L'importanza della materia – Un'alternativa al sovrasfruttamento. Edizioni Ambiente, Milano

Talamo C, Migliore M (2017) Le utilità dell'inutile. Economia circolare e strategie di riciclo dei rifiuti pre-consumo per il settore edilizio. Maggioli Editore, Sant'Arcangelo di Romagna

United Nations Environment Programme (2019) Global resources outlook 2019. Natural resources for the future we want

United States Department of Agriculture (2019) Rice outlook. Economic Research Service, RCS-19D | April 11, 2019

Wei K, Chen M, Zhou X, Dai Z (2015) Development and performance evaluation of a new thermal insulation material from rice straw using high frequency hot-pressing. Energy Build 87:116–122

Natural Language Processing for Information and Project Management

Giuseppe Martino Di Giuda, Mirko Locatelli, Marco Schievano, Laura Pellegrini, Giulia Pattini, Paolo Ettore Giana and Elena Seghezzi

Abstract The purpose of the paper is to investigate the state of the art of textual translation theories, methods, and tools into formal and numerical requirements to support information modelling and project management process. Natural Language Processing helps translate text requirements into numerical terms that are necessary for the application and success of information modelling and management in a data-driven process.

Keywords Project and information management · Natural language processing · Data-driven process · Artificial Intelligence

1 Introduction

The purpose of the article is to investigate the state of the art about theories, methods, and tools for translating documents into numerical requirements and possible applications in the field of information modeling and project management. Information modeling requires a precise and comprehensive definition of the initial requirements; the computational definition of the initial requirements is fundamental for the good application of information modeling and management. Natural language processing (NLP) can be applied to translate text into numerical data. In order to understand the proposed methodology and its possible applications, it is necessary to explain basic theories and the most recent developments and applications of NLP. We present case studies of applications in various fields focusing on possible applications in the AECO sector, with a focus on possible data-driven management of the construction process. A discussion on possible applications and further developments is also presented.

G. M. Di Giuda (✉) · M. Locatelli · M. Schievano · L. Pellegrini · G. Pattini · P. E. Giana · E. Seghezzi
Architecture, Built Environment and Construction Engineering—ABC Department, Politecnico di Milano, Milan, Italy
e-mail: giuseppe.digiuda@polimi.it

B. Daniotti et al. (eds.), *Digital Transformation of the Design, Construction and Management Processes of the Built Environment*, Research for Development,
https://doi.org/10.1007/978-3-030-33570-0_9

2 Natural Language Processing (NLP): Rule Based, Statistical and Deep NLP

Natural Language Processing (NLP) is an interdisciplinary field involving humanistic, statistical–mathematical, and computer skills. The aim of NLP is to process languages using computers. The human language can be defined as natural because it is ambiguous and changeable. On the contrary, machine language is defined as formal because it is unambiguous and internationally recognized. The NLP must deal optimally with the ambiguity, imprecision, and lack of data inherent in natural language. NLP tasks may take two main approaches: a machine learning (ML)-based approach (i) or a rule-based/statistical approach (ii). An ML-based approach uses ML algorithms for text processing (Pradhan et al. 2004), whereas a rule-based approach uses manually coded rules (Soysal et al. 2010). NLP approaches could be either shallow or deep. Shallow NLP conducts partial analysis of a sentence or extracts partial, specific information from a sentence. Deep NLP aims at full-sentence analysis toward capturing the entire meaning of a sentence (Zouaq 2011).

A table comparing the different approaches of the NLP has been created in order to identify the most suitable one for the numerical translation of the requirements (Table 1).

As shown below, the statistical approach to the NLP seems to have more advantages than traditional/manual rule based. However, among the various NLP techniques, those based on Artificial Neural Networks seem to have more chance of success.

ML-based methods efficiently manage the sparsity and non-structuring of learning data. Among the techniques of ML for NLP, those based on ANN ensure computational processing of text documents better, because they can respect the complexity, articulation, and multidimensionality of human natural language. By translating text into machine language, text information can be managed and used with methods, techniques, and tools typical of project and information management. In summary, NLP, based on ANN, transforms text documents into structured information resources (Callison-Bourne and Osborne 2003).

Despite the numerous advantages of the ANN compared to the rule based and statistics NLP, it offers few explanations on the relations found between the data and the connection among neurons with the output data. The phenomenon is called black-box effect which makes difficult to explain what is learned from the net (Paliwal and Kumar 2011; Waziri et al. 2017).

The latest efforts in the field of AI aim at overcoming the black box effect and the production of explainable AI model called the XAI model (Gunning 2017) (Fig. 1).

Table 1 Advantages and disadvantages of the various NLP approaches

NLP approach	Advantages	Disadvantages
Rule based	It produces outputs in relation to the data entered	Long time to implement and test a prototype
		It produces outputs in relation to the data entered
		Manual system creation, time-consuming, and expensive creation process
Statistical based	Rapid prototyping and testing	Robust system (Junqua and Van Noord 2001): it will always produce outputs, regardless of the inputs entered
	Robust system (Junqua and Van Noord 2001): it will always produce outputs, regardless of the inputs entered	
	Automated system creation, cheap creation process	
Artificial Neural Network based	Rapid prototyping and testing	ANNs operate as a black box (Jain and Pathak 2014)
	Robust system: ANNs can manage noisy and ambiguous data, prone to error or incomplete (Pradhan et al. 2004; Sivanandam and Deepa 2006; Rumelhart 1986)	
	Structure similar to the human brain, ANNs are very effective for solving a complex problem	
	ANNs have the ability to overcome the lack of data, they "Fill in the blank". This makes it particularly suitable for use in the NLP (Waziri et al. 2017)	

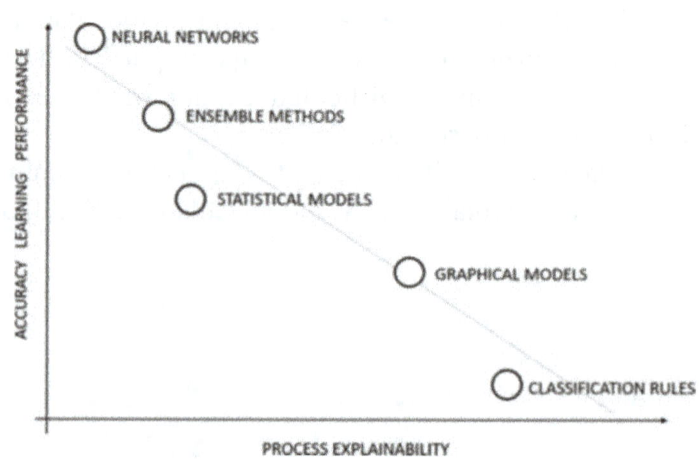

Fig. 1 Explainability and performance qualitative graph of AI learning method (Gunning 2017)

3 NLP Application for Project Management in AECO Sector

The increasing complexity and size of projects in the AEC sector make it difficult to identify and verify initial requirements expressed in natural language. As a result, there are numerous errors and shortcomings in the definition of requirements at the early stages of concept and design, with inefficiencies in terms of time, cost and quality (Frenette and Kyriakidis 2016). The use of NLP can help the project manager and the client to express project requirements in alphanumerical and quantifiable terms avoiding misunderstanding and increasing the project's chances of success. The use of NLP in the early stages of defining project requirements can be therefore considered a risk mitigation technique.

ANN approach to NLP allows to define requirements and predictions in order to monitor the progress of the project. This, in turn, leads to an optimization of the definition of requirements and the assessment of the project progress.

Alphanumeric translation of text documents into formal and structured data is a prerequisite for information modeling. The initial phase of defining requirements is fundamental for the good application and effectiveness of the information modeling method. The NLP for requirements definition can be integrated with the information modeling and management of the preliminary phase of the design and construction process (Zhang and El-Gohary 2015).

NLP approach can reduce time and cost in defining requirements by avoiding errors, loss of information and ambiguities in defining project objectives (Zhang and El-Gohary 2015).

The NLP approach can support Automated Compliance Checking in a BIM-based process (Zhang and El-Gohary 2015). Construction projects must comply with a variety of standards code. The manual conformity control process is time-consuming, costly, and error-prone (Han et al. 1998; Nguyen 2005). Automated Compliance Checking (ACC) and NLP should reduce Code Checking time, costs, and errors (Salama and El-Gohary 2013; Tan et al. 2010).

Some authors (Zhang and El-Gohary 2015) point out that, in addition, the proposed method has other potential benefits:

- allowing potential non-compliance cases to be identified in advance, which could save significant time and cost due to changes and/or rework (Ding et al. 2006),
- promoting the adoption of Building Information Modelling (BIM) and increase the cumulative benefits of adopting BIM as BIM would allow ACC (Pocas Martins and Abrantes 2010),
- enabling more efficient integration of stakeholder inputs into the design and exploration of what-if design scenarios. As a result, experimenting different design options and checking for compliance would be more efficient in terms of time (Frenette and Kyriakidis 2016),
- reducing violations of regulations, due to easier and more frequent Compliance Checking (CC) (Zhong et al. 2012).

The survey carried out shows that NLP can support the automatic identification of the limits, according to the regulations and support Automated Code Checking. This method applied to the BIM model can lead to a substantial reduction in time and errors (Zhang and El-Gohary 2015).

4 An Application: NLP for Risk Management

This paragraph proposes several applications of NLP in a specific construction related field. Risk management was chosen, as it reflects several declinations of NLP use.

Referring to safety risk management, NLP is applied to Case Base Reasoning. (CBR). CBR is an important approach in safety risk management of construction projects. It emphasizes that previous knowledge and experience of accidents and risks are extremely valuable and could help to avoid similar risks in new situations. CBR allows the creation and increasing size of Databases on construction accidents. These documents are written in natural language; as a consequence, retrieving information quickly and accurately from the database is still the main challenge. In order to improve the efficiency and performance of data recovery, a Natural Language Processing approach is proposed (Zou et al. 2017).

NLP gives the advantage to ensure an easier recovery of CBR knowledge and cases.

As regard to construction risk management, NLP approach is used to analyze bid document.

The bidding process takes place in the early stages of a construction project. Bidders should fully understand the uncertainties of the project before making decisions. The risk of uncertainty of construction projects, in turn, is determined by the content of the bid document. If the information provided in the bid notice is not accurate and clear, the uncertainty of the projects increases (Lee and Yi 2017).

NLP approach is proposed to predict risks in the bidding process of construction projects, by analyzing the uncertainty of the bidding document and using it as a factor to predict the bidding risk of a project. The model for forecasting tender risk was conducted using the pre-bid clarification information. Text mining was carried out on pre-bid documents, which are in an unstructured text data format, and the results were used as the main influencing factors for the risk forecasting models (Lee and Yi 2017).

In the field of safety and pre-bid risk mitigation, NLP is a useful method and tool to support the project manager in project risk mitigation.

The last application of NLP refers to procurement risk management.

As the size and complexity of construction projects significantly increased, the number of disputes among the parties involved during construction work is constantly increasing. To avoid such disputes, participants need to be sure of their contractual positions and rights. Most international construction projects require contract management teams to examine all possible contract risks during tendering periods. However, it is very difficult to review a large number of contracts in a short period of time. NLP approach is used to propose a model of automatic extraction of poisonous clauses (Lee et al. 2019).

The NLP approach proposed is, therefore, effective in automatically reading and extracting poisonous clauses from construction contracts. The algorithm, suitably modified, can be used to read and extract data from documents other than construction contracts.

5 Conclusions and Further Developments: NLP for Requirements Engineering

Given the importance of the correct and complete definition of the initial project requirements for the successful application of project and information management, the NLP approach to support the translation of the requirements into numerical and alphanumerical terms could help reducing the gap between expected and actual quality. The presented case studies show that, methods, techniques, and application cases of NLP have already been tested. However, no case of systemic application to construction project has been found since the early stages of planning. The use of NLP for the definition of initial requirements would have a greater impact if applied at an early stage. The NLP approach could be used for the definition of the initial requirements of a public client helping the public actor to define requirements on a numerical and alphanumerical basis and not on a simple text basis. Through the alphanumeric requirements translation, the public client would be the main actor in a Data-driven construction process, increasing its ability to understand, manage and direct the outcome of the design. The numerical translation of the requirements would make the entire process computable, and digitally manageable, which would see the client as able to contribute directly to the design process (Ciribini 2016). NLP supports requirements engineering by transforming the classic qualitative demand based on text data into a computational demand based on formal and structured data. During a data-driven process, the monitoring of the objectives to be achieved can be more effective and immediate, reducing the risk of overcoming time and costs and not reaching the expected quality level.

Possible further development of requirements translated into a computational form, through NLP, is the possibility of being readable and digitally managed. In this way, the design and construction process can be managed digitally from the earliest stages. Text translated into informal language data using NLP approach can be used to train an Artificial Intelligence. The A.I. needs a huge quantity of data to be trained. The amount of data needed for training A.I. can be an application limit. The AI, if properly instructed, could support the evaluation phase of the project to verify the degree of compliance of the project with the requests coming from the translation of text document. The system structured in this way would be configured as a decision support system for assessing the progress of the project. The monitoring of the project progress would be automatically managed by the AI to avoid errors and costs in economic and temporal terms.

References

Callison-Bourne C, Osborne M (2003) Statistical natural language processing. In: Farghaly A (ed) A handbook for language engineers. CSLI Publications

Ciribini ALC (2016) Information modelling management, BIM e digitalizzazione dell'ambiente costruito. Grafill S.r.l, Palermo

Ding L, Drogemuller R, Rosenman M, Marchant D, Gero J (2006) Automating code checking for building designs Design check. Clients driving innovation: moving ideas into practice. CRC for Construction Innovation, Brisbane, Australia, pp 1–16

Frenette M, Kyriakidis J (2016) Rethink requirements—The Natural language processing approach. Project management.com, 12 July. https://www.projectmanagement.com/webinars/334316/Rethink-Requirements—The-Natural-Language-Processing-Approach

Gunning D (2017) Explainable artificial intelligence (xai). Defense Advanced Research Projects Agency (DARPA)

Han CS, Kunz JC, Law KH (1998) Client/server framework for online building code checking. J Comput Civ Eng 12(4):181–194. 10.1061/(ASCE)0887-3801(1998)

Jain M, Pathak KK (2014) Applications of artificial neural networks in construction engineering and management: a review. Int J Eng Technol Manag Appl Sci 2(3):134–142

Junqua J-C, Van Noord G (2001) Robustness in language and speech technology, vol. 17 of Text, speech and language technology. Kluwer Academic Publishers

Lee JH, Yi JS (2017) Predicting project's uncertainty risk in the bidding process by integrating unstructured text data and structured numerical data using text mining. Appl Sci 7(11):1141. https://doi.org/10.3390/app7111141

Lee JH, Yi JS, Son JW (2019) Development of automatic-extraction model of poisonous clauses in international construction contracts using rule-based NLP. J Comput Civ Eng 33(3):04019003. https://doi.org/10.1061/(ASCE)CP.1943-5487.0000807

Nguyen T (2005) Integrating building code compliance checking into a 3D CAD system. In: Proceedings of international conference on computing in civil engineering, ASCE, Reston, VA, pp 1–12

Paliwal M, Kumar UA (2011) Assessing the contribution of variables in feed forward neural network. Appl Soft Comput 11(4):3690–3696

Pocas Martins JP, Abrantes V (2010) Automated code-checking as a driver of BIM adoption. J Hous Sci 34(4):286–294

Pradhan S, Ward W, Hacioglu K, Martin JH, Jurafsky D (2004) Shallow semantic parsing using support vector machines. In: Proceedings of the NAACL-HLT, association for computational linguistics, East Stroudsburg, PA, pp 233–240

Rumelhart DE (1986) Learning internal representations by error propagation. Parallel Distrib Process 1:318–362

Salama D, El-Gohary N (2013) Semantic text classification for supporting automated compliance checking in construction. J Comput Civ Eng: 04014106. https://doi.org/10.1061/(asce)cp.1943-5487.0000301

Sivanandam SN, Deepa SN (2006) Introduction to neural networks using Matlab 6.0. Tata McGraw-Hill Education, Columbus, United States

Soysal E, Cicekli I, Baykal N (2010) Design and evaluation of an ontology based information extraction system for radiological reports. Comput Biol Med 40(11–12):900–911

Tan X, Hammad A, Fazio P (2010) Automated code compliance checking for building envelope design. J Comput Civ Eng 24(2):203–211. 10.1061/(ASCE)0887-3801(2010)

Waziri BS, Bala K, Bustani SA (2017) Artificial neural networks in construction engineering and management. Int J Arch Eng Constr 6(1). https://doi.org/10.7492/IJAEC.2017.006

Zhang J, El-Gohary NM (2015) Automated information transformation for automated regulatory compliance checking in construction. J Comput Civ Eng 29(4):B4015001. https://doi.org/10.1061/(ASCE)CP.1943-5487.0000427

Zhong BT, Ding LY, Luo HB, Zhou Y, Hu YZ, Hu HM (2012) Ontology-based semantic modeling of regulation constraint for automated construction quality compliance checking. Autom Constr 28:58–70

Zou Y, Kiviniemi A, Jones SW (2017) Retrieving similar cases for construction project risk management using natural language processing techniques. Autom Constr 80(September 2016): 66–76. https://doi.org/10.1016/j.autcon.2017.04.003

Zouaq A (2011) An overview of shallow and deep natural language processing for ontology learning. In: Ontology learning and knowledge discovery using the web: challenges and recent advances. IGI Global, Hershey, PA, pp 16–38

Structuring General Information Specifications for Contracts in Accordance with the UNI 11337:2017 Standard

Claudio Mirarchi, Sonia Lupica Spagnolo, Bruno Daniotti and Alberto Pavan

Abstract Information Specifications are an important tool that allows the Customer or the Commissioning Body to define both the general and the specific rules and strategic information requirements, as well as the general guidelines for the formulation of the Pre-Contract BIM Execution Plan, by the Competitors, and then of the Contract BIM Execution Plan, by the Contractor. In Italy, in public contracts, the commissioning bodies "*can request for new works as well as for recovery or redevelopment interventions or variants, especially for complex jobs, the use of specific electronic methods and tools*" (Legislative Decree No. 50/2016, article 23, paragraph 13) "*such as modelling for building sector and infrastructures*" (Legislative Decree No. 50/2016, article 23, paragraph 1, letter h). These tools use interoperable platforms by means of open, non-proprietary formats (IFC, XML; etc.), in order not to limit competition between technology suppliers and involve specific projects among designers. The use of electronic methods and tools can only be requested from commissioning bodies equipped with suitably trained personnel. The subsequent implementing decree (Infrastructure and Transport, Ministerial Decree No. 560 of 12/01/2017) reaffirms this possibility in article 5 and introduces in article 6 a progressive obligation from 1 January 2019 for complex works depending on the tender amount (starting from > = 100 ml€). From 1 January 2025, the obligation will be extended to all tender amounts. The stringency of such methods and tools depends on the extent of the contract, but some Commissioning Bodies are moving towards applying them voluntarily. For this reason, research activity was conducted to test the possibility of defining the structure of the Standard General Information Specifications, drafted according to the UNI 11337-5-6: 2017, valid for contracts for the Inter-regional Superintendency for Public Works in the Lombardy and Emilia Romagna regions of the Ministry of Infrastructure and Transport. Starting from the analysis of works-only contracts based on the traditional, non-digital (non-BIM) design and through the validation of a case study.

C. Mirarchi (✉) · S. Lupica Spagnolo · B. Daniotti · A. Pavan
Architecture, Built Environment and Construction Engineering—ABC Department, Politecnico di Milano, Milan, Italy
e-mail: claudio.mirarchi@polimi.it

B. Daniotti et al. (eds.), *Digital Transformation of the Design, Construction and Management Processes of the Built Environment*, Research for Development,
https://doi.org/10.1007/978-3-030-33570-0_10

Keywords EIR · BEP · Public procurement · Information management · Interoperability · UNI 11337

1 Introduction

The definition of a Building Information Modelling (BIM) process requires the clear identification of the Client/Commissioning Body information requirements (BSI 2013; UNI 2017c; ISO 2018). Hence, for the purposes of the contract's information flow, in addition to the production of the legal and contractual documents (as mentioned in the tender notice), an assessment will also be conducted on the development of all "multidimensional object-oriented models", (Ministerial Decree No. 560 of 12/01/2017, article 4, paragraph 1), hereinafter: "graphic models" (UNI 2017a), necessary to fulfil the information requirements defined by the Client.

The Employer Information Requirement (EIR) allows the specification of the needs and information requirements of the Client/Commissioning Body for the execution of a public contract for works, supplies or services.

Through the Pre-Contract BIM Execution Plan (Pre-Contract BEP), it is possible to explain and provide specific information on the information management offered by the Competitors in response to the needs and in compliance with the requirements of the Client/Commissioning Body detailed in the EIR.

The following Contract BIM Execution Plan (Contract BEP) defines instead the (operational) planning of the information management to be implemented by the Contractor in response to the needs and in compliance with the requirements of the Client/Commissioning Body detailed in the EIR. Figure 1 presents the awarding process from the call up to the definition of the Contract BEP.

The EIR is of a supplementary nature, on purely informational issues, compared to the other tender documents and contracts.

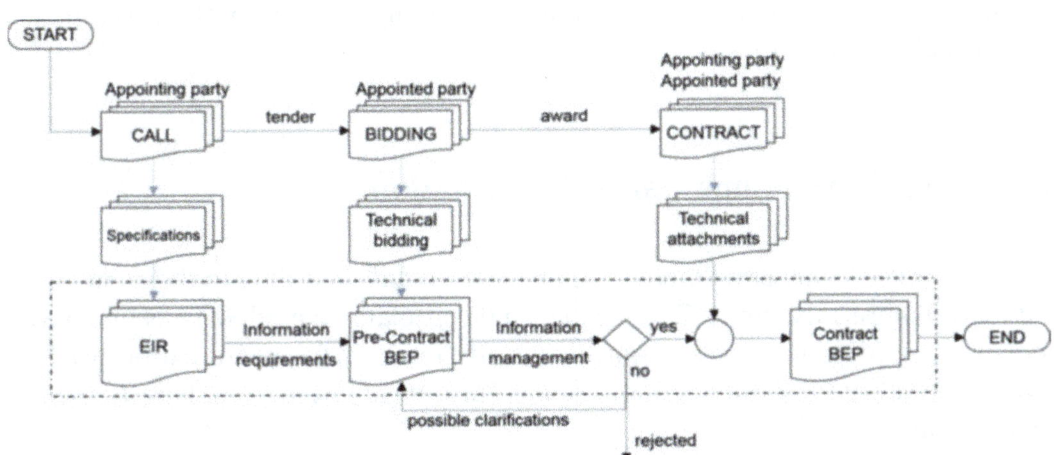

Fig. 1 Award information flow

The EIR, the Pre-Contract BEP and the Contract BEP are contractual documents and may refer to the consensual UNI 11337:2017 standard (parts 1, 3, 4, 5 and 6) (UNI 2017a, b, c, d) for each technical principle.

The EIR devised stipulate that in the Pre-Contract BIM Execution Plan, and consequently in the Contract BIM Execution Plan, the Commissioning Body may require some direct prior experience in the use of *BIM methods and tools* deemed the most significant by the Competitor with regard to the specific intervention and stage of the process. The Pre-Contract BIM Execution Plan, and consequently the Contract BIM Execution Plan, must provide a specific section in which the hardware and software tools that the Successful Bidder intends to use for the following contract in the information modelling activities are explained. These two points are useful for the identification of the Competitors background helping the Client in the evaluation of their capability in performing the works.

For the Contract for which the EIR was drafted, the Client/Commissioning Body should provide the Contractor with specific digital sharing and collaboration spaces that are accessible via the Internet, for data storage, models and digital documentation. In general, these requirements are based on the definition of a Common Data Environment (CDE) that can be related to a Digital Platform for Works Management to promote the coordination of different works by the Commissioning Body. At the time of entering into the contract, the technical specifications and the rules of access to these digital spaces for sharing and collaboration will be defined.

In the Pre-Contract BIM Execution Plan, and consequently in the Contract BIM Execution Plan, other useful supplementary solutions may be proposed for the purpose of improving transparency, consistency and information management of the Contract for which it is drafted.

In addition to these general criteria, the EIR may contain several specifications defined according to the needs and the scope of each specific work. Nevertheless, the identification of a general structure can facilitate the understanding of the requirements from both the Client and the Competitors and promote the standardisation of the information flows in the Client organisation paving the way for the development of data analysis workflows.

2 Models Development and Coordination

In a BIM process, the term "model" can be interpreted according to different scales and may comprise the entire CDE or the single specialistic model. Thus, the development of building information models usually requires the coordination between several specialistic information models that must make reference at least to the following points:

- identification of the common system of reference coordinates;
- development of individual graphic models (UNI 2017a) for all the disciplines involved in the intervention;

- definition of an aggregate graphical model (UNI 2017a) of all the specialistic models;
- coordination and control of all specialistic models by the analysis of interferences (clash detection; (UNI 2017c)) and inconsistencies (code checking; (UNI 2017c));
- creation of information management rules for the contract (Pre-Contract BIM Execution Plan, Contract BIM Execution Plan and related documents) according to the needs and information requirements of the client explained in the EIR;
- creation and management of libraries of (digital) objects necessary for modelling (UNI 2017a);

In the Pre-Contract BIM Execution Plan, and consequently in the Contract BIM Execution Plan, other useful supplementary solutions may be proposed for the purpose of improving transparency, consistency and information management of the Contract.

3 Organization of Graphic Models

Without prejudice to any contrary overriding indication contained in the specific tender documents, for the purpose of information modelling, individual graphic models for each discipline involved in the intervention are required. The following list provide an example of possible specialistic models.

- urban planning and topography—GIS (including terrain modelling, urbanisations, underground utilities and contour viability);
- architecture (including finishes, access for disabled persons and fixed furnishings);
- structures (including steel structures);
- mechanical, plumbing, sanitary and air-conditioning installations;
- electrical installations (including production from renewable sources) and data;
- special installations (including, for example, vertical and horizontal transport, fire prevention systems, etc.);
- energy efficiency and acoustics;
- workers' operational safety (working stage).

In the Pre-Contract BIM Execution Plan, and consequently in the Contract BIM Execution Plan, the Competitors may propose a different organization of the models to optimise, e.g. the model dimensions, the collaboration workflows, the output organization, etc.

4 Specifications for the Use of Objects in Digital Models

In order for the graphic models to be developed by applying modelling criteria allowing them to be read, queried, and subsequently reworked easily, in all cases in which this is possible, the rules for a correct parameterisation of the elements by introducing the appropriate geometric constraints must be abided by. Below are some reference rules that exemplify the general principles described above with reference to the use of digital objects in the development of information models:

- all objects used in the graphic models should be associated with the natural level of localization (according to the specific discipline and regulations);
- all the vertical elements (walls, pillars, etc.) must be modelled as discrete elements in their vertical development according to a subdivision that is consistent with the decomposition of the Working Breakdown Structure (WBS) applied to the specific class of objects;
- all the structural elements are to be bound to the axes associated with them;
- all the plant machines positioned on the ceiling must be associated with the reference level of the discipline/underlying environment.

The definition of specific rules to develop the model allows easier control of the models and can limit the presence of data quality issues that can affect the effectiveness and efficiency of the entire process (Mirarchi and Pavan 2019).

5 Documentation

Without prejudice to any contrary overriding indication contained in the specific tender documents, the graphic documents must be extracted from their corresponding graphic models in case of information modelling. Representations of non-modelled knots and details are allowed solely for the purpose of completing the relevant information on the LOD (see the following sections) of some objects (where precise modelling is not necessary, in order to fulfil the objectives and uses of the envisaged model: safety, interferences, inconsistencies).

The MEP models must allow for extrapolation of both the real-scale representation of the elements (according to the reference LOD), comprising pipes and ducts, and the representation as a single-line diagram for the required set representations.

For non-graphic documents (UNI 2017a), each piece of data contained on or extractable from the models must be able to be extracted without any duplications.

The Contractor must ensure data consistency between the models and the documentation (graphs, documents, multimedia; (UNI 2017a)) of each discipline.

In any case, the Successful Bidder will have to ensure the coordination (and consistency) produced by the data for each document even though not extracted from models.

6 Data Sharing and Collaboration

The EIR must consider the processes and tools for data sharing and collaboration. Hence, it must contain the requirements related to the definition of the Common Data Environment (CDE) that should be provided by the Client/Commissioning Body according to the need of organising multiple works and promote standardised processes in its organization.

The Contractor shall deposit (uploading into the CDE) each graphic model and digital document concerning the specific work by the scheduled end of each level of processing/modelling laid down in the EIR. The uploading stages must be defined according to an information delivery plan that considers the verification and validation processes.

Thus, in order to ensure maximum efficiency in sharing the data, the following information must always be identifiable (according to UNI (2017b)):

- "*State of definition*" (L0; L1; L2; L3);
- "*State of approval*" (A0, L1, L2, L3).

The Contractor should specify in its Pre-Contract BIM Execution Plan how it intends to pursue this requirement even regardless of the CDE made available by the Client/Commissioning Body.

The modelling flow proposed in the EIR is synthesised in Figs. 2 and 3 with specific reference to the post-awarding phase (before the effective start of the works) and the works execution phase until the closeout (As-Built Work In Progress and Final As-Built).

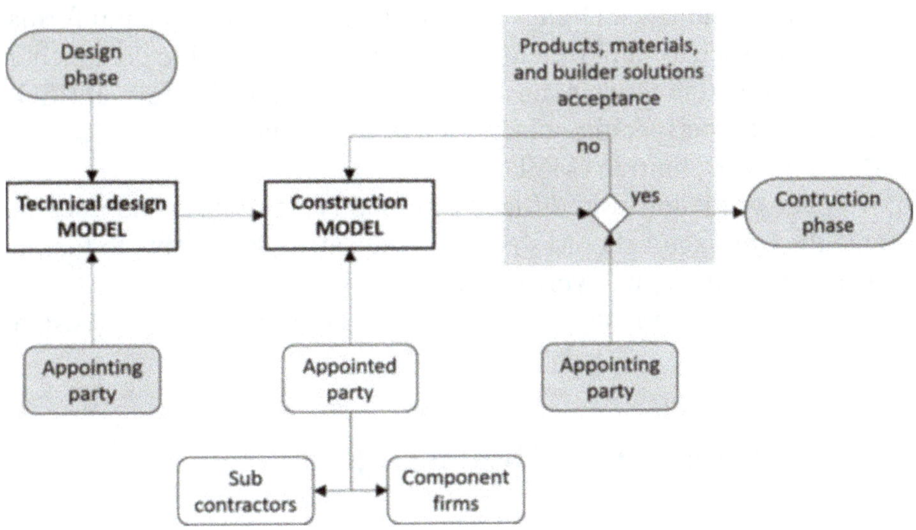

Fig. 2 Modelling flow in the post-awarding phase before the start of the works

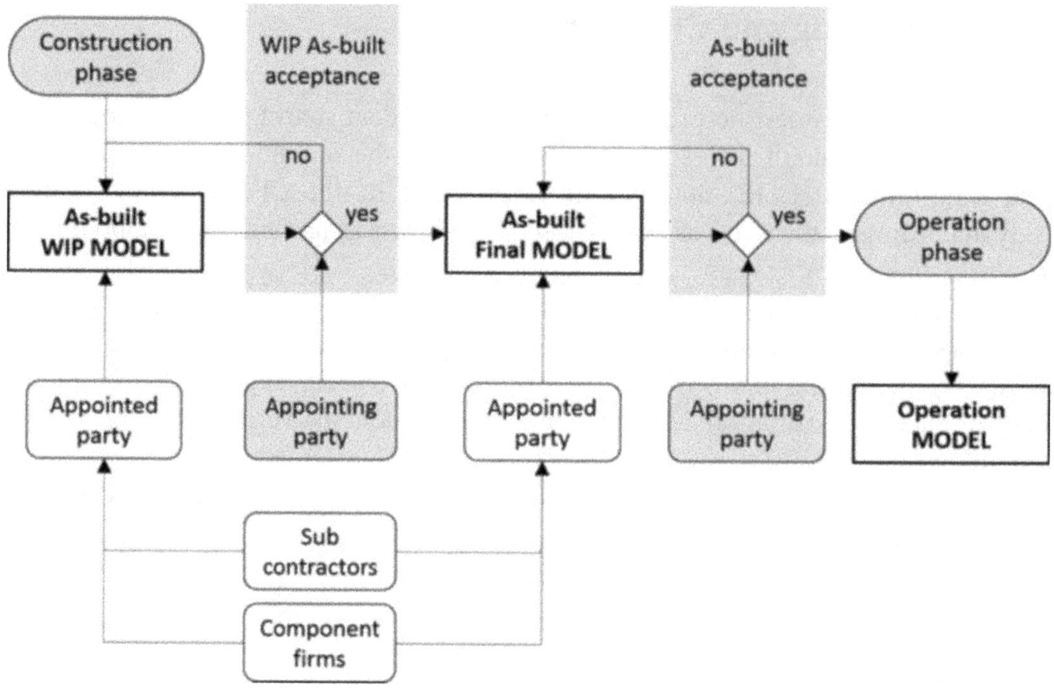

Fig. 3 Modelling flow in the work execution phase until the closeout

7 Data Exchange Format

The BIM process should promote the use of open formats to avoid any limitation to the Competitors in the bidding phase. This principle must be applied to both digital models and documents. Digital documents should be in open pdf format or in other open formats such as xml, rtf, etc., depending on the specific document and its intended use. Information models must be shared using the IFC format (ISO 2013) defined according to the specific objectives and use of the model, i.e. according to specific Model View Definition (MVD).

Moreover, due to the intrinsic difficulty in modifying models and documents in open format, the Contractor should share also the native documents to facilitate the re-use of the models in the life cycle of the good.

Open and proprietary formats should always be coordinated and consistent.

8 Level of Development of Digital Objects

In the Pre-Contract BIM Execution Plan, and consequently in the Contract BIM Execution Plan, the LOD reference scale (Level of Development of Digital Objects; (UNI 2017b)) must be defined.

Each discipline will have to ensure that the level of development of digital objects/elements used in its document/model is aimed at achieving the objectives

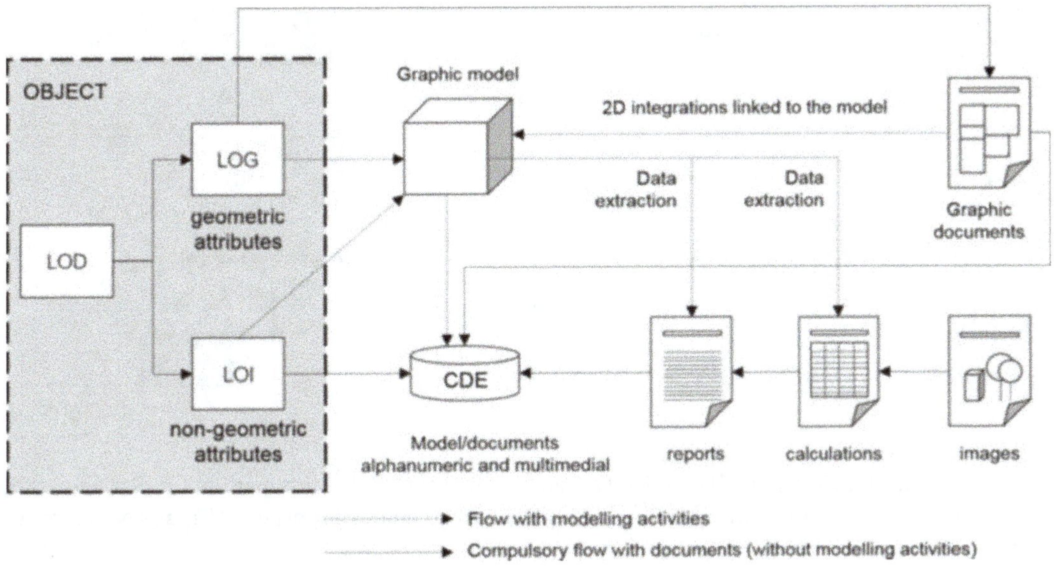

Fig. 4 Diagram of LODs; geometric and non-geometric information attributes

defined by the law, set out in the EIR, or proposed by the Contractor, for the model/document itself. The minimum LOD of the objects/elements relating to each document/model will still have to meet every legal requirement for the phase of the process it relates.

The structure of the LOD[1] should guarantee the completeness and consistency of the information through the use of mutually interconnected graphic or geometric attributes (LOG; (UNI 2017b)) and non-graphic or informational ones (LOI; (UNI 2017b)). Figure 4 proposes an example of integration between objects, models and documents in the CDE highlighting the graphic and non-graphic nature of LOD).

As a matter of principle, we consider the following development levels of objects/elements as roughly consistent (those typical of the design level are in bold):

1. Survey:
   - LOD B-**F/G** (UNI 2017b);
   - LOD, LOI 2-**5/6** (BSI 2013; NBS 2015);
   - LOD 200-**500** ((BIMForum 2016) Part I and Part II).
2. Construction and Security:
   - LOD **E** ((UNI 2017b) paragraph 5 and annex I);
   - LOD, LOI **5** (BSI, 2013; NBS 2015);
   - LOD **400** ((BIMForum 2016) Part I and Part II).

[1]NB: the LOD of each digital object is made up of the geometric and non-geometric attributes defined in the model, in the graphic documents related to it, in the documentary (reports and calculations) or multimedia (images) texts, supplementary notes, jointly filed in the CDE and/or on the construction supervisor Digital Platform. This condition is valid provided that all the geometric and non-geometric attributes external to the graphic models are nevertheless unequivocally associated with the object or the objects for which the LOD is being assessed.

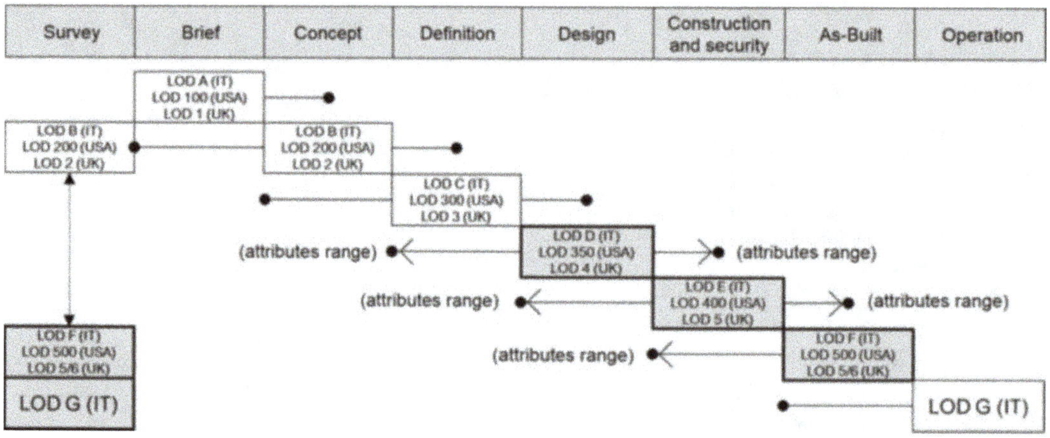

Fig. 5 Comparative diagram of some LOD scales (national [IT] and international [USA and UK]) and definition range of the corresponding informational attributes in respect of the process stages of Italian public law

3. As-Built:

   - LOD **F** ((UNI 2017b) paragraph 5 and annex I);
   - LOD, LOI **5-6** (BSI 2013; NBS, 2015);
   - LOD **500** ((BIMForum 2016) Part I and Part II)

Figure 5 proposes a compared vision of the LOD scales according to UK, USA and Italian standards.

Some objects need to be defined according to both a 3D and a 2D graphic representation due to the need of transfer the technical information to the specialists. For example, digital objects representing the MEP devices must, in any event, contain the two-dimensional graphic representation as prescribed by the law (for example, the IEC symbols, UNI-CIG symbols, etc.).

9 Conclusion

This paper presents the research conducted in order to define a general structure of the Employer Information Requirements (EIR) for the Inter-regional Superintendency for Public Works in the Lombardy and Emilia Romagna regions of the Ministry of Infrastructure and Transport. In the light of the relevant state of the art at international level, as well as the regulations in force in the national context, a structure capable of being adapted to various types of contract has been developed, in such a way as to allow a standardisation of the same to optimise its use both by the Commissioning Body and by Competitors. Moreover, the identification of a general structure that can be adapted according to the specific works paves the way for the process standardisation of the Commissioning Body.

The specifications described herein have been validated on a first case study and this has allowed the Authors to fine-tune the specifications according to the real needs of both the Commissioning Body and the Competitors. Further application of the proposed structure by other Commissioning Bodies and/or in other works would enable a validation, also considering the potential peculiarities of other Public Clients.

References

BIMForum (2016) Level of development specification

BSI (2013) PAS 1192-2:2013, Specification for information management for the capital/delivery phase of construction projects using building information modelling. The British Standards Institution, UK

ISO (2013) ISO 16739:2013-Industry Foundation Classes (IFC) for data sharing in the construction and facility management industries

ISO (2018) ISO 19650-1:2018—Organization of information about construction works—Information management using building information modelling—Part 1: concepts and principles

Mirarchi C, Pavan A. (2019) Building information models are dirty. In: 2019 European conference on computing in construction. Chania, Greece

NBS (2015) BIM toolkit. Available at: https://toolkit.thenbs.com/ (Accessed: 1 April 2017)

UNI (2017a) UNI 11337-1-Building and civil engineering works—Digital management of the informative processes—Part 1: Models, documents and informative objects for products and processes. Italy

UNI (2017b) UNI 11337-4-Building and civil engineering works—Digital management of the informative processes—Part 4: Evolution and development of information within models, documents and objects. Italy

UNI (2017c) UNI 11337-5-Building and civil engineering works—Digital management of the informative processes—Part 5: Informative flows in the digital processes. Italy

UNI (2017d) UNI 11337-6-Building and civil engineering works—Digital management of the informative processes—Part 6: Guidance to redaction the informative specific information. Italy

Design Stage

Introduction

Bruno Daniotti, Marco Gianinetto

The part II presents an overview of various research activities developed in order to support designers and clients in the first design stages, in the framework of progressive digital transformation and BIM introduction.

A new design approach is changing the traditional one with the introduction fo Building Information Modelling, allowing then to anticipate choices, preventing mistakes, pathologies and useless loss of time and economic troubles.

Some key topics are here discussed:

BIM platform for clash detection and code checking: in this part are presented innovative tools for a Clash detection as the verification of geometric interferences and can be a coordination problem in 3D BIM and Code checking as the verification of the compliance of the digital model with the corresponding regulation.

New design approaches: taking into account the architectural design activity two aspects are explored which requires more consideration from a methodological perspective; design optimization, and architectural simulation.

Digital parametric design is exploited to help defining, controlling and assessing architectural works and to develop sustainable and innovative surfaces for adaptive acoustics; new awareness on the topic of architectural acoustics design has pushed towards finding new integrated methodologies able to deliver tailored solutions for the design with sound, which embed performance criteria early in the design phase by means of simulation and computational techniques.

Smart city policies and related support tools: Smart City policies have attracted significant funding over the last few years. However, little evidence is available of their impact on urban economic performance. In this paper it's studied the urban growth and innovation impact of Smart City policies.

Finally, specific studies are presented in order to investigate the changes in architecture and engineering firms due to digitalization: the main objective of this research project has been to understand the pro-cess-oriented and organizational changes that the adoption of digital technol-ogies bring on, as well as the new forms of process and organization associat-ed with the digital transformation of architectural and engineering firms.

Clash Detection and Code Checking BIM Platform for the Italian Market

Caterina Trebbi, Michelangelo Cianciulli, Francesco Matarazzo, Claudio Mirarchi, Guido Cianciulli and Alberto Pavan

Abstract Development of experimental products for the BIM environment, and plug-ins are integrated with a collaborative platform to create, manage and validate BIM models in a common data environment.

Keywords ACDat · CDE · Clash detection · Code checking · IFC · IDM · MVD · Rule checking · Model checking · mvdXML · Bsi · BuildingSMART · BCF

1 Introduction

The term BIM[1] means a process of creating and managing a digital model containing all the information relating to a construction. It is essential to create a single environment in which it is possible to manage all data and information centrally throughout the life cycle of a building. The UK voluntary standard BS 1192-1:2007 (BS 2007) originally theorized this environment as Common Data Environment (CDE). The Italian voluntary technical regulation UNI 11337 introduces this environment in the process of digitization of the construction sector as *Ambiente di Condivisione Dati* (ACDat), outlining its characteristics, objectives and advantages. An *IT infrastructure for data collection and management* (UNI 11337-5:2017), where all those accredited can share information, in accordance with specified procedures. The platform ensures transparency and a reduction in errors. All actors involved in the process can relate, sharing data and design, construction or maintenance choices. In a BIM process, it is also essential to coordinate data and information contained in different

[1]Building Information Modeling

C. Trebbi (✉) · C. Mirarchi · A. Pavan
Architecture, Built Environment and Construction Engineering—ABC Department, Politecnico di Milano, Milano, Italy
e-mail: caterina.trebbi@polimi.it

M. Cianciulli · F. Matarazzo · G. Cianciulli
ACCA Software, Bagnoli Irpino, Italy

B. Daniotti et al. (eds.), *Digital Transformation of the Design, Construction and Management Processes of the Built Environment*, Research for Development,
https://doi.org/10.1007/978-3-030-33570-0_11

BIM models: coordination in the model, between the models and towards the relevant regulations.

Model checking is now an integral part and key element in information modeling and management (Ciribini et al. 2014). In standard design processes, only 5–10% of the project information content is systematically checked (BuildingSmart Finland 2012). The model checking, however, allows an automatic validation of 40–60% of the project (IBIDEM). Part 5 of UNI 11337 defines these activities as *Clash detection* and *Code checking* and plans to carry out these activities within the *Coordination Levels* (UNI 11337-4:2017, UNI 11337-5:2017).

Clash detection means the verification of geometric interferences and can be a coordination problem in 3D BIM (Akponeware and Zulfikar 2017). Two objects can collide because they occupy the same physical space (*hard clash*): a pipe that passes through a beam. On the other hand, two objects generate interference because they do not collide, but they are too close together (*clearance clash* or *soft clash*): space for maintenance of a plant element is not enough. Conflicts can also develop over time (4D or workflow clash): two objects designed to overlap during assembly.

Code checking is the verification of the compliance of the digital model with the corresponding regulation. Using specific software to make these checks allows you to speed up times and limit errors, improving the efficiency of building design (Greenwood et al. 2010). Automated rule checking has been identified as potentially providing significant value to the AEC (Solihin and Eastman 2015). It is essential to verify the conformity of the models with regulatory or technical requirements. The kitchen must have a minimum useful area of five square meters (building code of the City of Milan). It may be necessary to verify compliance with the requirements of *Employer Information Requirements* (*E.I.R.*) or of *BIM Execution Plan* (*B.E.P.*) (PAS 1192-2:2013). Particular modeling criteria may be required. It may be useful to verify the correspondence with custom requirements defined, for example by the client in the contract documents. The market does not appreciate balconies with a depth less than 1.40 m.

It is essential to identify such interferences and inconsistencies, to share them with the specialists concerned and, if necessary, to proceed with their solution in a coordinated manner. This way the potential flaws are detected in advance and a reliable performance in the following applications is guaranteed through an information model achieving fully coordinated information management (Ciribini et al. 2014). *usBIM.platform* is an ACDat cloud that integrates with several plug-ins for the management of the process that extends from the design to the realization and subsequent management of the item. Specifically, the Politecnico di Milano is providing the development of *Code checking* (*usBIM.code*) and *Clash detection* (*usBIM.clash*) plug-ins.

2 Methods

We have studied the software and information systems currently on the market (Table 1).

These are, for most of their activities, highly evolved documentary archives. Most of the computer solutions proposed are overseas and are related to software houses that produce graphic modeling systems.

Based on the guidelines of national (UNI 11337-4, 5:2017) and international (PAS 1192) regulations, we have carried out various hypotheses of structuring the CDE and WORKFLOW for the verification of the design documentation (Fig. 1).

After the analysis phase, we tested the platform by setting up several case studies. We have assumed and structured the working folders, the associated permissions, set the workflows for the definition of protocols and procedures for the approval of documents through a GATE construction, special folders with a GATE area where documents pending for approval are placed. In addition, we have developed a file encoding system with the platform's #TagBIM system. Thanks to the tests we worked out how to integrate, on this platform, additional model checking systems in order to completely manage the BIM process. Furthermore, the tests conducted demonstrated the possibility of implementing a number of useful functions for coordination, such as the implementation of *Markups* and *Issues* (visible annotations on the shared model to a specific user, containing information for the management of the problem). We have shared the implementation proposals with the software house.

We have carried out an accurate study of the administrative rules and procedures in order to determine the parameters necessary for code checking and the telematics validation of the digital model of the building. In the first phase we focused on Italian regulations (Milan, Naples and Rome), and then on international regulations: Berlin, Madrid, London and Paris. For each country, we looked for the guidance regulations relating to each discipline (building, hygiene, urban planning, architectural barriers and fire prevention). Once we had defined and analyzed the standards, we selected the articles that could be computable by the machine. Articles were defined by verifiable parameters without needing the aid of a human analysis (such as an

Table 1 Analyzed software systems

BIM360 (*Autodesk*)	BIMx (*NEMETSCHECK*)	TRIMBLE CONNECT (*Trimble*)
ProjectWise (*Bentley*)	Aconex (*Oracle Aconex*)	usBIM.platform (*ACCA software*)
BIMPLUS (*ALLPLAN*)	STR Vision Teamwork (str)	

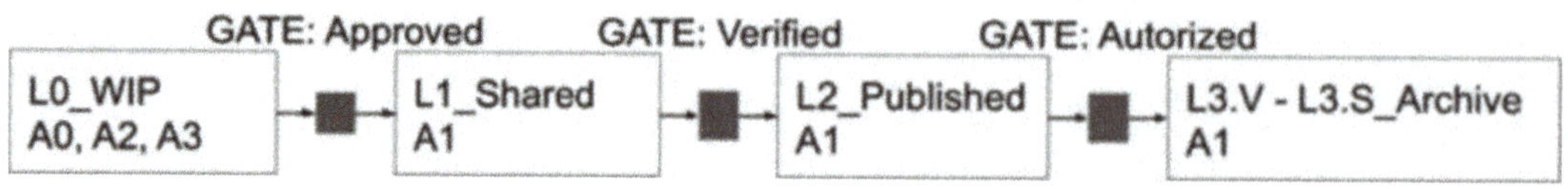

Fig. 1 Structuring hypothesis

analysis conducted by a municipal technician). Each of them was "disarticulated" into a list. It contains the subject, the number and the full text of the prescription; the measurable parameter quoted in the prescription (surface, volume, etc.) and its unit of measurement. Moreover, in the last two columns, the indication for the respect of the article is identified along with the definition of the parameter under consideration (Table 2).

We have made a comparison between different cities in order to determine code checking rules that are valid for different countries. This analysis showed that, for different building regulations, there are many similarities but also many differences.

Table 2 Table of disarticulations: building code of the City of Milan and D. 14 June 1989 on the elimination of architectural barriers

Min. Room Size	97—The usable area [...] bedroom for 1 person: 8 mq;	Surf. (m^2)	$S_{ulp} \geq 8$ mq	$\mathbf{S_{ulp}}$: Usable area – bedroom for 1 person; **Usable floor area** = floor area without surrounding walls
Inatural and Direct Lighting	105.1—[...] 1/10 of the floor surface [...] The distance between the window and the furthest point [...] 2.5 the height from the ground of the highest point of the window surface	Surf. (m^2)	S_{wtot}-S_c-(2/3)$S_a \geq$ 1/10 $\frac{1}{10}$ S_{loc} $D_{loc} \leq 2.5\ H_w$	$\mathbf{S_{wtot}}$: Total window area $\mathbf{S_c}$: Window surface less than 0.60 m from the floor level $\mathbf{S_a}$: Window area calculated by Art.105.4 $\mathbf{S_{loc}}$: Floor surface $\mathbf{D_{loc}}$: Local depth $\mathbf{H_w}$: Highest point of the window surface
Bathrooms	8.1.6—[...] The axle of the W.C. cup or bidet must be placed at a minimum distance of 40 cm from the lateral wall; The front edge at 75–80 cm from the rear wall and the top floor at 45–50 cm from the floor	Length (m)	$D_{l\text{-}bidet}$, $D_{l\text{-}wc} \geq$ 0.40 $0.75 \leq D_{a\text{-}bidet}$, $D_{a\text{-}wc} \leq 0.85$ $0.45 \leq D_{s\text{-}wc}$, $D_{s\text{-}bidet} \leq 0.50$	$\mathbf{D_{l\text{-}bidet}}$, $\mathbf{D_{l\text{-}wc}}$: Distance between side wall and bidet/WC cup axis $\mathbf{D_{a\text{-}bidet}}$, $\mathbf{D_{a\text{-}wc}}$: Distance between front edge of bidet/WC and rear wall $\mathbf{D_{s\text{-}bidet}}$, $\mathbf{D_{s\text{-}wc}}$: Distance between bidet/toilet top and floor

This is a fundamental aspect because all of this could result in the creation of a different information delivery manual (ISO 29481-1:2016; Berard and Karlshoej 2012; Karlshoej 2011) and a different model view definition (BuildingSMART international) with the aim of code checking models for different building regulations. The idea was therefore to develop BIM tools that allow, starting from a generic IFC model now produced by all BIM authoring software, to enrich it, to make it conform to another view (MVD) as required by regulatory organizations. In this way, the regulatory organization itself is interested in investing in the development of such BIM tools, as well as the software houses concerned. Anyone can make a significant contribution at any time to the development of these tools because an open format is used.

The developed system is *usBIM.code*. It consists of two separate modules that perform different functions: *CODEmaker* and *CODEcontroller*.

CODEmaker is the BIM tool that allows you to digitize, in IFC, the missing re-requested information. The system works on an IFC file generated by any BIM authoring software; it enriches it with the addition of graphical objects and information needed for *Code checking*; it saves an IFC containing all the additional data. The objects and information needed for the checks are specific to the various building regulations, and has emerged from the previous research phase. In this way, any software can use the new information to perform code checking. *CODEcontroller* is the BIM tool that uses the input information for code checking and related analysis. We have paid particular attention to the issue of the inclusion of information in the IFC file, which is an important issue discussed with the software house. As the analysis showed, building regulations often do not explain clearly which surfaces should be included in the calculation of aero-illuminating surfaces. If a window presents an opaque section, it is cautious to exclude it from the calculation of the illuminating surface. Allowing the user the possibility to define precisely the area of the window that contributes to the lighting allows the system to make a more precise and cautious calculation.

Once we had received the beta version of the *usBIM.code* program we performed tests. We have developed models of both residential and mixed-use buildings (offices, shops and residences), using both Revit (Autodesk) and Edificius (ACCA software) to test their compatibility. After completing the tests, we discussed the results with the software company for the necessary improvements (Fig. 2).

The ACCA software system for interference verification is *usBIM.clash*. The plug-in can detect any *hard clash*, *hard clash* with tolerance and *clearance clash* present on the IFC model (or between models). In order to perform its verification the program executes computational geometry algorithms considering all the elements one by one that are part of the selection made. Using specific data structures, the system processes a list for each element considered. The list contains the remaining elements of the model that could give rise to a clash. This analysis, quick but approximate, considers the selected verification settings. For example, in order to verify whether two objects collide, before making a more precise but also computationally more onerous verification, the respective bounding boxes are calculated and it is verified whether these are not disjointed. In order to achieve a better level of performance, the

Fig. 2 Test model (IFC2X3)

algorithms make considerations of this kind over several phases, for all three types of interference (hard clash, hard clash with tolerance and clearance clash). Then, after the system has populated the list with the elements that at a rough analysis generate interference with the current element, we proceed with progressively more precise and increasingly expensive assessments. When a check fails it means that the two elements considered do not generate interference and therefore you can delete the element from the list and continue repeating it with the next. If you repeat the process for all the elements in the list and the list is not empty, there is some interference. In the algorithms used, attention has been paid to the management of calculation errors due to the finished precision of computers when dealing with decimal numbers.

Once the beta program arrived from the software house, we conducted tests on simplified models, specifically modeled in both Revit and Edificius in order to verify and perfect the three different verification algorithms. We first verified the interference on entities of the same model, then on entities of different models. We performed clash detection with both *usBIM.clash* and *Navisworks Manage 2018*, in order to compare the results (Fig. 3).

We simulated the verification activities required by UNI 11337 in the coordination levels of a BIM process. Tests have shown the need to implement a number of features to better manage the resolution of the interference. For example, this includes the possibility of assigning an interference to a specific user by adding notes or creating interference groups (e.g. according to the user who must solve them or the elements that generate it). To make communication between process actors more efficient, we have suggested a standard report for collision transmission, containing all the useful data which can be exported. The user can set the content of the report by filtering the information they consider to be useful.

REPORT– NAME: *[Clash_ARC VS STR]*		
SETTINGS: Collisions: *[yes]* – Tollerance value: *[0.01 m]* – Entity same model: *[yes]*		
SUMMARY: N° collisions: *[35]* - New: *[35]* - Active: *[0]* - Examined: *[0]* - Approved: *[0]* - Fixed: *[0]*		
COLLISIONS:	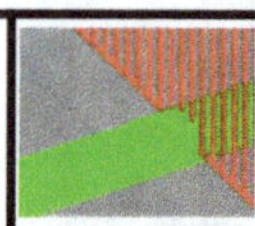	Name: / Collisione 1/ - Description: / Intersezione/ - Creation date: / 01/01/2019/ Distance: / -0.5 m/ Status: / Attivo/ – Assigned to: / Mario Rossi/
	Element 1:	
	ID: *[1434]* - Layer (Floor): *[GF – Ground Floor]*	
	Path: [File -> *Single templates.ifc* ->0 ... > Basic-Generic wall - 30 cm/	
	Element 2:	
	ID: *[1224]* - Layer (Floor): *[GF – Ground Floor]*	
	Path: *[File ->Single templates.ifc ->0 …>Railing]*	
	Comment 1: By *[Claudio Bianchi]* - Assigned to: *[Mario Rossi]* – On date: *[02/01/2019]* Text of the comment: *[Controlling the railing …]*	

Fig. 3 Standard interference verification report

3 Results

The *usBIM.platform* system allows you to customize the data sharing environment according to your project needs. The use of *GATE* folders, in combination with the definition of associated *permissions* and the definition of *workflows*, allows for a correct application of procedures/processes of revision and validation of documentation. Unlike most commercially available CDEs, procedures can be freely structured within the platform according to the needs of the individual project. The system supports open formats (BCF, IFC, etc.). Unlike the main solutions proposed by the market, it is not only a "container of information"; but a management system for the informative content of a project throughout its life cycle. The IFC project model can be displayed in the cloud without having BIM authoring software and can be federated with other project models.

Being able to perform model checking tests in the sharing platform leads to a new way of working based on sharing information and transparency of processes. Creating standard sets of rules according to the reference standard contributes to the optimization of verification processes. Most of the solutions in commerce today are foreign, concentrated on an international market, far from national legislative requirements. *usBIM.code*, currently under development, verifies the correspondence of projects to local legislation and/or specific design needs quickly, without compromising the quality of assessment. You can choose the technical or legislative rule to verify and import the IFC model file into the software. The system will analyze and highlight verification data and calculate missing data. If necessary, the user will enter the missing data into a graphical form. In the section labeled "diagnostic" it will be possible to visualize a list of the requirements that are not respected and the objects (or parts of the project) that are not compliant. Checks can also be carried out on models developed by BIM authoring software with insufficient information. The insertion of missing data is done on an open format (IFC) to ensure the process is intelligible.

The problems associated with the exchange of information, currently investigated (Belsky et al. 2016; Bloch and Sacks 2018; Sacks et al. 2017), are thus overcome. The implemented information also remains stored in a new IFC file which can also be read by other software.

The actors taking part in the process are varied; each of them elaborates its own BIM model using a specific software of BIM authoring according to its needs. It should be stressed that in a real process not all detected interferences should be considered in the same way. Some, for example, are "false positives" because they are easily manageable on the construction site. Other collisions, on the other hand, should not be neglected and should be resolved at the design phase. *usBIM.clash* works with models in IFC format; it can verify interferences both between entities belonging to the same model and between entities belonging to different digital models. It is a simple and immediate system, unlike most clash detection software. IFC models for which verification is to be conducted are imported into the software; through the appropriate commands, the criteria for selecting the entities on which to process conflict verifications are defined. These criteria may be recorded and used subsequently to repeat the verifications on the updated models. The type of clash to be identified must be defined: *hard clash*, *hard clash* with tolerance or *clearance clash*. The system then produces a table with all interferences. By selecting the interference, the system provides a view of the model entities involved in the collision. In addition, each interference reports a set of information useful for its management. You can export an interference report in different formats (pdf, html, xml, csv, xlsx). In the future, the software will also be able to export in BCF standard form.

4 Conclusion

BIM not only results in three-dimensional modeling, but also, and above all, in effective information management, in collaboration and coordination between professionals, in process automation. With this study we have looked into new collaborative procedural practices in order to understand how to better exploit this new way of working. The presence on the CDE of all information relating to the intervention, being complete and up to date, allows for an informed management of the whole process. Every activity must be traced allowing for maximum control over operations; the data must be organized and catalogued clearly; the roles and work processes of those involved in the process must be defined; the platform structure must be adapted to requirements (B.E.P., E.I.R., etc.) or internal needs.

We theorized a system where, in a single environment, in addition to sharing information it is possible to use model checking tools. In a traditional design process, process assessments are carried out manually and on a sample basis. This makes the process slow and fertile to the proliferation of errors. The tests we have carried out confirm that delegating massive project assessments to a machine allows users to optimize time and costs. The study, through the analysis of rules and administrative procedures, has allowed us to define the parameters necessary for the automatic validation of digital models. The goal is to create software capable of adapting to different needs; to create a proven set of rules specific to different regulations or needs.

The analyses and the work carried out confirm that the path is still long but the goal is attainable. The advantages that can be obtained are numerous, starting from the design phase, all the way up to the realization phase and, consequently, also to the management and maintenance phases.

References

Akponeware AO, Zulfikar AA (2017) Clash detection or clash avoidance? an investigation into coordination problems in 3D BIM. Buildings 7(4):75. https://doi.org/10.3390/buildings7030075

Belsky M, Sacks R, Brilakis I (2016) Semantic enrichment for building information modeling. Computer-Aided Civil Infrastruct Eng 31:261–274. https://doi.org/10.1111/mice.12128

Bloch T, Sacks R (2018) Comparing machine learning and rule-based inferencing for semantic enrichment of BIM models. Autom in Constr 91:256–272. https://doi.org/10.1016/j.autcon.2018.03.018

Berard O, Karlshoej J (2012) Information delivery manuals to integrate building product information into design. Electron J Inform Technol Constr 17:64–74

BS 1192:2007, Collaborative production of architectural, engineering and construction information – Code of practice

BuildingSMART International, Model View Definition (MVD)—An Introduction. https://technical.buildingsmart.org/standards/mvd/

Finland BuildingSmart (2012) COBIM series 06—quality assurance. Common BIM Requir 1:1–27. https://doi.org/10.1073/pnas.0703993104

Ciribini ALC, Ventura SMB, Marzia B (2014) Informative content validation is the key to success in a BIM-based project. Territorio Italia 1:87–111. https://doi.org/10.14609/Ti_2_15_1e

de Farias TM, Roxin A, Nicolle C (2018) A rule-based methodology to extract building model views. Autom Constr 92:214–229. https://doi.org/10.1016/j.autcon.2018.03.035

Greenwood D, Lockley S, Malsane S, Matthews J (2010) Automated compliance checking using building information models. In: Construction, building and real estate research conference

Khemlani L (2015) Automating code compliance in AEC, AECbytes Feature

Karlshoej J (2011) Information delivery manuals. http://iug.buildingsmart.org/idms

Nawari N (2012) The challenge of computerizing building codes in a BIM environment. American Society of Civil Engineers (ASCE), pp 285–292. https://doi.org/10.1061/9780784412343.0036

Sacks R, Ma L, Yosef R, Borrmann A, Daum S, Kattel U (2017) Semantic enrichment for building information modeling: procedure for compiling inference rules and operators for complex geometry. J Comput Civil Eng 31:04017062. https://doi.org/10.1061/(asce)cp.1943-5487.0000705

Solihin W, Eastman C (2015) Classification of rules for automated BIM rule checking development. Autom in Constr 53:69–82. https://doi.org/10.1016/j.autcon.2015.03.003

Zhang S, Teizer J, Lee JK, Eastman CM, Venugopal M (2013) Building Information Modeling (BIM) and safety: automatic safety checking of construction models and schedules. Autom Constr 29:183–195. https://doi.org/10.1016/j.autcon.2012.05.006

Standards and Laws

UNI 11337:2017-4 - Building and civil engineering works – Digital management of the informative processes - Part 4: Evolution and development of information within models, document and objects. UNI - Ente Italiano di Normazione
UNI 11337:2017-5 -Part 5: Informative flows in the digital processes
PAS 1192-2:2013 - Specification for Information Management for the capital/delivery phase of construction projects using Building Information Modelling
ISO 29481-1:2016 - Information delivery manual – Part 1: Methodology and format
Italy:
D.Lgs 18/04/2016, 50 - Codice dei contratti pubblici
D.M. 02/04/1968, 1444 - Limiti inderogabili di densità edilizia, di altezza, di distanza fra i fabbricati e rapporti massimi tra gli spazi destinati agli insediamenti residenziali e produttivi e spazi pubblici o riservati alle attività collettive, al verde pubblico o a parcheggi, da osservare ai fini della formazione dei nuovi strumenti urbanistici o della revisione di quelli esistenti
G.U. 16/11/ 2016, S. G. 268 - Quadro delle definizioni uniformi
Codice Civile Italiano
L. 17/08/1942, 1150 - Legge urbanistica
L. 24/03/1989, 122 - Disposizioni in materia di parcheggi, programma triennale per le aree urbane maggiormente popolate, nonché modificazioni di alcune norme del testo unico sulla disciplina della circolazione stradale
D.M. 16/05/1987, 246 - Norme di sicurezza antincendi per gli edifici di civile abitazione
D..M. 14/06/1989, 236 - Prescrizioni tecniche necessarie a garantire l'accessibilità, l'adattabilità e la visitabilità degli edifici privati e di edilizia residenziale pubblica, ai fini del superamento e dell'eliminazione delle barriere architettoniche
D.M. 05/07/1975 - Modificazioni alle istruzioni ministeriali 20/06/1896, relativamente all'altezza minima ed ai requisiti igienico-sanitari principali dei locali di abitazione
Milan:
Piano di Governo di Territorio - Piano delle Regole - Norme Tecniche di Attuazione del Comune di Milano
Regolamento Edilizio del Comune di Milano
Regolamento Locale di Igiene della Regione Lombardia
L.R. 20/02/1989, 6 - Norme sull'eliminazione delle barriere architettoniche e prescrizioni tecniche di attuazione
Rome:
Piano Regolatore Generale - Norme Tecniche di Attuazione del Comune di Roma
Regolamento Generale Edilizio del Comune di Roma
Naples:
Norme Tecniche di Attuazione del Comune di Napoli
Regolamento Edilizio del Comune di Napoli
L.R. 20/03/1982, 14—Indirizzi programmatici e direttive fondamentali relative all' esercizio delle funzioni delegate in materia di urbanistica

Digital Culture for Optimization

Samir Al-Azri

Abstract Humanity proceeds along the path of hybridization with the technological tool. The centre of gravity of the creative process moves within the computational man-tool ecosystem, with capital consequences: the instrument, like an "organ", guides our way of thinking, designing, and planning. In this context two aspects of the process of architectural design will be explored in this chapter which requires more consideration from a methodological perspective: design optimization and architectural simulation. The optimization stage is related to the process of realizing the objective of architecture, inherently prescribing the outcome. Representation and simulation, on the other hand, are important as a tie between the concept (or an idea of the architect) and the materialization of that concept, that is, a link between the sensory and non-sensory. Both aspects are important but highlight a different side of the process.

Keywords Topology optimization · Digital simulation · Computation in nature · Adaptive topology

1 Introduction

Architecture, like music or other forms of creative works, is based on the experience that is evoked by the sensory capabilities of the user. Unlike music which disappears with the sound, the building has the power to enforce its dominance by its permanence (Drewniak 2009). Not only does the building not disappear, but it has the opportunity of lasting and being experienced by different generations.

Architecture is relative by nature. It is undefined, unmeasurable, and is dependent on the perspective of the viewer. A simple task of trying to research a definition would yield countless philosophical terms and quotes from renown architects, all providing their own account or perspective on the matter. Such subjectivism has

S. Al-Azri (✉)
Architecture, Built Environment and Construction Engineering—ABC Department, Politecnico Di Milano, Milan, Italy
e-mail: samir.alazri@polimi.it

B. Daniotti et al. (eds.), *Digital Transformation of the Design, Construction and Management Processes of the Built Environment*, Research for Development,
https://doi.org/10.1007/978-3-030-33570-0_12

created a professional community that is dispersed more than it is united. The meaning of architecture spans a large spectrum from Hassan Fathy's "architecture of the poor" to Frank Gehry's extravaganza. Within this fragmentation, the main discourse about architecture today examines whether it is an art form, a process/practice or a final product (Fisher 2015). Architectural practice and research in recent years have touched upon completely new fields of study. The technological advancements and scientific discoveries from different fields drive more architectural processes. Focusing on the process/practice discussion, it is imperative to consider the role of the advanced technologies we have available today and how the design culture is shaped by the digital revolution.

In this context two aspects of the process of architectural design will be explored in this chapter which requires more consideration from a methodological perspective: design optimization and architectural simulation. The optimization stage is related to the process of realizing the objective of architecture, inherently prescribing the outcome. Representation and simulation, on the other hand, are important as a tie between the concept (or an idea of the architect) and the materialization of that concept, that is, a link between the sensory and non-sensory. Both aspects are important but highlight a different side of the process.

2 Optimization

Optimization is the process of finding the best and optimal solution to a given problem—an idea that can be traced back to the Ancient Greeks, but only gained momentum in application with Newton's development of modern calculus (Kiranyaz 2014). A renowned optimization problem is the travelling salesman problem, where a salesman must visit all the cities in the United States but should figure out the path that would minimize the travelling distance without visiting the same city twice. This problem is still in use today to benchmark optimization algorithms. The reasoning of optimization is based on the existence of an optimum solution in which the designer can strive to achieve. From that premise, the designer should be able to explore the large solution space to identify that one ideal solution. This process is usually in the later design phases and involves committing an extensive amount of resources to optimize it and usually requires expertise from the designer.

Nature, as we know it, is the result of billions of years of development and adaptation. It features panoply of organisms that inherit this long process of evolution including ourselves. However, as Alan Watts contemplates in his book "*The book on the taboo against knowing who you are*", we as humans suffer from a hallucination, from a false and distorted sensation of our own existence as living organisms. We have been brought up to believe that we came into this world rather than we came out of it. This simple ideological twist alone might draw our very distinction from every other organism. When nature was conceived as something alien and hostile, we decided to conquer it rather than cooperate with it (Watts 1973). We marked a unique geological era which is called by the new science "the era of Anthropocene".

It seems that we have been successful at conquering the world but at the cost of jeopardizing our own existence.

Despite different outlooks on architecture, finding solutions for the conditions brought about by ourselves, now loom over as an oppressive imperative for the future practice (Luebkeman 2015). Architects can no longer indulge themselves with designing beautiful buildings without considering their environmental impacts. For the first time, we can sense the urgency of proactively participating in more sustainable practice. We are left with no other choice other than cooperating with nature to guarantee our survival as a species. More attention has been drawn to natural sciences and its subdisciplines in recent years in response. The most cutting-edge institutions/studios around the globe are increasingly committing themselves to the study of nature and its potential pertinence to architecture.

A new "design culture" as mentioned by Neri Oxman is emerging, thanks to the advancements of technology. It is distinguished by the underlying approach to the very act of creating. Products and buildings, for the most part of the history of craftsmanship, have been created by assembling different parts. However, now at the intersection of computational design, additive manufacturing, materials engineering and synthetic biology, we can grow products and even our buildings just as nature. Instead of having distinct parts delivering different purposes, the output of the new process is a product with varying properties, gradually transitioning its functionality over different areas exactly the same way that the trunk of a tree transitions into its branches, leaves and so forth (Oxman 2012).

The "Silk Pavilion" project developed by the Mediated Matter Group at MIT Media Lab capitalizes on the inherent intelligence of silkworm to optimally produce silk fibres of variable properties, according to the functions needed. These fibres are optimized for a wide range of different conditions including, but not limited to, mechanical properties such as strength and toughness. This research explores the possibility of merging digital and biological fabrication to deliver a holistic and sustainable design approach in the production of non-woven fibre-based constructions (Oxman et al. 2014).

In another example, a project carried out by the students of The Bartlett School of Architecture explores the embedded computational capabilities of Slime Mold or Physarum polycephalum for generating urban design proposals for Liwa Oasis in UAE. Physarum polycephalum is a single biological cell with countless cell nuclei. This brainless organism can stretch itself to any shape in multiple branches called plasmodium searching for resources. Upon facing a source of food, it leaves traces as a memory, otherwise eventually fading away in the absence of food. The outcome of this rather heuristic process is an optimized network distributed between the sources. This lends a sophisticated network optimization model to be adopted for scenarios such as Liwa where future developments heavily rely on the informed use of the existing resources (UCL 2015).

There is a fundamental issue in the optimized design process, which is the emphasis on a certain design objective and disregarding the rest. For an optimization process to converge into a single solution, a certain function would have to be minimized or maximized. This implies that all significant design outcomes are quantitative. An example would be the optimization of the energy consumed in a building. A building, however, is more than the measure of its performative quantifiable properties (the sum is always larger than the parts). However, there are many other factors that render a building design meaningful. This goal-oriented approach has given rise to many factions each concerned with what they consider as the most important aspect of a building design. Performance-driven designs, for example, emerged recently as a response to the energy crisis and a solution to climate change. It is not to say that their goals are not legitimate, however, it can be argued that other design characteristics are just as important.

Genetic algorithms are prominent in the design optimization scene today. However, we can apply the concept of "survival of the fittest" to the buildings as well. Only the outstanding buildings were worth being preserved; the rest were demolished with time. The architects of the past surely built many buildings that were not considered worth keeping or documenting, and that was probably most of the cases. Nevertheless, we are accustomed to looking back and applauding past generations for their work, neglecting the fact that they made mistakes just the way we do today. Architecture as a concept is going through an optimization process within itself and we are converging to a better built environment with time.

The ecological paradigm and computational design thinking in architecture have led to novel methodologies. Finding inspirations from nature, which is commonly given the term "bio-mimicry", has a long history in the architectural practice. However, it has been for the most part, limited to mimicry of the forms and shapes. This approach, despite exceptions of good examples, is a rather thin abstraction of what nature can offer in terms of efficiency and potential solutions for complex problems.

On the positive side, more architects are either involving themselves with in-depth study of natural processes or resort to close collaboration of specialists in other fields including biology and genetics, finding solutions for future challenges. An increasing number of inter-disciplinary studios are emerging. There is a shift in the perception of nature from a mere source of aesthetical inspiration, towards a repertoire of the intelligence which is developed through billions of years of evolution (Pawlyn 2011). The most cutting-edge institutions and research labs are striving to decipher complex natural systems and implementing their knowledge of their processes. These attempts include translating the optimized behaviours of particular organisms into tailor-made solutions for particular design problems, modifying certain natural organisms to find optimum functionality for certain purposes or directly applying the organisms within the design process.

3 Simulation in Architecture

Architecture is realized in a dimension between what we envision and what exists. Considering the mind as the source of concepts and the place where ideas are processed, and the building as the final product, there lies a medium in between through

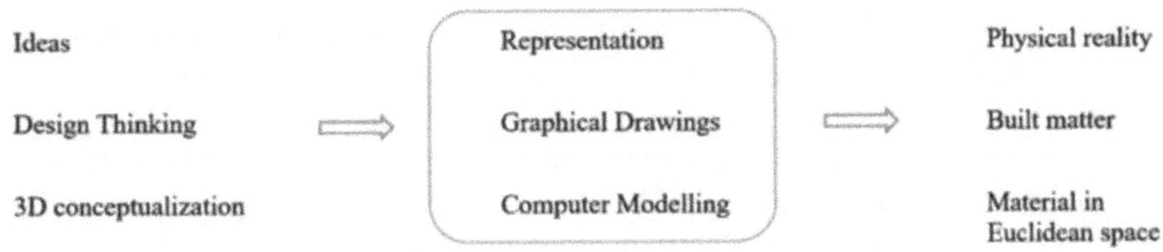

Fig. 1 The medium between concept and realization

which information needs to be communicated. This intermediate bridge between the two realities was previously referred to as architectural representation (Fig. 1).

Representation in architecture is the accumulation of information found in geometry. The representation can take many forms, from drawings to physical models, and the amount of information determinable depends on the viewing perspective. A computer model, for instance, can be rotated and viewed from an infinite number of positions. However, the information transferred through the screen is that of a still image at a specific timeframe. Therefore, we can think of representation as the collection of information from all perspectives, and hence all images.

The idea of images, and information stored within, incites one to question the understanding of geometry. Starting with the Greeks and the introduction of the Euclidean geometry, the ability to deduct measures from drawings constantly puzzled the mathematicians. This was only rationalized during the Renaissance period with the works on perspectives, making it a revolutionary period for architectural graphics. However, curves remained the most difficult objects to represent, and circles were used to estimate the curvature at different points of a curve.

> *There are no solids in the universe. There's not even a suggestion of a solid. There are no absolute continuums. There are no surfaces. There are no straight lines.—Buckminster Fuller*

Simple geometries can be transformed and rationalized into complex geometries through morphing, be it addition or subtraction. Similar works can be observed in the works of Philibert Delorme and his shell models. This approach of understanding geometry raises the question of the reality of the models and objects we are building. It is understood that a curve is nothing but an estimation created from many smaller circles that are connected through their tangent vectors, and no perfect sphere can be created. On the other hand, we rest assured that we can build a structure that is perfectly straight and proclaiming that is. The tools used are levels that check the altitude difference between two points, but since the earth is not flat, how straight are these "straight line"? In our microscale compared to the world, such an approximation is accepted. However, the models and calculations used to ensure a structure is stable, are perfect geometries. Is this an acceptable deduction? Or are we dependent on the confidence of our experimental realism?

> If process drives outcome, we may not know where we are going, but we will know we want to be there—Bruce Mau.

It is suggested by Maurice Conti in his Ted Talk that we are on the cusp of a new age of our activities as a species. After the four major eras that define how we have been working, namely, hunter-gatherer, agricultural, industrial and digital age, we are now at the dawn of a new era called "augmented age". For the first time in history, the tools developed by humankind will augment it through its daily actions. Unlike how tools worked in the past, merely in a passive form that needed direct exertion of intention and authority from its user, new tools are being developed to actively contribute and even think (Conti 2017).

Simulation has always been a key element in the process of design. Every design undergoes some sort of simulation, even at the most elementary level. An architect will simulate, at least in a mental simulation, the way the building occupants will interact in a space. This simulation will be driven by certain design drivers or objectives, while the constraints will depend on gained knowledge through experience, maybe construction techniques or local codes. This simulation is limited to the designers computing capabilities and has been historically outsourced to physical models to try and comprehend the complexity at hand. Great examples include working with strings and weights, and soap films, by Gaudi and Frei Otto, respectively.

With the advancement of computing powers in the digital age, the interaction between human and technology evolved to an unprecedented level. Technology has become an extension of human activities, leveraging their abilities in all aspects. Transhumanism is a common term now from magnetic implants to augmented reality. In the design field, simulations have come a long way as technological tools. Mathematical algorithms have been developed that make digital simulations of physical phenomenon's accessible to designers and engineers. Interestingly, the design process did not adopt these tools as design drivers at first and instead used these models mostly in the advanced stages to validate or analyze the performance of a proposal. This shift allocated the design solution generation responsibilities to the designer and limited the simulation process to selection or enhancements. Naturally, the solution space explored was limited and the designs produced by human capabilities were streamlined to compliment a familiar manufacturing process.

The challenges on the horizon are increasingly growing complex for humans to handle. The problems lying ahead such as climate change are far more complex for us to analyze and organize by ourselves. However, our capabilities are being augmented by emerging technologies. The most advanced computers lend us unprecedented powers of analyzing millions of data points only in a matter of seconds (Conti 2017). Novel manufacturing technologies are capable of producing almost anything. Simulation in this regard finds a vital role. And how we go about designing as architects is subject to fundamental disruption.

Simulation as a by-product of the advancements of technologies has been turning into an active collaborator for designers. It is not only a model to validate the feasibility of a design per se, but also a contribution to an informed design process. The computational simulations shed light on completely new facets of our propositions, giving us real-time models of the consequences of our decisions. Therefore, the design outcomes as mentioned by Mau will somewhat become unpredictable but in a rather positive way.

As previously mentioned, more designers and researchers are turning to nature to find solutions for their complex design problems. In an attempt of adopting the ingenuity of many different organisms, the research can sometimes lead to a team of computational scientists trying to develop algorithms or models that simulate the favoured behaviours and patterns being played out by the particular organisms. These models will, later on, be exploited as simulation tools that will allow them to implement the intelligence of nature into their design process. For example, in the case of Liwa Oasis, in order to perform the same network optimization process at the local scale, the biological algorithm of slime mould was digitally simulated (UCL 2015). In the case of "The Silk Pavilion", in order to understand the structural complexity of a cocoon, advanced imaging techniques were carried out using magnometer motion sensing to capture the movements and the patterns of the silkworm during the production of the cocoon. The silkworm motion-tracked during the three days yielded a point cloud model with high levels of precision to be investigated (Oxman et al. 2014). Eventually, the investigations led to a more informed design of the silk scaffold.

The simulations provided us with means of understanding and visualizing forces and energy. This process allowed us to build towers and skyscrapers cladded with glass that are consuming significantly more energy for heating and cooling. Consequently, a climatic crisis evolved and we are currently trying to simulate our way out of this calamity. How much of the technology contributed to today's environmental challenges and can we trust it not to create more? Or will it be the saviour and enhance our well being?

4 Case Study: Adaptive Topology

To examine the influence of the digital tools on designers, an experiment was conducted in the form of a workshop. A group of 10 students were introduced to a new process of simulation in the form of topology optimization. This optimization process would allow their designs to be remodelled in a topology that reduces material while complying structurally. The students did not have prior knowledge of the process and were introduced for this first time during the workshop.

The design task in hand was to design a chair for the Salone Del Mobile 2018. Initially, the students were given the freedom of interpreting their own design processes and the only given requirement was that the chair was to be fabricated through additive manufacturing. Upon completion of their first concept designs, the students were then introduced to topology optimization and the available simulation tools; Ameba through Rhino/Grasshopper and Inspire by Altair. The students were then given the options of either starting the design from scratch and using the new simulation tool as a generative design process, or remodel their existing designs to reduce the amount of material needed for fabrication.

The results revealed that all the students attempted improving their designs with the topology optimization process, but none steered away from their initial concept/ideas. Reducing the amount of material was a sustainable task, which they agreed was necessary and were willing to undertake. However, they did not compromise on their initial form and design (Fig. 2). In addition, none of the participants was tempted to restart their design process with the new simulation tools. This observation questions the idea of a design agency in the new digital era and focuses on the designer's continuous demand for autonomy. The continuous disruptions caused by explosive innovations in different fields raise the question of authority in design. Designers may well have to adapt themselves to the present conditions that necessitate collaboration from many different sources. Despite the predictions about the future, arguing that technology would take over in different sectors, we might as well hope for better future scenarios that will be played out by the productive collaboration of humans, nature and technology.

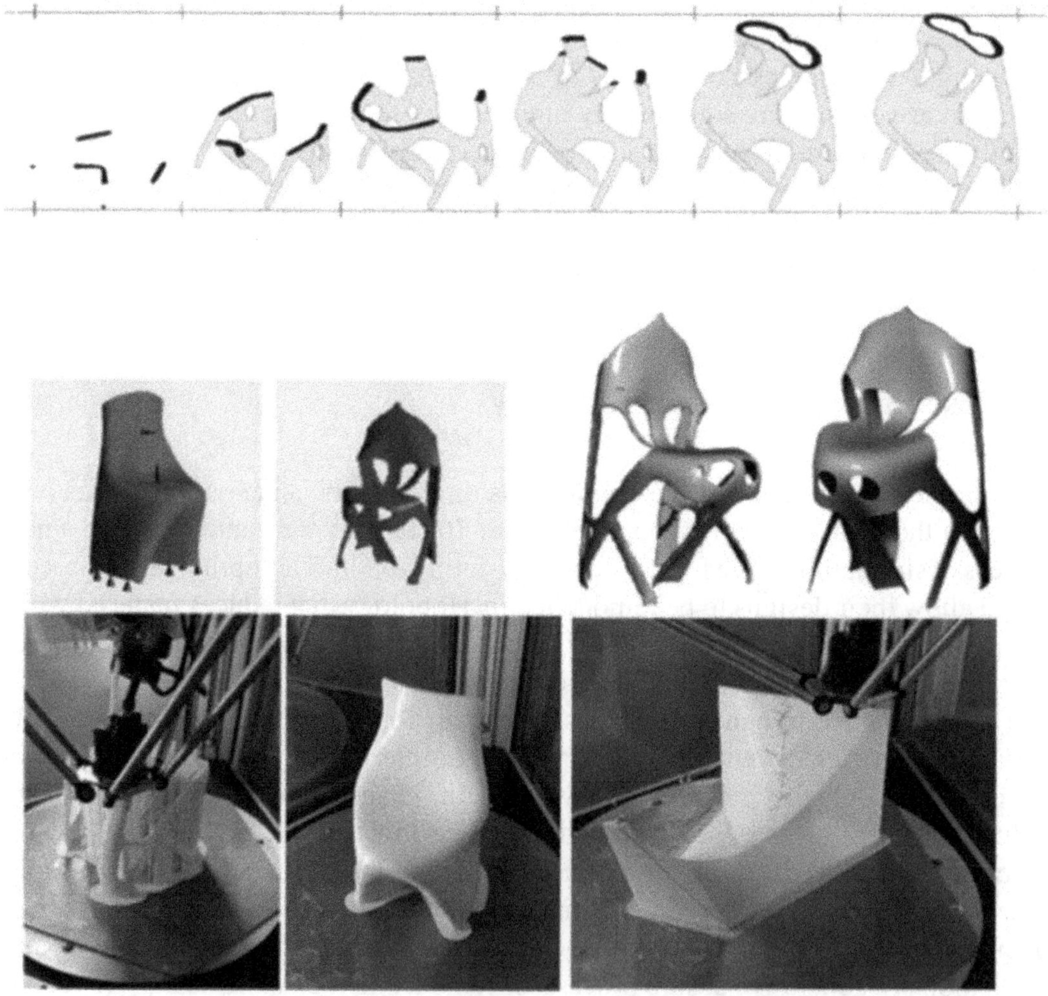

Fig. 2 Design and production of "Adaptive Topologies" workshop at Politecnico di Milano. *Image Credits* Iris Koni, Elena Manandise, Zeynep Kalaycioglu

References

Architectural Record (n.d.) Accessed from https://www.architecturalrecord.com/ext/resources/archives/features/critique/2012/images/1204-Commentary-slideshow/Perspective-Drawing-3.jpg

Conti M (2017) The incredible inventions of intuitive AI. [video] Available at: https://www.ted.com/talks/maurice_conti_the_incredible_inventions_of_intuitive_ai/discussion?langua

Drewniak T (2009) Philosophy of architecture and architectonics of philosophy. Filosofía UIS 8(2):11–31

Fisher S (2015) Philosophy of architecture. Accessed from Stanford Encyclopedia of Philosophy: https://plato.stanford.edu/entries/architecture/

Kiranyaz S (2014) Multidimensional particle swarm optimization for machine learning and pattern recognition. Springer, Berlin

Luebkeman C (2015) 2050: designing our tomorrow. Wiley, New York

Mumford L (2015) Mass-production and the modern house. Accessed from Architectural Record: https://www.architecturalrecord.com/ext/resources/archives/inTheCause/onTheState/0311mumford.pdf

Oxman N (2012) Material computation. Wiley, New York

Oxman N et al (2014). Silk pavilion: a case study. In: Fiber-based digital fabrication. gta Verlag

Pawlyn M. (2011). Biomimicry. In: Architecture. RIBA Publishing

Stasinopoulos TN (2015) The need for architectural materialism. Plea, Bologna

The Bartlett School of Architecture, UCL. (2015). MArch Urban Design (UD) 2015. The Bartlett School of Architecture, UCL

Watts A (1973) The book; on the taboo against knowing who you are. Vintage Books

Performance-Based Design Approach for Tailored Acoustic Surfaces

Andrea Giglio, Ingrid Paoletti and Maia Zheliazkova

Abstract The acoustics of architecture has a large influence on the physical and psychological state of people, impacting on their communication, concentration and behaviour. Nevertheless, the sound issues of the interior of common spaces are often faced in the last stage of the design process, as an accurate acoustic design requires a high level of expertise. Owners, companies, public administration, and so on face this issue by selecting off-the-shelf acoustic panels or furniture, which might not always respond to the acoustic requirements, specific for each space. This leads to unsatisfying acoustic conditions, consequently increasing the initial investment. Nowadays, a new awareness on the topic of architectural acoustics design has pushed towards finding new integrated methodologies able to deliver tailored solutions for the design with sound, which embed performance criteria early in the design phase by means of simulation and computational techniques. This paper describes a workflow developed in the frame of the research project: "EcoAcustica. Sustainable and innovative surfaces for adaptive acoustics", aimed at creating acoustic surfaces and customised to improve the interior acoustics in a global geometry scale. The design of the new ABC Department Digital Fabrication Lab at Politecnico di Milano is used as a case study for the proposed methodology.

Keywords Design stage · Simulation · Applied research · Architectural acoustics · Performance-based design

1 Introduction

EcoAcustica project emerges from the collaboration between the innovative composite materials system, developed by Woodskin® and TecnoSugheri, the main importer

A. Giglio (✉) · I. Paoletti · M. Zheliazkova
Architecture, Built Environment and Construction Engineering—ABC Department, Politecnico di Milano, Milan, Italy
e-mail: andrea.giglio@polimi.it

B. Daniotti et al. (eds.), *Digital Transformation of the Design, Construction and Management Processes of the Built Environment*, Research for Development,
https://doi.org/10.1007/978-3-030-33570-0_13

and manufacturer of the eco-sustainable thermoacoustic CORKPAN panels, supported by the scientific expertise of ABC Department at Politecnico di Milano in advanced technological research for architecture.

Implementing innovative computational design and fabrication methods, the project develops a customisable system aimed at providing high-performance acoustics for interiors and tailored according to their functional and spatial characteristics. The control of the acoustic conditions is achieved at two levels: a global level of the overall geometry for controlling early reflections and sound scattering, and on local material level through sandwich composites differentiation and surface treatment for tuning sound absorption.

The origami-like structure of the Woodskin® technology allows in achieving this degree of adaptivity/flexibility. The system combines flexible textile sandwiched between two layers of solid material (mainly wood). Once glued together, in the solid flat surfaces, cuts are created that allow them to be flexed away from the rigid form, thus creating a desired three-dimensional shape.

EcoAcustica project fosters the contamination between academically led research and the craftsmanship of small and medium enterprises (SMEs), and works towards the introduction of innovation and excellence in the supply chain within the Industry 4.0 research agenda. This effective collaboration enables the synergy between the two SMEs involved, creating an environment for applying in an industrial context an innovative process for design and fabrication of fully customisable acoustic panels.

2 Room Acoustics Modelling Techniques

The quality of acoustics in an enclosed space is affected by the geometrical and material properties of its surfaces (Cox and D'Antonio 2017). When the sound wave strikes a surface it can be absorbed, transmitted or reflected. The achievement of high-comfort conditions for humans depends on the control of these phenomena combined with considerations on the functional programme of the space.

Every functional programme has highly specific sound requirements: the sound conditions for speech are not the same as those for music events and vice versa. The main value used universally to guarantee proper sound conditions is the reverberation time (T_{60}), defined as the time required for the sound in an enclosed space to decay by 60 dB (ISO 3382-1). This in turn defines the amount of absorptive surfaces needed in the room. For example, for proper speech intelligibility, the reverberation time required is of about 1 s, while for symphonic music it is 2 s.

2.1 Simulating Acoustic Phenomena

Various room acoustics modelling techniques and computer models have been developed over the last few decades (Rindel 1995). Nowadays, highly accurate predictions are achievable, thanks to custom software that can be more easily adapted to the designers' needs than the classic standalone ones.

The most important room acoustic techniques are: ray tracing, image-based methods, combined methods and beam tracing.

The ray tracing method creates a dense spread of rays, which are reflected around a room and tested for intersection with a spherical detector (Kulowski 1984; Lehnert 1993). This one is in part passed by the beam tracing method that considers ray with volume using conical (Van Maerck and Martin 1993) or triangular beams (Lewers 1993). The image method considers the rays calculating images of the sound sources in reflecting walls (Lee and Lee 1988). In this way, it can provide more accurate information at a point detector (Allen and Berkley 1979). The most used method combines the potentialities of ray tracing and image methods to get room acoustical impulse responses (Vorlander 1989).

Through the application of the finite element method (FEM), the original problem of determining field variable distributions in a continuum domain is approximately transformed into a problem of determining the field variables at some discrete (nodal) positions within each element (Gladwell 1965). The method is usually used for propagation of sound vibration in the structures (Jiang et al. 2011).

The boundary element method (BEM) is derived through the discretisation of an integral equation that is mathematically equivalent to the original partial differential equation. It is used mainly in applications such as the sound output of a loudspeaker, the noise from a radiating source such as an engine and the interior acoustic modes of an enclosure (Hargreaves and Cox 2010).

2.2 *Integrated Approaches for Architectural Acoustics*

Although the room acoustics modelling techniques have originally been developed for the acoustic prediction and design of spaces for music (theatres, concert halls, auditoria etc.), the problems are equally challenging in other typologies of rooms and, thanks to the computational processes, the same methods can be adopted for them.

The main problem is that the reverberation time cannot be calculated only by the classical equations of Sabine and Eyring because the rooms can be very irregular, the diffusion of sound can be uneven and very different from the simple assumption of a diffuse sound field, and the sound absorption can be distributed irregularly over the surfaces. Another problem is that the reverberation time is not the most complete parameter describing the complex acoustical conditions of a room. Parameters, such as speech transmission index (STI) for workrooms and open plan offices, sound pressure level (SPL), clarity (C_{50}) or definition (D_{50}) for auditoria, should also be considered in the acoustic design of those spaces.

Nowadays, 3D computer modelling and numerically controlled fabrication processes have introduced new territories for the design of architectural acoustics. With 3D modelling programmes such as Rhino3D and Grasshopper incorporating

NURBS, geometries can be tailored to meet aesthetic, functional and performance objectives. These processes also allow embedding acoustic simulation in the initial design stages. Computational and optimisation design tools are taking a stronger role in geometrical acoustics. Mathematically defined geometries can be manipulated and exchanged between the modelling and analysis programmes, allowing for an easier shaping of rooms and surfaces (Bassuet et al. 2014).

Computational design processes applied to architectural acoustics help:

- To balance the relation between the shape of the space and sound comfort;
- To assist designers' creativity and explore a wide range of options but at the same time narrowing the field of possible solutions to only the most fitted ones;
- To check and confirm the validity of a particular design, and potentially aim towards more deterministic solutions, reducing the risk of wasting time or material;
- To define a novel performance-driven aesthetics;
- To develop techniques coupling in an auditory model a set of subjective criteria or spatial parameters, used for design optimisation.

3 Methodology

3.1 Performance-Driven Design for Global Form-Finding of Acoustic Surfaces

In response to these points, the paper describes an approach to the generation of acoustic surfaces with tailored performance for absorption and diffusion, employing simulation and evaluation tools. FEM analysis of the sound pressure level in the time and frequency domains is used for the identification of potentially problematic acoustic zones and frequency ranges, and accurate placement of acoustically absorptive surfaces, followed by ray-tracing as a means for formal definition of these surfaces on the global scale of the panel. These two performance criteria are used as the main design generation drivers of two geometric configurations: a ceiling and a wall partition, and are implemented in a parametric 3D design environment—Grasshopper for Rhino3D. Focusing on maximising the absorption by means of orienting the analysed geometries such as they intersect as many rays as possible, while minimising the overall surface area, here multi-objective optimisation algorithms are used for the generation of possible options. The goal is applying this performance-based strategy for the design of acoustic surfaces and solutions that are tailored not only to fit a specific acoustic condition requirement but also to minimise production cost and waste.

The objective of the EcoAcustica project towards an efficient manufacturing process is addressed by using the flexible Wood-Skin technological system for the generation of developable surfaces with planar triangulated facets. Such geometries allow to be produced from a single flat sheet by numerically controlled milling procedures,

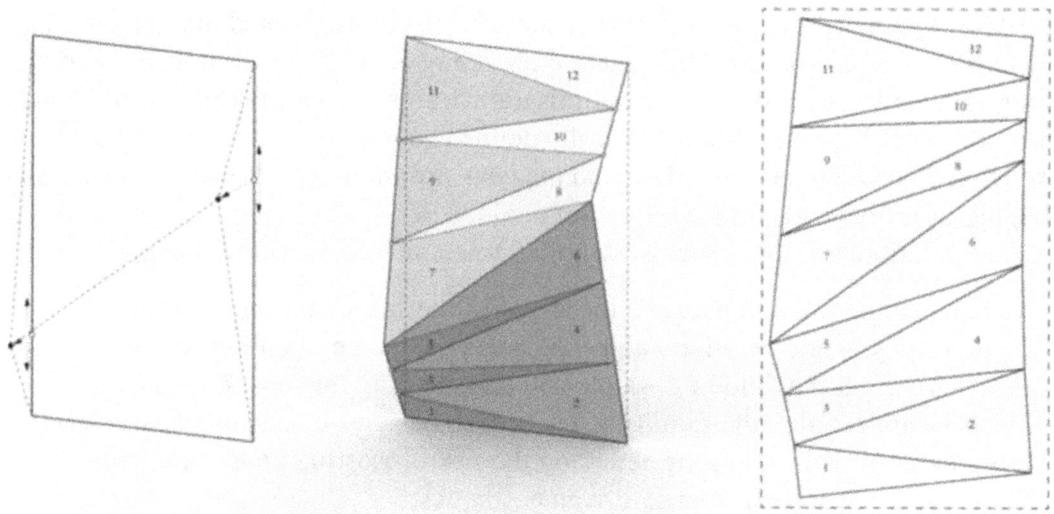

Fig. 1 Design logic of the developable surfaces

which guarantees reduced material waste and a facilitated assembly process that eliminates complex joinery (Fig. 1).

3.2 Empirical Studies on Materials and Absorption Coefficients

The integration of the ICB cork as the absorptive component of the acoustic system plays a central role in the EcoAcustica project. Four main types of new base composites were designed, so as to allow the differentiation of acoustic properties. Each composite targets a specific behaviour (absorption, diffusion and combination of absorption and diffusion), combining an acoustically reflective material, like okoume or plywood, with absorptive one—ICB cork, the most effective in mid- to high-frequency range. Following the fabrication logic of the Wood-Skin system, the combination of various composites within the same element allows for the versatility of the overall system. The absorptive properties of the panels are also enhanced with the possibility of varying the air gap behind that the Wood-Skin system allows. Based on preliminary studies of the behaviour of these material composites, prototype panels have been developed and tested according to the ISO 354:2003 standard. These empirical data on the sandwich composition and absorption coefficients are used in the simulation phase.

4 Case Study

4.1 Case Study Selection and Criteria

This paper applies the proposed methodology for the acoustics optimisation of the new laboratory of Department ABC at Polytechnic of Milan. The space, originally used as a classroom, has been designed to accommodate a new lab for advanced digital fabrication and prototyping for architecture. It hosts the following functions:

- Two fabrication areas with numerically controlled machinery—an area for robotic fabrication and a space for FDM 3D printers and a CNC machine (Fig. 2);
- A work and study space, dedicated for students and researchers;
- Offices, a meeting room and services.

The diversity of the functions present in the laboratory makes it a challenging acoustic environment, as naturally each of these activities requires a specific sonic treatment. The research in this paper focuses on the work and study area, as the one where a lack of proper acoustic conditions could lead to an uncomfortable environment for focused work or even simple conversation or discussion.

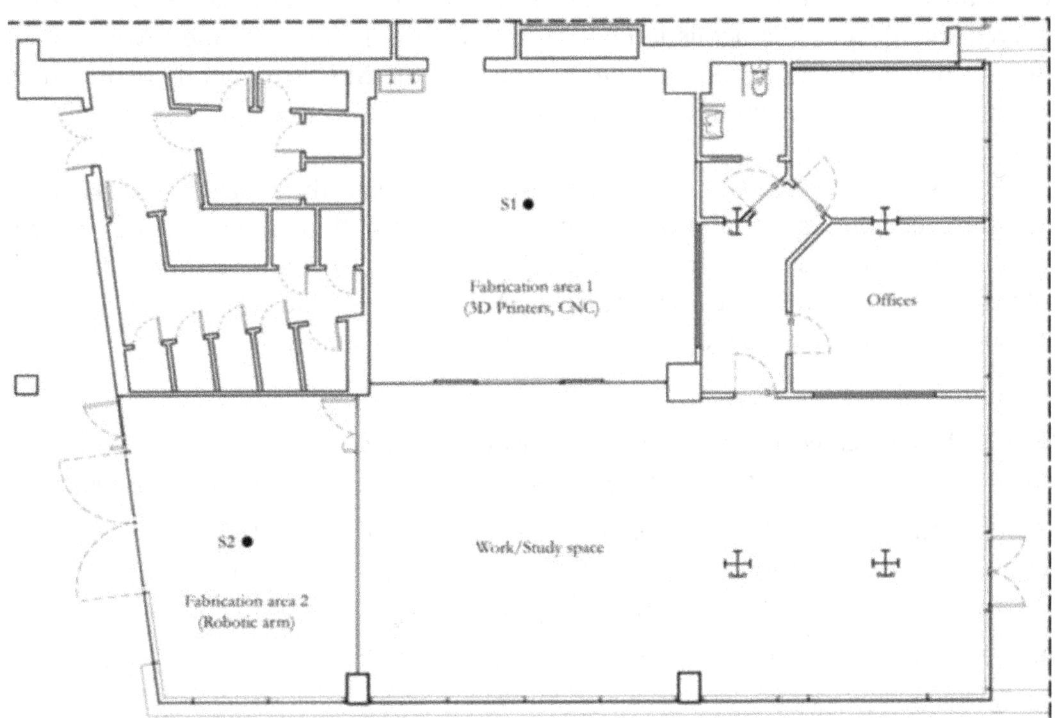

Fig. 2 Plan of the future Digital Fabrication Lab and the main sound sources

4.2 *Simulation and Analysis of the State-of-the-Art Conditions*

For the acoustic analysis and evaluation of the proposed interior design, simulations were performed with FEA software (Comsol), using the ray acoustics interface in order to understand the high-frequency (reverberant) behaviour of the space. The simulations were made considering the worst conditions, that is, with doors opened towards the work and study area, and with both sound sources active at the same time and employed in full capacity. The maximum sound power level of the robotic arm was hypothesised at 75 dB (0.01 W), considering a mill as an end effector, while the value for the sound power level for the CNC and the 3D printers, set as a logarithmic sum, was specified to 105 dB (0.03 W).

Two scenarios were considered for the simulations:

- Scenario 0: no acoustic treatment
- Scenario 1: standard acoustical dropped ceiling (200 mm construction depth and 50 mm mineral wool) (Table 1).

The goal of the acoustic performance analysis was to obtain information about the sound distribution in the space, through reverberation time (T_{60}) and time-dependent sound pressure level (SPL), calculated in full octave frequency bands. The results showed that despite the acoustic ceiling in Scenario 1 improves the reverberation time, from 2.75 to 1.2 s, compared to Scenario 0, the SPL levels in the work/study area in the speech-relevant frequency band are still significantly high—average 75 dB (Fig. 3).

The results were used in the next phase to delineate the areas of intervention, providing the boundary conditions for the spatial distribution of the acoustic surfaces and their geometric characteristics. The objective was to decrease the overall SPL values in the work/study area with at least 10–15 dB (ISO 3382-3), and decrease the reverberation time, employing less surface area than Scenario 1.

Table 1 Absorption coefficients of the materials used for simulation

Materials	125 Hz	250 Hz	500 Hz	1000 Hz	2000 Hz	4000 Hz
Wall (plaster)	0.30	0.02	0.03	0.04	0.05	0.07
Glass	0.30	0.20	0.10	0.07	0.05	0.02
Steel columns	0.13	0.09	0.08	0.09	0.11	0.11
Rough concrete	0.02	0.03	0.03	0.03	0.04	0.07
Floor (resin)	0.05	0.04	0.03	0.08	0.04	0.08
Acoustical ceiling	*0.55*	*0.80*	*0.75*	*0.65*	*0.60*	*0.50*

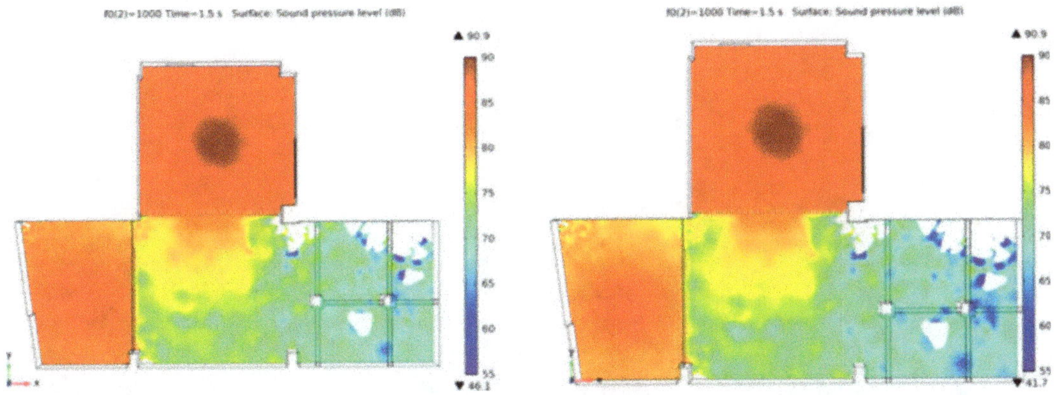

Fig. 3 Sound pressure levels for 1000 Hz, 1.5 s of Scenario 0 (left) and Scenario 1 (right)

5 Surfaces for Tailored Performance

Based on the analysis of the state-of-the-art conditions, two acoustic treatment strategies were defined:

- Scenario 2: an optimised solution on the ceiling
- Scenario 3: an optimised self-supporting partition.

Scenario 2 focuses on providing larger area of absorption to lower the reverberation time of the overall space. Scenario 3 instead uses a single partition, aimed at blocking first- and second-order reflections, preventing sound distribution towards the rest of the space. The Rayshoot method (Snail plug-in) was used for the generation of rays from the two sound sources and for calculating the intersection points with the analysed geometries. Multi-objective evolutionary optimisation was performed with the Octopus plug-into generate a diversity of optimised trade-off solutions, ranging between the extremes of each target.

The constraints, used for the geometry generation for Scenario 2 and for determining the boundary conditions of the optimisation process, include setting the displacement of the centre point of the panels in Z to 1 m, as well as limiting the number of tessellations to four per each face. Owing to the higher sound power level, and the closer proximity to the study area, the absorption from sound source 1 was prioritised (Fig. 4). The optimisation produced a range of candidate solutions, and the one with most optimal values for all the three objectives was selected.

Scenario 3 follows similar form-finding logic and the same objectives of Scenario 2, but the position of the partition within the space was added as an additional constraint and an opportunity to be moved to match changes in the functional programme (Fig. 5). The boundaries of the possible locations were derived from the SPL maps, resulting from the initial simulation. The selected candidate provides maximum ray intersection values from both the sources.

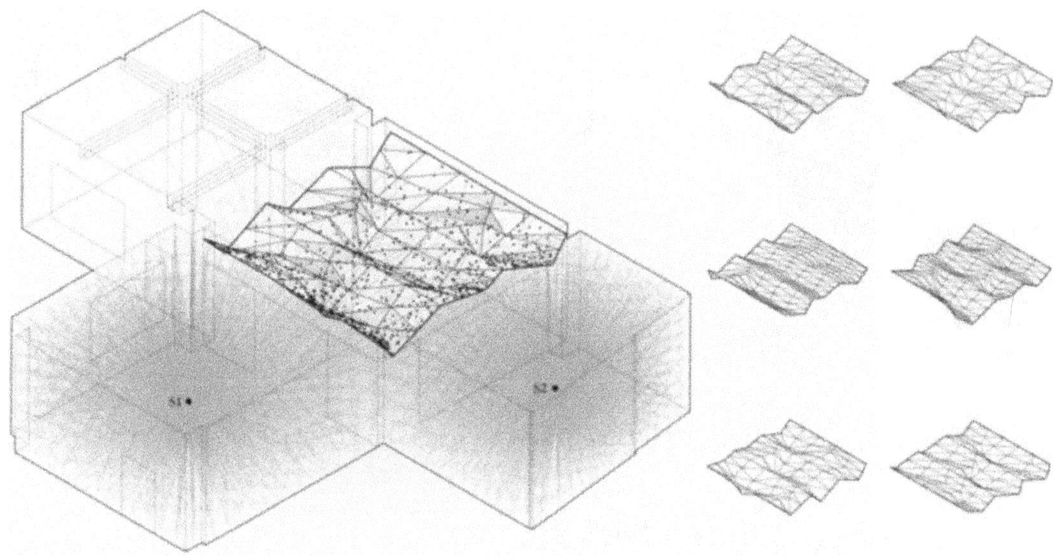

Fig. 4 Form-finding Scenario 2: ceiling (left) and candidate solutions (right)

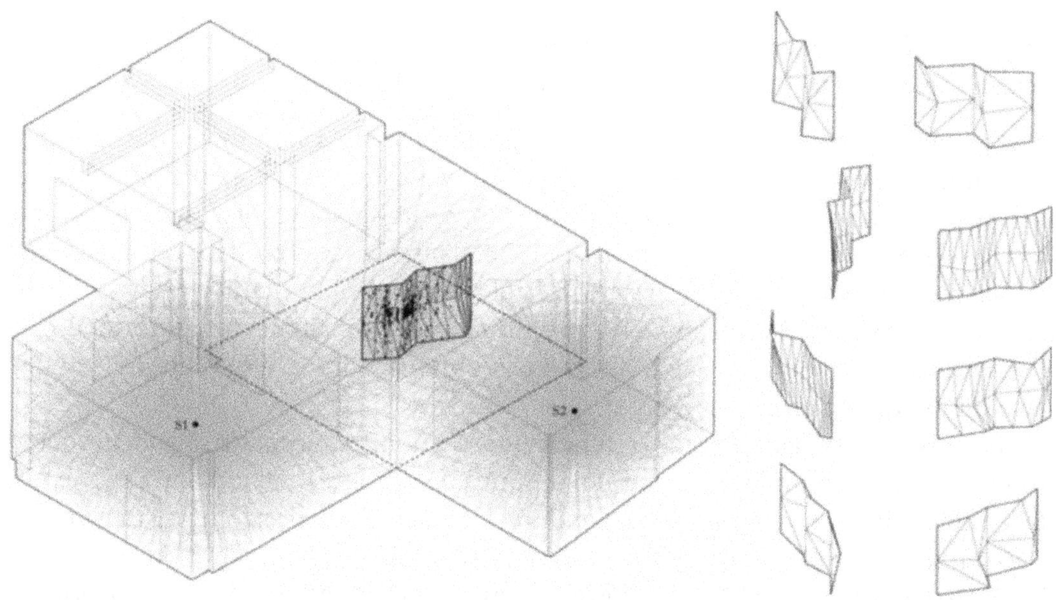

Fig. 5 Form-finding Scenario 2: wall (left) and candidate solutions (right)

6 Simulation and Results

The absorption values used for the analysis of the acoustic performance of the selected candidates were taken from the reverberation room tests of the prototypes, developed prior. The α values are: 125 Hz—0.55; 250 Hz—0.80; 500 Hz—0.75; 1000 Hz—0.65; 2000 Hz—0.60; 4000 Hz—0.50.

The simulation results for Scenario 1 showed that orienting the surfaces provides significant lowering of the overall SPL values within the work/study area, compared

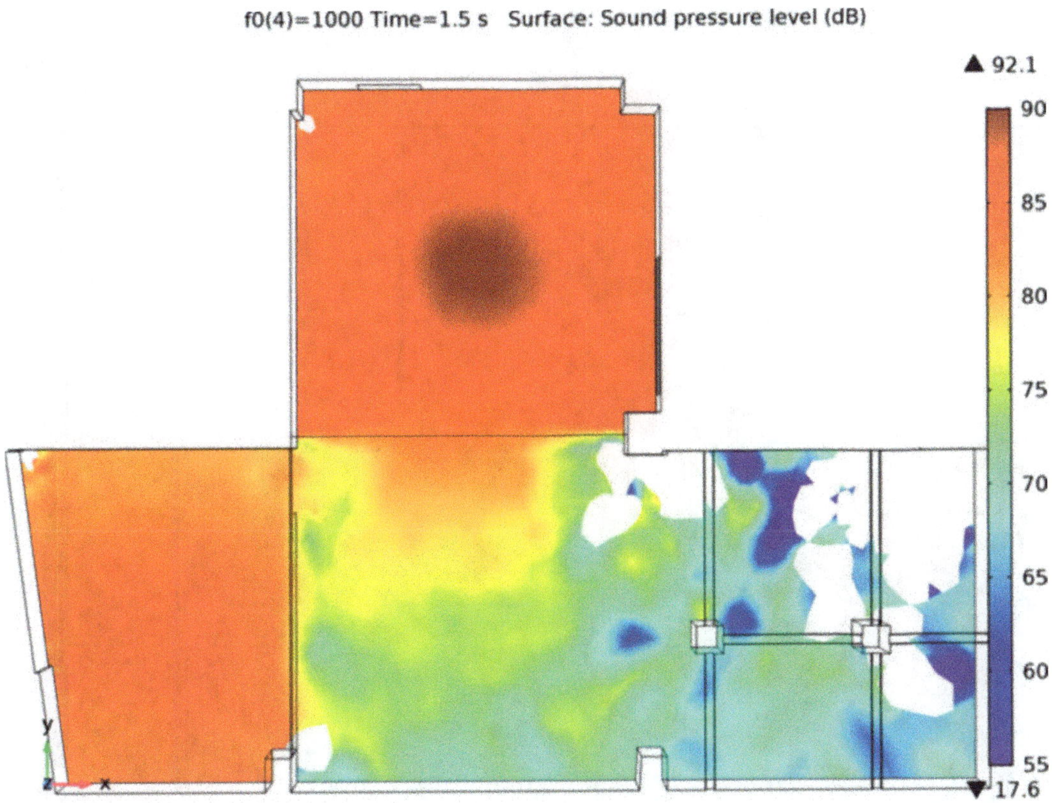

Fig. 6 SPL at 1000 Hz, 1.5 s for Scenario 2: ceiling

to Scenario 1 (Fig. 6). The reverberation time, however, was not affected much, and Scenario 1 provided shorter T_{60} values. This can be attributed to the fact that the overall surface area of the ceiling in Scenario 2 is much lower.

The results are similar in Scenario 3, showing lower SPL values within the work/study area than Scenario 1 (Fig. 7). This shows that the optimised position of the partition in the space influences early reflections, blocking the sound closer to the source. However, the higher SPL values, compared to Scenario 2, evidence that more absorption surface area is needed for the solution to be fully effective.

7 Conclusions and Discussion

The results of the acoustic analysis of the proposed solutions clearly demonstrate that the system, both in ceiling and wall configuration, contributes to creating an improved acoustic environment in the work/study area, compared to standard solutions with dropped acoustic ceiling. The outcomes show that following the proposed methodology for performance-driven form generation in the global scale, it is possible to achieve lower sound pressure levels by means of geometrical optimisation aimed at orienting the absorptive surfaces such as to attenuate sound closer to the

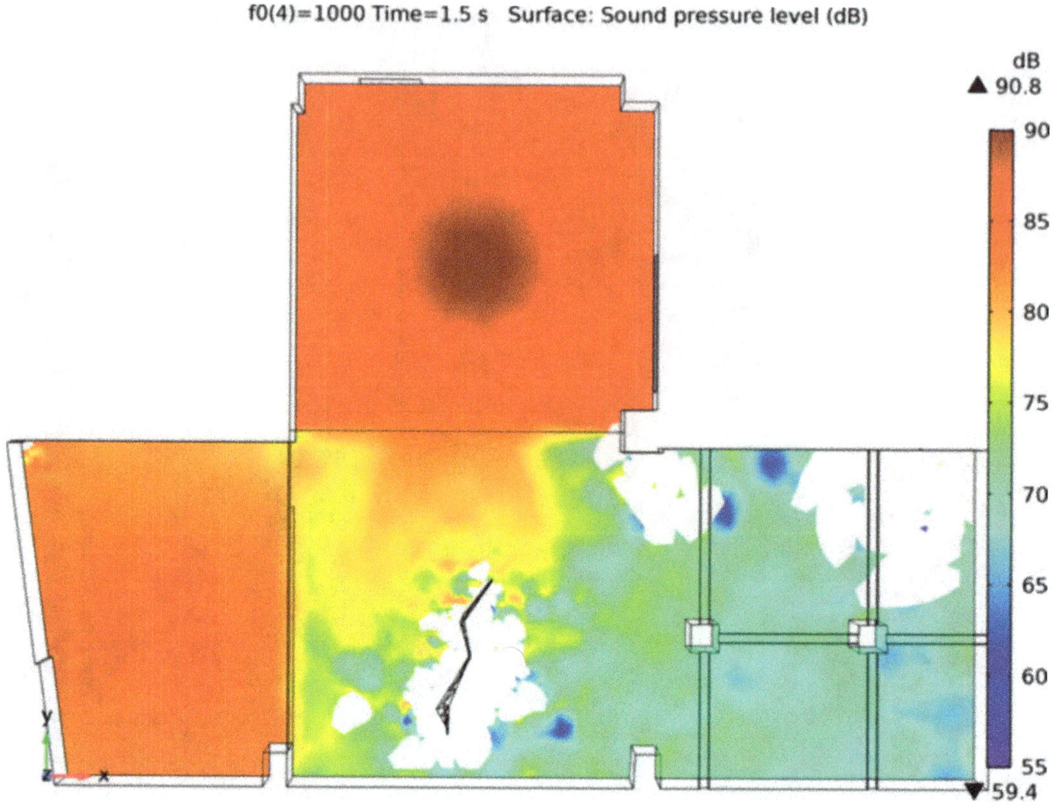

Fig. 7 SPL at 1000 Hz, 1.5 s for Scenario 3: wall

source and impact early reflections. Employing less surface area consequently would also decrease the cost of the overall acoustic treatment.

The innovative aspect of the work is featured in the employment of developable surfaces, which allow the fabrication of geometrically complex panels from a single flat sheet. Moreover, the customised sandwich compositions combined in the Wood-Skin technological system offer an additional layer of tailored performance, where the type and thicknesses of the composing materials could be chosen precisely according to the performance targets and functional programme.

The methodology allows the design of acoustic surfaces which not only could be tailored for an optimised absorption but with their inherent corrugations contribute for more uniform sound distribution and increased diffuse sound field, thus limiting the zones with high levels of SPL. The performance criteria will be implemented in the next developments of this project as an additional design target. Future work also includes creation of a custom fabrication process that allows the toolpath optimisation of the openings within the panels, such as the milling time is significantly shortened, leading to lower production cost. Full-scale prototypes will be developed and consequently tested, according to the ISO standards, with the results implemented in a holistic computational process.

References

Allen JB, Berkley DA (1979) Image method for efficiently simulating small-room acoustics. J Acoust Soc Am 65(04):943–950. Published by The Acoustical Society of America (ASA)

Bassuet A, Rife D, Dellatorre L (2014) Computational and optimization design in geometric acoustics. Build Acoust 21(1):075–086

Cox TJ, D'Antonio P (2017) Acoustic absorbers and diffusers. In: Theory, design and application. CRC Press

Gladwell GML (1965) A finite element method for acoustics. In: Proceedings of fifth international conference on acoustics

Hargreaves JA, Cox TJ (2010) A transient boundary element method for acoustic scattering from mixed regular and thin rigid bodies. Acta Acust U Acust 95(4):678–689

Jiang J, Huang K, Zhao Y (2011) Calculating room acoustic parameters by finite element method. In: International conference on consumer electronics, communications and networks, CECNet, Proceedings

Kulowski A (1984) Algorithmic representation of the ray tracing technique. Appl Acoust 18:449–469

Lee H, Lee B (1988) An efficient algorithm for image method technology. Appl Acoust 24:87–115

Lehnert H (1993) Systematic errors of the ray-tracing algorithm. Appl Acoust 38(2–4):207–221

Lewers T (1993) A combined beam tracing and radiant exchange computer model of room acoustics. Appl Acoust 38(2–4):207–221

Rindel JH (1995) Computer simulation techniques for acoustical design of rooms. Acoust Aust 23(3–81)

Van Maerck D, Martin J (1993) The prediction of echograms and impulse responses within the Epidaure software. Appl Acoust 38(2–4):93

Vorlander M (1989) Simulation of transient and steady state sound propagation in rooms using a new combined ray tracing/image source algorithm

Standards and Laws

BS EN ISO 354:2003 Acoustics—Measurement of sound absorption in a reverberation room

ISO 3382-1 Acoustics—Measurement of room acoustic parameters—Part 1: Performance spaces

Do Smart City Policies Work?

Andrea Caragliu and Chiara Del Bo

Abstract Smart City policies have attracted significant funding over the last few years. However, only less evidence is available of their impact on urban economic performance. In this paper, we look at the urban growth and innovation impact of Smart City policies, exploiting a dataset collected for these analyses comprising data on Smart City characteristics of 309 European metro areas, Smart City policy intensity, along with the urban growth and innovation outputs. Economic growth is measured as real GDP increases, while innovation is captured by patent applications to the European Patent Office, both measures being calculated between 2008 and 2013. Patent counts include technologically narrower classes, namely high-tech, ICT and specific Smart City technology patent applications. Instrumental variables and propensity score matching estimates suggest that cities engaging in Smart City policies more than the EU average tend to grow faster and patent more intensively.

Keywords Smart city · Program evaluation · Instrumental variables · Propensity score matching

JEL Classification Codes R11 · R12 · H43

This paper presents an original synthesis of the work carried out by the authors. Particularly, empirical analyses have been originally described in Caragliu and Del Bo (2018, 2019).

A. Caragliu (✉)
Architecture, Built Environment and Construction Engineering—ABC Department, Politecnico di Milano, Milan, Italy
e-mail: andrea.caragliu@polimi.it

C. Del Bo
Department of Economics, Management and Quantitative Methods—DEMM, Università degli Studi di Milano, Milan, Italy
e-mail: chiara.delbo@unimi.it

B. Daniotti et al. (eds.), *Digital Transformation of the Design, Construction and Management Processes of the Built Environment*, Research for Development,
https://doi.org/10.1007/978-3-030-33570-0_14

1 Introduction

The literature on Smart Cities has achieved vast academic and policy success, with a fast acceleration of scientific production over the past few years. The pervasive presence of information and communication technologies (henceforth, ICTs) in the present-day cities and the involvement of urban dwellers in collecting, sharing and exploiting information collected by sensors has elicited a debate cutting across different disciplines (Fig. 1), with a major focus on engineering, computer science and telecommunication studies.

This literature has also been criticized on the grounds of the heterogeneity in the type of scientific output, which often comprises gray literature and unpublished manuscripts as shown in Table 1.

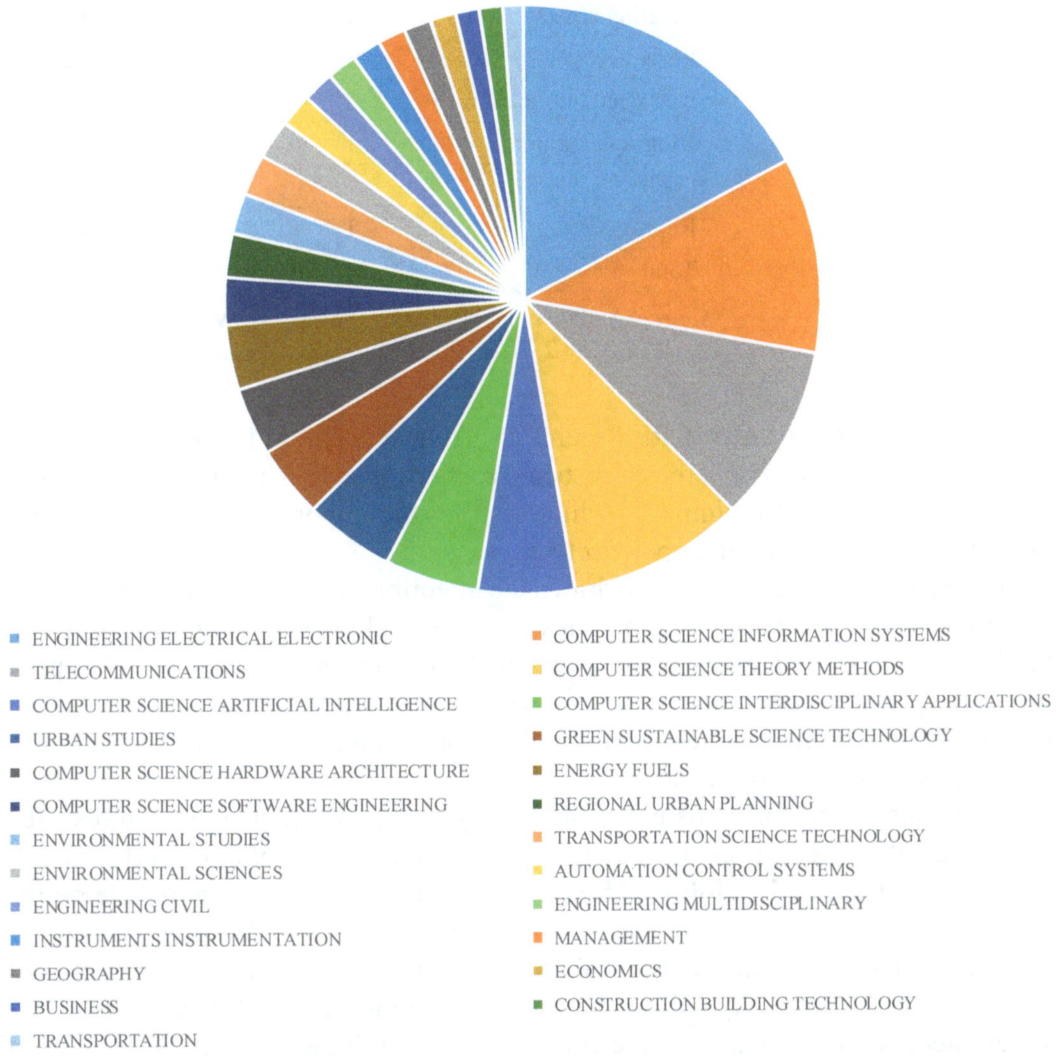

Fig. 1 Scientific outputs on Smart Cities divided by scientific discipline as of May 2019. *Source* Author's elaboration on the basis of Web of Science (WoS) raw data. Search made on May 2, 2019. Search strategy: «Smart cit*»

Table 1 Typologies of scientific outputs on Smart Cities as of May 2019

Document types	Record count (May 2019)	%
Proceedings paper	6490	49.34%
Article	5347	40.65%
Book chapter	612	4.65%
Editorial material	301	2.29%
Review	243	1.85%
Book review	62	0.47%
Book	34	0.26%
News item	25	0.19%
Correction	10	0.08%
Letter	8	0.06%
Data paper	7	0.05%
Meeting abstract	7	0.05%
Early access	3	0.02%
Retracted publication	2	0.02%
Biographical item	1	0.01%
Poetry	1	0.01%
Reprint	1	0.01%
Total	13,154	1

Source Author's elaboration on the basis of WoS raw data. Search made on May 2, 2019. Search strategy: «Smart cit*»

Yet, the interest in Smart Cities seems far from vanishing. In fact, if anything, research on Smart Cities is increasingly being carried out across all major universities, with a growth of scientific impact,[1] which testifies the importance of the topic from both an academic and a policy perspective.

Consequently, Smart Cities have drawn the attention of countless policymakers at all administrative levels, shaping policies aimed at making cities *smarter*. Relevant funding has been provided by bodies such as the European Commission (European innovation partnership on smart cities and communities; henceforth, SCC), European Investment Bank, and countless States and regional Authorities. Yet, the landscape of policy assessment is rather scant.

Within this framework, this paper offers a synthesis of the quantitative research we carried out on the impact of Smart City policies on urban economic performance. The paper is structured as follows. Section 2 briefly summarizes the debate on the definition of Smart Cities, culminating with the one provided in Caragliu et al. (2011) which underlies all the empirical contributions discussed later. While in Sect. 3 a brief recap of the way Smart City policies have been implemented and translated into data is presented. Section 4 discusses the empirical methodologies for policy assessment adopted in the empirical estimates, which are synthesized in Sect. 5, and the dataset collected for our analyses. Finally, Sect. 6 draws a number of conclusions.

[1]Works satisfying the WoS search strategy used for Fig. 1 and Table 1 suggest that roughly 50% of all (12,395) WoS entries have been published between 2017 and 2019.

2 Defining Smart Cities

Since the dawn of Smart City literature, a heated debate has raged even on the very definition of what a true Smart City really is (Hollands 2008). Vallianatos (2015) extends the timeline of "smart cities" and "big data" efforts by a considerable amount—all the way back to the late 1960s. "*Beginning in the late 1960s and through most of the 1970s, the little-known Community Analysis Bureau used computer databases, cluster analysis, and infrared aerial photography to gather data, produce reports on neighborhood demographics and housing quality, and help direct resources to ward off blight and tackle poverty*" (*ibid.*).

Over time, several definitions of the concept of Smart City have emerged, each differing in terms of the main *smart* characteristic deemed as the most relevant to define the very notion of urban smartness. Initial conceptualizations revolved around ICTs as the main pillar around which a city should build its smart pathway.

Earlier definitions related to the concept include the "*wired city*" (Dutton et al. 1987), whereby the focus is on networking the urban space per se and the "*intelligent city*" (Komninos 2009), which expands this idea by considering also the cognitive element of a digital city and the relationship between individual cognitive skills and the urban information system.

Linking smartness to the availability, development and use of ICT is based on the idea that, in order to succeed, cities and urban dwellers must be interconnected. The focus on ICTs and data availability is also related to a policy shift from mainly top-down to increasingly more bottom-up approaches, paving the way for greater importance of citizens' direct participation in the urban governance and their links with city officials.

This latter aspect is related to the importance of e-government in the creation of a Smart City (Deakin and Al Waer 2011). This in turn suggests an important role for the human capital component of cities and might also give rise to concerns related to the availability, management and privacy issues linked to big data (Batty 2012).

The vision of ICT-centered smartness eventually spurred several academic projects and research, not without critique against the strong business orientation of this concept. In fact, ICTs have been from the very inception of this concept considered at the roots of urban smartness, mostly because of the vast empirical evidence supporting ICT-led development (Caragliu 2013). This attracted the interest of large corporations, aiming at profiting from the interest of urban administrations in such technologies (Vanolo 2014).

Despite the relevant academic success and the significant investments in Smart City projects at all administrative levels, however, only a handful of works provide a clear-cut definition of what a Smart City really is, particularly by extending their characterization beyond the pure technological content, thus including contextual elements.

Because context elements are expected to co-explain the success of Smart City policies, only two notable definitions are mentioned (in chronological order), which truly go beyond ICTs as a means to define urban smartness, highlighting the most important elements.

Giffinger et al. (2007) provide a classification of European medium-size cities according to six axes (Smart people, Smart governance, Smart environment, Smart economy, Smart mobility and Smart living). Their definition reads as follows: "*A Smart City is a city well performing in a forward-looking way in these six characteristics, built on the 'smart' combination of endowments and activities of self-decisive, independent and aware citizens*" (Giffinger et al. 2007, p. 13).

Caragliu et al. (2011) build on the classification by Giffinger and co-authors, providing a comprehensive and operational definition of urban smartness. In this case, cities are identified as smart when "*investments in human and social capital and traditional (transport) and modern (ICT) communication infrastructure fuel sustainable economic growth and a high quality of life, with a wise management of natural resources, through participatory governance*".

This definition presents two main advantages.

- It is inspired by an urban production function approach whereby urban smartness is defined as a precondition to urban economic performance;
- The definition decomposes the concept along six dimensions, which can be individually measured, using data from official sources. Therefore, this definition has been among the first to be empirically verified (Caragliu and Del Bo 2012).

In the rest of the paper, this last definition represents the foundations of all empirical estimates.

3 Delimiting Smart City Policies

While academics have actively participated in the debate about the definition of the concept of Smart City, they have relatively neglected the policy appraisal side.

Despite the non-negligible funding available at all spatial scales, albeit chiefly from the European Union through the SCC initiative,[2] to date insufficient attention has been paid to a careful analysis of both the economic rationale for Smart City policies, as well as their potential growth-enhancing effects on cities.

Two major issues seem relevant for this analysis. On the one hand, Smart City policies must show some feature that makes this specific object of policy different from other axes of intervention. In other words, the economic rationale for Smart urban policies should be clarified. On the other hand, once the nature of such policies has been defined, their expected impact on urban growth should be discussed, with an eye on a possible empirical strategy to appraise the impact.

[2]It is worth stressing that this study focuses on European cities. Asian and US cities have experienced rather different paths towards urban smartness. In the USA, for instance, public–private partnerships (fostered among other entities by SmartAmerica, a White House Presidential Innovation Fellow project) foresee heavy involvement of private actors.

On the first front, Smart Cities can be defined as the result of the interplay among the six axes of the Giffinger et al. definition. Vast theoretical work and empirical verifications are available on the role of human (Berry and Glaeser 2005) and social (Glaeser and Sacerdote 1999) capital; transport (Duranton and Turner 2012) and ICT (Basu et al. 2003) infrastructure; quality of life (Lenzi and Perucca 2018) and participatory governance (Rodriguez-Bolívar 2018) in urban development. Smart Cities are urban areas benefitting from a simultaneous engagement in all these six axes.

Recent evidence shows that Smart City policies are undertaken by urban areas that already score high in one or more of the axes of the definition used in this paper (Neirotti et al. 2014). Context factors, thus, not only enhance urban smartness, they also empower local dwellers and increase chances of success for Smart City policies.

From an economic perspective, our research has identified three main microfoundations for a positive policy impact to be expected:

- Smart City policies make cities more productive (Chourabi et al. 2012). For instance, Smart City policies have often stimulated the widespread availability of knowledge and information, especially in terms of big data (Kitchin 2015).
- Smart City policies enhance citizens' participation (Bakici et al. 2013). Participation of various social groups in the construction of Smart Cities is one of the most notable elements differentiating this concept from other similar notions found in the literature.
- Smart City policies offer increased business opportunities for local firms (Lee et al. 2014).

While some qualitative work on Smart City policy effectiveness has been carried out, a grand overview of the empirical association between Smart City policies in a cross-section of cities and urban performance is mostly absent. This translates into the following research question:

RQ. What is the economic impact of adopting Smart City policies on urban growth and innovation?

4 Methods and Data for Policy Impact Assessment

The research question of this paper faces a number of relevant empirical issues.

The two most relevant problems are related to the nature of Smart City policies impact (Do Smart City policies directly foster urban performance?), and the potentially relevant issue of endogeneity (Do Smart City policies foster urban performance, or is there selection bias?).

While there is very limited evidence of a direct causal impact of smart urban features on economic performance, individually taken, each axis of the adopted definition of urban smartness has been found to be positively associated with economic performance. Our work hinges on this finding (and tests it empirically) in order to identify the causal link between this type of policy and urban economic performance. The natural candidates for this type of econometric exercise are the instrumental variable (IV) and propensity score matching (PSM) estimators.[3]

As for the data employed in these analyses, in order to measure Smart City policy intensity, five main data sources on policy intensity have been analyzed:

- Cities implementing smart policies in the list prepared by the European Parliament (2014);
- Member cities of the Eurocities network;
- Cities participating in Framework Programme 7 Smart City initiatives;
- Cities actively cooperating with IBM for the deployment of Smart urban technologies;
- Member cities of the Lighthouse, ERRIN, and ICLEI networks.

Additionally, data have been collected to calculate the mean urban smartness indicator for a sample of 309 EU cities based on the six dimensions of the definition in Caragliu et al. (2011). All the remaining data are collected at the EU metro area level, apart from the indicator of urban quality of institutions (Charron et al. 2015). This latter portion of the data is collected at NUTS2 level, and the value of each NUTS2 region is assigned to the metro area located in the region.

5 Empirical Results

Table 2 presents the empirical estimates discussed in Caragliu and Del Bo (2018). The estimates are presented in five columns, whereby additional control variables are added to highlight possible multicollinearity issues in the models tested. Column (5) is the preferred specification and suggests that a 16% increase in Smart City policy intensity is associated with a 1% faster GDP growth. Our findings suggest a positive and statistically significant association between investing in Smart City policies and urban GDP growth. These estimates also suggest that this association is causal: Smart City policies foster economic performance, thus ruling out reverse causality.

Figure 2 graphically represents the main findings found in Caragliu and Del Bo (2019). This is obtained as follows. First, all patent applications to the European Patent Office from 2012 onwards have been geo-referenced and assigned to an EU metro area. Each patent application is assigned to one or more International Patent Class (IPC) codes, reflecting the nature of the technology characterizing the patent.[4]

Next, three sub-classes of patent applications have been identified, namely high-tech patent applications, ICT patent applications and Smart City patent applications

[3]Owing to space limitations, most technical details are only briefly summarized here. For further analysis, the readers are invited to look into Caragliu and Del Bo (2018, 2019) where they will find a more thorough discussion regarding the technical details of the empirical estimates.

[4]When more IPC classes are available for each patent, the first has been used as the one characterizing the patent.

Table 2 Smart City policies and urban GDP growth

Dependent variable	Metro area GDP growth rate, 2008–2013				
Model	(1)	(2)	(3)	(4)	(5)
Constant term	0.08*** (0.00)	0.10*** (0.00)	0.24*** (0.02)	0.09** (0.04)	0.02 (0.03)
Initial per capita GDP	−0.02*** (0.00)	−0.02*** (0.00)	−0.03*** (0.00)	−0.04*** (0.00)	−0.01 (0.00)
Intensity of Smart City	0.11***	0 23***	0.22***	0 24***	0.16***
Policies	(0.00)	(0.00)	(0.00)	(0.06)	(0.06)
Population density	–	−0.01*** (0.00)	−0.01*** (0.00)	−0.01*** (0.00)	−0.01*** (0.00)
R&D expenditure	–	–	0.03*** (0.00)	0.02*** (0.00)	0.02*** (0.00)
Quality of local institutions	–	–	–	0.04*** (0.01)	0.04*** (0.01)
Dummy New Member States					0.05*** (0.01)
Number of obs.	309	309	309	309	309
R^2	0.26	0.28	0.28	0.29	0.32
Joint F test	51.42***	30.94***	46.43***	40.32***	56.52***
Estimation method	IV	IV	IV	IV	IV
Variable insturmented	Intensity of Smart City policies				
Instruments used	Urban smartness; dummy, equal to 1 if the city is the Country capital				
Underidentification test (Kleibergen-Paap rk LM statistic)	46.13***	34.17***	34.03***	30.65***	21 41***
Weak identification test (Cragg-Donald Wald F statistic)	50.47***	33.23***	32.61***	31.11***	19 12***
Hansen J statistic (overidentification test of all instruments)	19.24***	6.33**	2.41	0.36	0.48

Source Caragliu and Del Bo (2018)
Note Heteroskedastic-robust standard errors in brackets
*, **, and *** indicate significance at the 90, 95, and 99%, respectively

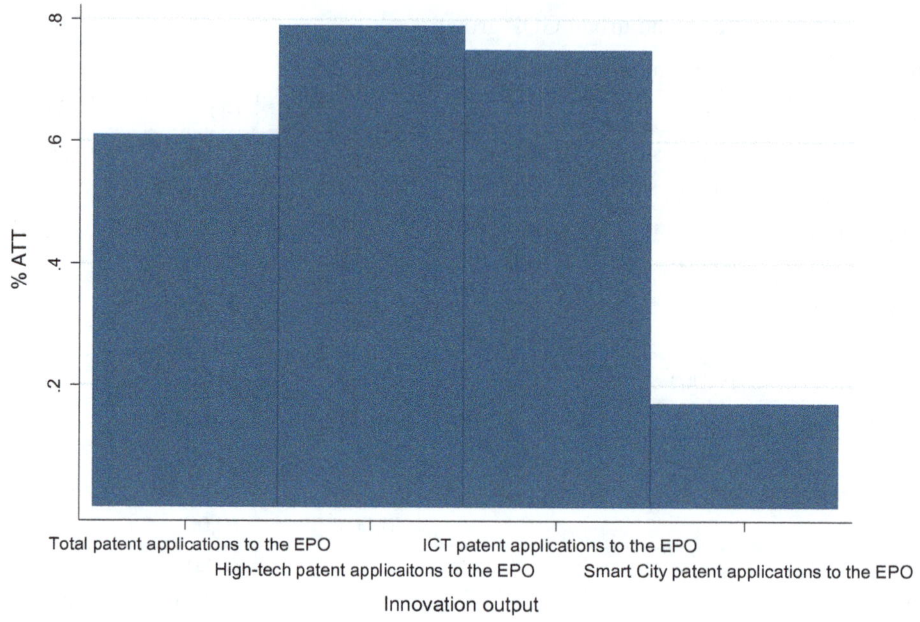

Fig. 2 Percentage average treatment effect on the treated cities. *Source* Caragliu and Del Bo (2019)

(IPO 2014). Finally, the sample of cities analyzed has been divided into those above and below the median intensity of Smart City policy intensity. PSM has been used to isolate the net impact of engaging in these policies on the innovation outcome described above.

Figure 2 suggests that the effects of Smart City policies are not limited to the most strictly defined IPC class of patents; rather, the impact is highest when looking at high-tech patents. Thus, an interesting spillover effect may be at play.

6 Conclusions and Ways Forward

This paper has provided a synthesis of the empirical work presented in Caragliu and Del Bo (2018, 2019) on the impact of Smart City policies on urban growth and innovation. Results suggest a statistically significant association between Smart City policies and the outcomes mentioned above; moreover, the use of advanced econometric techniques allows for inference in the causality direction of this link.

Yet, many more research questions relating to the economics of Smart Cities remain open.

First of all, the existence of a direct link between Smart urban features, and the possible synergic role they may play in stimulating economic growth, is yet to be inspected. Ideally, this exercise would require longer time spans in the data, in order to uncover possible long run effects that the database collected for this paper cannot capture.

Moreover, a sound conceptual classification of existing Smart City policies could also be beneficial. The work here summarized is restricted by the limited availability of data concerning Smart City policy intensity, which could be overcome by a more formal definition of what makes a true Smart City for the policy provided.

Lastly, many criticisms have been discussed on the unequal effects that Smart City policies may have; for instance, benefitting the wealthy and educated, while causing a digital divide in the poor and uneducated segments of the population. This debate would greatly benefit from a sound empirical assessment of the impact of these policies on the intra-urban and intra-national distribution of resources.

References

Bakici T, Almirall E, Wareham J (2013) A smart city initiative: the case of Barcelona. J Knowl Econ 4(2):135–148

Basu S, Fernald JG, Oulton N, Srinivasan S (2003) The case of the missing productivity growth, or does information technology explain why productivity accelerated in the United States but not in the United Kingdom? NBER Macroecon Annual 18:9–63

Batty M (2012) Building a science of cities. Cities 29:S9–S16

Berry CR, Glaeser EL (2005) The divergence of human capital levels across cities. Papers Reg Sci 84(3):407–444

Caragliu A (2013) Dynamics of knowledge diffusion: the ICT sector in L ombardy. Reg Sci Policy Pract 5(4):453–473

Caragliu A, Del Bo C (2012) Smartness and European urban performance: assessing the local impacts of smart urban attributes. Innov Eur J Soc Sci Res 25(2):97–113

Caragliu A, Del Bo C (2018) The economics of smart city policies. Sci Reg 17(1):81–104

Caragliu A, Del Bo CF (2019) Smart innovative cities: the impact of Smart City policies on urban innovation. Technol Forecast Soc Chang 142:373–383

Caragliu A, Del Bo C, Nijkamp P (2011) Smart cities in Europe. J Urban Technol 18(2):65–82

Charron N, Dijkstra L, Lapuente V (2015) Mapping the regional divide in Europe: a measure for assessing quality of government in 206 European regions. Soc Indic Res 122(2):315–346

Chourabi H, Nam T, Walker S, Gil-Garcia JR, Mellouli S, Nahon K, Scholl HJ (2012) Understanding smart cities: an integrative framework. In: 2012 45th Hawaii international conference on system sciences, pp 2289–2297. IEEE

Deakin M, Al Waer H (2011) From intelligent to smart cities. Intell Build Int 3(3):140–152

Duranton G, Turner M (2012) Urban growth and transportation. Rev Econ Stud 79(4):1407–1440

Dutton WH, Blumler JG, Kramer KL (May 1987) Continuity and change in conceptions of the wired city. In: Wired cities: shaping the future of communications, pp 3–26. GK Hall & Co

Giffinger R, Fertner C, Kramar H, Kalasek R, Pichler-Milanović N, Meijers E (2007) Smart cities. Ranking of European medium-sized cities. Centre of Regional Science of Vienna. www.smart-cities.eu. Accessed 21 Dec 2014

Glaeser EL, Sacerdote B (1999) Why is there more crime in cities? J Polit Econ 107(S6):S225–S258

Hollands RG (2008) Will the real smart city please stand up? City 12(3):303–320

Intellectual Property Office (2014) Eight great technologies. The Internet of Things: a patent overview. https://www.gov.uk/government/publications/new-eight-great-technologies-internet-of-things. Accessed 21 Nov 2017

Kitchin R (2015) Making sense of smart cities: addressing present shortcomings. Cambr J Reg Econ Soc 8(1):131–136

Komninos N (2009) Intelligent cities: towards interactive and global innovation environments. Int J Innov Reg Dev 1(4):337–355

Neirotti P, De Marco A, Cagliano AC, Mangano G, Scorrano F (2014) Current trends in Smart City initiatives: some stylized facts. Cities 38:25–36

Lee JH, Hancock MG, Hu MC (2014) Towards an effective framework for building smart cities: lessons from Seoul and San Francisco. Technol Forecast Soc Chang 89:80–99

Lenzi C, Perucca G (2018) Are urbanized areas source of life satisfaction? Evidence from EU regions. Papers Reg Sci 97:S105–S122

Rodriguez-Bolívar PM (2018) Governance models and outcomes to foster public value creation in smart cities. Sci Reg 17(1):57–80

Vallianatos M (2015) Uncovering the early history of 'Big Data'and 'Smart City'in Los Angeles. Boom Calif. https://boomcalifornia.com/2015/06/16/uncovering-the-early-history-of-big-data-and-the-smart-city-in-la/. Accessed 2 May 2019

Vanolo A (2014) Smartmentality: the Smart City as disciplinary strategy. Urban Stud 51(5):883–898

Digital Design and Wooden Architecture for Arte Sella Land Art Park

Marco Imperadori, Marco Clozza, Andrea Vanossi and Federica Brunone

Abstract Digital design is increasingly sinking the construction sector, shaping and validating architecture according to various criteria and introducing the wood industry to the 4.0 approach. Within the study entitled "Architecture at Arte Sella", parametric design, structural validations and CNC procedures are exploited to help define, control and assess several architectural woodworks, created with famous designers. This contribution describes the design and construction experiences of Atsushi Kitagawara (2017) and Kengo Kuma (2018–2019), the two masterpieces installed in the land art park of Arte Sella (Trento, Italy) and developed, thanks to the Politecnico di Milano team, from design to mock-ups, testing and construction.

Keywords Diffuse lattice structure · Parametric design · FEM-based structural analysis · CNC-based processes · Wooden land art

1 Introduction

Digital design is increasingly sinking the construction sector, shaping and validating architecture according to various criteria (Vanossi 2018; Tedeschi 2014; Vanossi and Imperadori 2013), and supporting the realizations at several stages and dimensions of the construction process (Eastman et al. 2008).

It has found, in particular, a ready stage for a 4.0 approach within the wood-based industry, thanks to computer-aided design (CAD) and computer-aided manufacturing (CAM) systems, already available for timber solutions, both within the design and

M. Imperadori (✉) · F. Brunone
Architecture, Built Environment and Construction Engineering—ABC Department, Politecnico di Milano, Milan, Italy
e-mail: marco.imperadori@polimi.it

M. Clozza
D3WOOD, Lecco, Italy

A. Vanossi
CMB, Carpi, Italy

B. Daniotti et al. (eds.), *Digital Transformation of the Design, Construction and Management Processes of the Built Environment*, Research for Development,
https://doi.org/10.1007/978-3-030-33570-0_15

architecture fields (Bianconi and Filippucci 2019b). Numerous experiences (Bianconi and Filippucci, 2019a; Imperadori et al. 2019; Kobayashi and O'Keefe 2019; Imperadori and Brunone 2018; Kuma 2018; Kobayashi 2017; Kuma and Daniell 2015; Menges et al. 2016; Vanossi et al. 2014, 2015), indeed, have demonstrated the wide spectrum of possible applications for digitalized processes to wooden realizations—artworks, land/urban art pieces, small-scale pavilions and so on—from computational design to digital fabrication, robotics and so on.

Within this field, the research project "Architecture at Arte Sella" has defined the opportunity to develop a continuous and iterative workflow, based on parametric design, structural validations and computer numerical control (CNC) procedures to define, control and assess several wooden installations, all characterized by an inner complexity of conception, production, assembly and structural validation. After a detailed description of the followed process methodology, this paper refers to, specifically, two design and construction experiences of masterpieces, first conceived for Arte Sella (Montibeller et al. 2017; http://www.artesella.it/en/), a mountain land art park near Trento, Italy. The first one is the *Forest Byobu* (2017) conceived in collaboration with Atsushi Kitagawara and AKA as a re-use project for the outstanding wooden structural lattice of the Japan Pavilion façade at EXPO Milano 2015 (Figs. 1, 2 and 3) (Kitagawara et al. 2019). The second one is *Kodama*, a wooden blades lattice defining a small polyhedral pavilion (Figs. 4, 5 and 6), by Kengo Kuma (Kuma et al. 2019). Both wooden structures have been developed, thanks to the team of Politecnico di Milano, collaborating with the most innovative and advanced research teams and industries of the fields, from design to mock-ups, testing, construction and maintenance, and leading to a methodology which has since then been replicated in several later experiences, between Italy (Lecco, Favara, and Ischia), Japan (Tokyo) and Taiwan.

2 Process Methodology

Since its kick-off in 2017, "Architecture for Arte Sella" has counted already the realization of several wooden artworks for this inspiring land art park in the valley of Borgo Valsugana (Montibeller et al. 2017; http://www.artesella.it/en/).

Reading into a cross-section of all these experiences let us trace a defined, shared and, therefore, validated method of design, assessment and realization, which has been applied to each project, applying CAD, CAM and BIM at different levels of development for the different stages of the process. Indeed, the methodology accounts for computational design of timber elements, joinery and three-dimensional structures from the very beginning of the design conception to structural validation, right up to digital fabrication through CNC-cutting processes, along with the management of construction issues, but always looking to the materiality of architecture as well and the need to assess all the stages in a hands-on fashion, via mock-ups and tests. The process thus far defined can be outlined as follows, considering each stage as

Fig. 1 Japan Pavilion wooden façade, conceived and realized for EXPO Milano 2015, thanks to an intense and iterative process between computational and parametric design, FEM-based structural analysis and digital fabrication with CNC machines, which allow both real scale mock-ups and final fast and precise assembling on site (*drawings and photo courtesy* Atsushi Kitagawara Architects and Galloppini Legnami)

Fig. 2 Forest Byobu design definition (*drawings credit* Galloppini Legnami and Politecnico di Milano) and assembling phases (*credit* Politecnico di Milano), up to the final complete structure, a geometric pattern contrasting the natural landscape of its location at Arte Sella, the Contemporary Mountain (*credits* for the first row, Ph. Giacomo Bianchi for Arte Sella; for the latter, Politecnico di Milano)

Fig. 3 The *Byobu* experiences. Above to the left, the *University Byobu* in Politecnico di Milano—Lecco campus (*source* Politecnico di Milano) and, to the right, the *Urban Byobu* at FARM Cultural Park, Favara (*source* Politecnico di Milano); in the middle, the *Kigumi Infinity Byobu* at Mori Museum of Tokyo (*credit* Ph. Shigeo Ogawa for Atsushi Kitagawara Architects); and finally the one on the island of Ischia (*source* Politecnico di Milano and PIDA Association)

Fig. 4 *Kodama Pavilion* in Arte Sella, from the first sketch by Kengo Kuma, to the FEM-based structural validation and parametric management of elements' dimensions, codifications and assembling phases; in the middle, the infinite number of mock-ups realized through digital fabrication and the assembling stages, performed by D3WOOD (*credit* Politecnico di Milano); lastly, Kengo Kuma land art masterpiece (*credit* Ph. Giacomo Bianchi for Arte Sella)

Fig. 5 The Taiwanese sample of Kodama Pavilion realized, accounting for the same digitalized processes of design and production, by Kengo Kuma Lab in 2018 (*credit* Ph. Takahiro Hirayama for Kengo Kuma Lab)

Fig. 6 *Kodama 2.0* is the Arte Sella pavilion refurbishment damaged during the devastating storm of October 2018: the structure has been (i) protected to retain the humidity content of wood elements before restoration; (ii) new pieces have been cut via CNC machine, (iii) burned and carved to render a Japanese haiku on the surface, and (iv) relocated to its original location, obtaining a new aesthetic for the wooden lattice (*credit* Politecnico di Milano and Arte Sella)

interconnected within an iterative workflow of conceptual hypothesis, parametric translation, virtual and/or real evaluation and validation:

- Design conception—beginning with the first sketches, the design concept passes through a parametric translation in order to optimize the singular elements of the construction as a module defined by parameters (e.g. dimension, mutual relation, and global proportion); the same is evaluated through real mock-ups at different scales, which are able to outline the real consistency of architecture, and can be elusive in virtual reality: space, as a relationship between solids and voids;
- Structural validation—finite elements methods (FEM)-based assessment procedures are carried out alongside structural tests on real prototypes, which are useful to assess the reliability of the virtual analysis of singular elements and joineries;
- Details and connections—the prototyping process, feasible exactly because of the CAM and CNC-based fabrication, allows for tests on a 1:1 scale of the construction details, such as wood-by-wood joinery and/or steel anchors to the ground, in terms of both structural stresses distribution and construction feasibility of the assembling processes (from the modular element to the whole timber structure);
- Production and assembly—at this final stage, BIM and parametric environments allow users to manage the CNC-based production, thanks to an open flux of data and information between the design and production files, where all the single pieces are unequivocally codified (minimizing the waste of resources); moreover, these methodologies help to set virtual assembling scenarios for the whole structure, in order to optimize construction (minimizing its timeframe).

This fil-rouge of digital design and fabrication has run throughout the experiences of Arte Sella, and indeed beyond them, since the same ones have been already replicated in several contests, attesting the reliability of this process methodology.

3 Case Studies

The outlined process has been applied on two different experiences (*Byobu* and *Kodama* pavilions), both started with the realization of the first samples (*Forest Byobu* and *Kodama*) located in Arte Sella and then continued with derivative projects (*Urban Byobu*, *University Byobu*, *Kigumi Infinity Museum Byobu*, and *Kigumi Infinity Ischia Byobu*; *Kodama pavilion* in Taiwan and *Kodama 2.0*, restoration project of the first original pavilion, after damages caused by a heavy storm), accounting for the methodology described above.

3.1 Wooden Byobu: Atsushi Kitagawara

Forest Byobu, and more generally the *Byobu* experiences all around Italy (and Japan), derives as a sequel from the EXPO Milano 2015 challenge to design and realize the wooden structure of the Japan Pavilion façade, tackled by Atsushi Kitaga wara Architects (www.kitagawara.co.jp), Ishimoto Architectural and Engineering Firm, and Ove Arup and Partners—for the design stage—and Galloppini Legnami—for the construction—with the collaboration of Politecnico di Milano. The project, indeed, had the initial aim to be preserved, thanks to the potential dismantling of the original façade lattice, a 115 × 115 mm glulam studs' pattern created by the multiple wood-by-wood connections of a modular element and resulting in a three-dimensional diffusive grid of solid joints and voids (with no metal connectors), as the Japanese aesthetic teaches. Here, parametric models helped in defining structure porosity, optimizing the studs' dimensions and proportions with the joints' distribution at the design stage; then the virtual model was used in order to assess the structural properties and distribution of loads along the timber lattice; finally, the data regarding single geometries were sent to the CNC machine for the production, both for initial prototypes and for the final realization (Fig. 1).

However, despite the intense endeavor behind its definition, the original structure has been dismantled, but the project has survived in the form of both the aesthetic potentials of its mock-ups and the replicability of its digital design/fabrication/assembling process (Kitagawara et al. 2019).

Forest Byobu, therefore, has been realized as a trace of the former façade, a 3.50 × 3.20 × 1.34 m sample of larch wood of the geometrical and rational wooden structure that went back to the forest, in contrast to its proper and natural landscape. The project followed the same design process; subsequently, all the larch timber elements were produced via CNC-based cutting processes, transferred to Arte Sella and assembled by Galloppini Legnami in just one day of operations (Fig. 2).

With the same dimensions and proportions, following an ideal connection of all the experiences, other samples of the so-defined *Byobu* have been realized, between Italy and Japan, once again thanks to the high replicability of the digital process behind the wooden structure. The last one has been the *Kigumi Infinity Ischia Byobu*, conceived for the seismic area of the island, left damaged since 2017 by a strong earthquake and still uninhabited. Here, the connections between the modular elements, repeated across the structure, and its following capability to resist as a whole to horizontal forces, being also self-bearing, aim to represent the possibility to find strength and resilience within the collaboration of singularities (Fig. 3).

3.2 Kodama: Kengo Kuma

Kodama pavilion at Arte Sella is a complex lattice structure of 335 massive larch blades realized out of the same dimensional proportions—58 mm thick, 300 mm wide and 1000 mm long—the repetition and mutual connection with no steel elements of which result in a polyhedral volume of solids and voids, according to Kuma's theory of particles and diffused structures (Kuma and Daniell 2015).

Behind this mutual relation of complexity and simplicity, there has been an intense endeavor (Fig. 4), which relied on, at first, virtual parametric models to vary the blades' proportions, check the final geometry and help in the FEM-based structural verifications; after that, the model facilitated the production of the wooden elements via CNC-cutting processes, for the real prototyping assessment, by Ri-Legno and D3WOOD. The latter has been essential in order to test and validate the nodes' structural behavior and the construction feasibility in joineries of assembling six panels. In the end, the so-validated geometry—as a whole and a sum of connections of modular elements—was processed into digital environments in order to manage the pieces' production, codification and final assembly at Arte Sella (Kuma et al. 2019).

The digitalization of all the stages, from the first design conception to production and construction, was made possible not only to communicate and transfer data and ideas but more generally to adapt the process to other experiences, such as the one in Taiwan. This version of the *Kodama*, indeed, followed the same path of conception/production/assembling, in direct contact with D3WOOD and the Politecnico di Milano team, as a result of a workshop led by Kengo Kuma Lab and its students of Tokyo University (Fig. 5).

Finally, this open access process has been adapted once again.

In October 2018, a strong storm devastated the lands of Arte Sella park, killing millions of trees and destroying lots of beautiful masterpieces. The original Kodama Pavilion was heavily damaged. The restoration challenge was accepted (and won) by the team of Politecnico di Milano, Arte Sella, D3WOOD and Ri-Legno. Inspired by the ancient Japanese *kintsugi* crafting technique, which embraces the damages as a natural process of life and exalts them as golden scars, the destroyed blades have been removed and replaced with new ones, and that results in a vibrant variation of colors among the already existing preserved pieces and the new ones (produced via CNC-cutting procedures by Ri-Legno). The latter, indeed, has been treated by D3WOOD with the *yakisugi* technique, burnt in order to resist decay and be visible as a new intervention. Two of them have been, finally, carved with the words of Kaya Shirao's haiku *Wind hisses through the treetops the music of Koto reaches the sky* (in both Italian and Japanese), which fix the event on the structure as a "golden scar" carved by Nature itself (Fig. 3).

In this last case, the process has been facilitated by the digitalized method in recognizing the new elements needed, their production via digital fabrication and manufacturing, and allocation on site for the *Kodama 2.0* experience.

4 Conclusions and Further Developments of the Research Project

The digitalization of the process, from design to assessment, production and construction, aims to gain better control over each phase and procedure, transferring the earlier ideas and sketches into an engineered knowledge. It is demonstrated by the real wooden structural prototypes, which have been able to go beyond numerical experimentations and become proper sculptures for the Arte Sella land art park, integrating creativity and engineering into art.

The new horizons of this research project include a new masterpiece by Ian Ritchie: *Levitas*. The grid-shell structure, conceived to float among the trees of Arte Sella mountain landscape, pushes the engineering challenge even further, introducing the field of reverse engineering, in order to control the complex geometries that the designer wants.

Acknowledgements The authors would like to credit and thank first the architects and designers who defined these three outstanding land art pieces for Arte Sella: Atsushi Kitagawara, Kengo Kuma and Ian Ritchie. A special acknowledgment goes also to Emanuele Montibeller Giacomo Bianchi and Floriano Tomio from Arte Sella—The Contemporary Mountain, for having joined with passion and enthusiasm every design challenge brought up by these experiences.
Finally, the authors would like to credit the design teams of designers, producers, researchers and students of Politecnico di Milano and thank them for their essential contribution to each realization. For the *Byobu experiences*: Kuwabara-san and Mayuko-san from AKA—Atsushi Kitagawara Architects, Atelier2—Arch. Valentina Gallotti, for the close collaboration, Galloppini Legnami, for the cooperation in developing details and the availability to make the structure reliable for re-use projects all-over Italy, PIDA association and arch. Giovannangelo De Angelis for the enthusiasm in realizing the last prototype as a symbolic art piece for their island, Gianluca Crippa, Chiara Re Depaolini, Roberta Simone and Mirko Borzone from the Politecnico di Milano. And for *KODAMA* and *KODAMA 2.0*, Toshiki Hirano, Prof. Jun Sato of the University of Tokyo and the Kengo Kuma lab team, the computational team of the Politecnico di Milano (Matteo Pedrana, Leandro Robutti and Fabrizio Miele), D3WOOD team (Marco and Claudio Clozza) and Ri-Legno Srl (Giulio Franceschini, Lavinia Sartori and Giorgio Franceschini) for the CNC work, Rothoblaas Srl and Elena Bonaldo from SFS for the material supply, Federica Iachelini and Clara Rinaldi of the Politecnico di Milano and Takahiro Hirayama, Masumi Ogawa and Ifan Yim, from the University of Tokyo.

References

Bianconi F, Filippucci M (eds) (2019a) Digital wood design. Lecture notes in civil engineering, vol 24. Springer, Cham

Bianconi F, Filippucci M (2019b) WOOD, CAD AND AI: digital modelling as place of convergence of natural and artificial intelligent to design timber architecture. In: Bianconi F, Filippucci M (eds) Digital wood design. Lecture notes in civil engineering, vol 24. Springer, Cham

Eastman C, Teicholz P, Sacks R, Liston K (2008) BIM handbook. A guide to building information modeling for owners, managers, designers, engineers, and contractors. Wiley & Sons, Inc., Hoboken, New Jersey

Imperadori M, Salvalai G, Vanossi A, Brunone F (2019) SMALL IS MORE. Wooden Pavilion as a path of research. In: Bianconi F, Filippucci M (eds) Digital wood design. Lecture notes in civil engineering, vol 24. Springer, Cham

Imperadori M, Brunone F (2018) Insegnare costruendo. Architettura temporanea tra ricerca e didattica. AGATHÒN 04 Int J Archit Art Des 21–28

Kitagawara A, Imperadori M, Kuwabara R, Brunone F, Matsukawa M (2019) Wooden Byobu. From architectural façade to sculpture. In: Bianconi F, Filippucci M (eds) Digital wood design. Lecture notes in civil engineering, vol 24. Springer, Cham

Kobayashi H, O'Keefe D (2019) Empathic architecture: digital fabrication and community participation. In: Bianconi F, Filippucci M (eds) Digital wood design. Lecture notes in civil engineering, vol 24. Springer, Cham

Kobayashi H (2017) The Veneer house experience: the role of architects in recovering community after disaster. In: Yan W, Galloway W (eds) Rethinking resilience, adaptation and transformation in a time of change. Springer, Cham

Kuma K, Imperadori M, Clozza M, Hirano T, Vanossi A, Brunone F (2019) KODAMA: a polyhedron sculpture in the forest at Arte Sella. In: Bianconi F, Filippucci M (eds) Digital wood design. Lecture notes in civil engineering, vol 24. Springer, Cham

Kuma K (2018) Ja 109 Kengo Kuma: a lab for materials. Shinkenchiku-sha Co Ltd., Tokyo

Kuma K, Daniell T (2015) Small architecture/natural architecture. Architectural Association Publications, London

Menges A, Schwinn T, Krieg OD (2016) Advancing wood architecture: a computational approach. Routledge, United Kingdom

Montibeller E, Tomaselli L, Bianchi G (2017) Arte Sella. The contemporary mountain. The new beginning. Silvana Ed., Milano

Tedeschi A (2014) AAD algorithms-aided design. Parametric strategies using grasshopper, Edizioni Le Penseur, Brienza

Vanossi A. (2018) Interaction and intersection between digital modelling and design in architecture: different approaches in parametric design. In: handbook of research on form and morphogenesis in modern architectural contexts, pp 152–174. Information Science Reference, Milano, February 2018

Vanossi A, Parello G, Imperadori M, Bernardo C, Liotta SJA, Ito Y, Occhipinti F (2015) Architecture for archeology: identifying new modular and flexible types of shelter adaptable to the diverse needs of archaeological sites. In: XII Forum Internazionale Le Vie dei Mercanti–proceedings, pp 457–466. La scuola di Pitagora, Anversa-Capri, June 2015

Vanossi A, Parello G, Imperadori M, Bernardo C, Liotta SJA, Ito Y (2014) BIM for archaeology, Use of BIM process and parametric model in a temporary shelter adaptable to the diverse needs of archaeological sites. In: PPC 2014 conference proceedings, pp 191–200, Monza, Mantova (Italy), May 2014

Vanossi A, Imperadori M (2013) BIM and optioneering in dry technology small-scale building. In: ICT, automation and the industry of the built environment, pp 53–65. Maggioli, Milano, March 2013

The Impact of Digitalization on Processes and Organizational Structures of Architecture and Engineering Firms

Cinzia Talamo and Marcella M. Bonanomi

Abstract The digitalization of the architecture, engineering, and construction (AEC) industry leads to new forms of process through which buildings are designed, constructed, and operated, and to new forms of organization through which professionals work and interact. Certain activities of the conventional building process disappear, while others appear, distribution of work is reviewed, and new relationships, roles, and responsibilities emerge. Although many architecture and engineering (A/E) firms claim that they have already undertaken a digital transformation, there is still little awareness of the new forms of process and organization associated with digitalization. This lack of knowledge about the process-oriented and organizational changes makes it difficult to establish a work environment within and between firms that is conducive to digital innovation. Given the above considerations, the main objective of this research project has been to understand the process-oriented and organizational changes that the adoption of digital technologies bring on, as well as the new forms of process and organization associated with the digital transformation of architectural and engineering firms. To achieve this, a case-study analysis of two A/E firms—one in Italy and one in Canada—has been performed.

1 Introduction

Productivity growth in the architecture, engineering, and construction (AEC) industry has stagnated globally in recent decades (Barbosa et al. 2017). The AEC industry has not been able to keep pace with the overall economic productivity growth. A recent report by McKinsey (Remes et al. 2018) finds a positive correlation between the productivity growth of an industry and its degree of digitalization. However,

C. Talamo (✉)
Architecture, Built Environment and Construction Engineering—ABC Department, Politecnico di Milano, Milan, Italy
e-mail: cinzia.talamo@polimi.it

M. M. Bonanomi
Chair of Innovative and Industrial Construction, ETH Zurich, Zurich, Switzerland

B. Daniotti et al. (eds.), *Digital Transformation of the Design, Construction and Management Processes of the Built Environment*, Research for Development,
https://doi.org/10.1007/978-3-030-33570-0_16

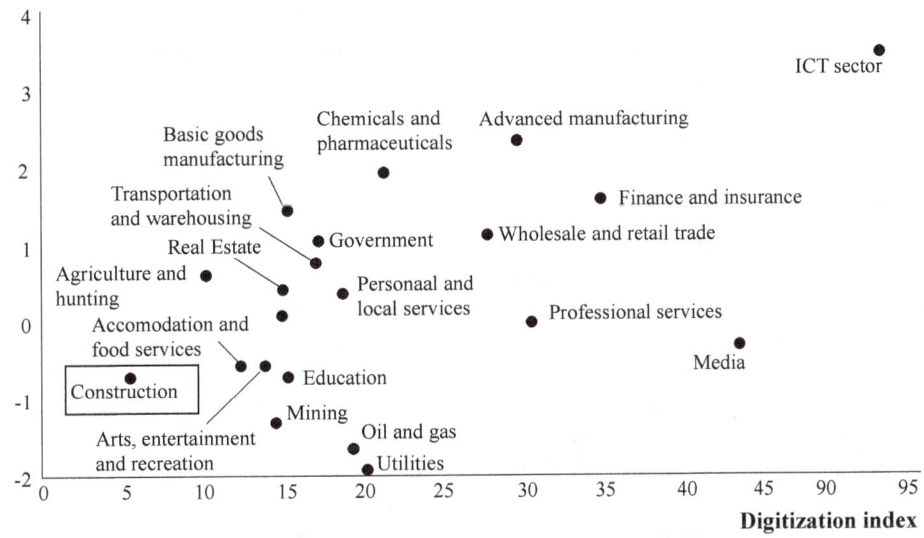

Fig. 1 Correlation between productivity growth and digitization index across industries in Europe: highlighted in the red box the construction industry (Remes et al. 2018)

according to the McKinsey 'digitization index', construction is among the least digitized sectors in the world: it comes second to last in the United States—right after the agriculture and hunting industry—and last in Europe (see Fig. 1).

Italy and Canada have not been an exception to this trend (Remes et al. 2018). Italy ranks behind other nations in digitalization,[1] ranking 15th out of 21 nations[2]—and behind countries such as Finland, Netherlands, and Ireland—in a recent study about building information modeling (BIM) adoption (Kassem and Succar 2017). According to the same study, Canada performs slightly better, ranking 11th. To investigate the level of BIM diffusion, this study overlays three BIM fields (technology, process, and policy) with as many BIM capability stages (modeling, collaboration, and integration) to generate nine diffusion areas[3] (Succar and Kassem 2017). The results show an irregular distribution of the diffusion rates across the 21 countries analyzed (see Fig. 2). For example, in the Netherlands, the United Kingdom, China, Finland, and South Korea, the diffusion is quite balanced across the nine diffusion areas. On the contrary, Italy and Canada, like Malaysia, Mexico, Russia, Spain, Switzerland, Qatar, and the UAE, show unbalanced diffusion rates. Additionally, some diffusion areas are missing in these countries. This is the case, for example,

[1]By 'digitalization', we mean the transformation of processes, organizational settings, and project delivery methods that need to be coupled with the adoption of digital technologies in order to gain the full benefits of the digitalization of the architecture, engineering, and construction industry.

[2]USA. United Kingdom, UAE, Switzerland, Spain, South Korea, Russia, Qatar, Portugal, New Zealand, Netherlands, Mexico, Malaysia, Italy, Ireland, Hong Kong, Finland, China, Canada, Brazil, Australia.

[3]Modeling technologies, modeling processes, modeling policies, collaboration technologies, collaboration processes, collaboration policies, integration technologies, integration processes, integration policies.

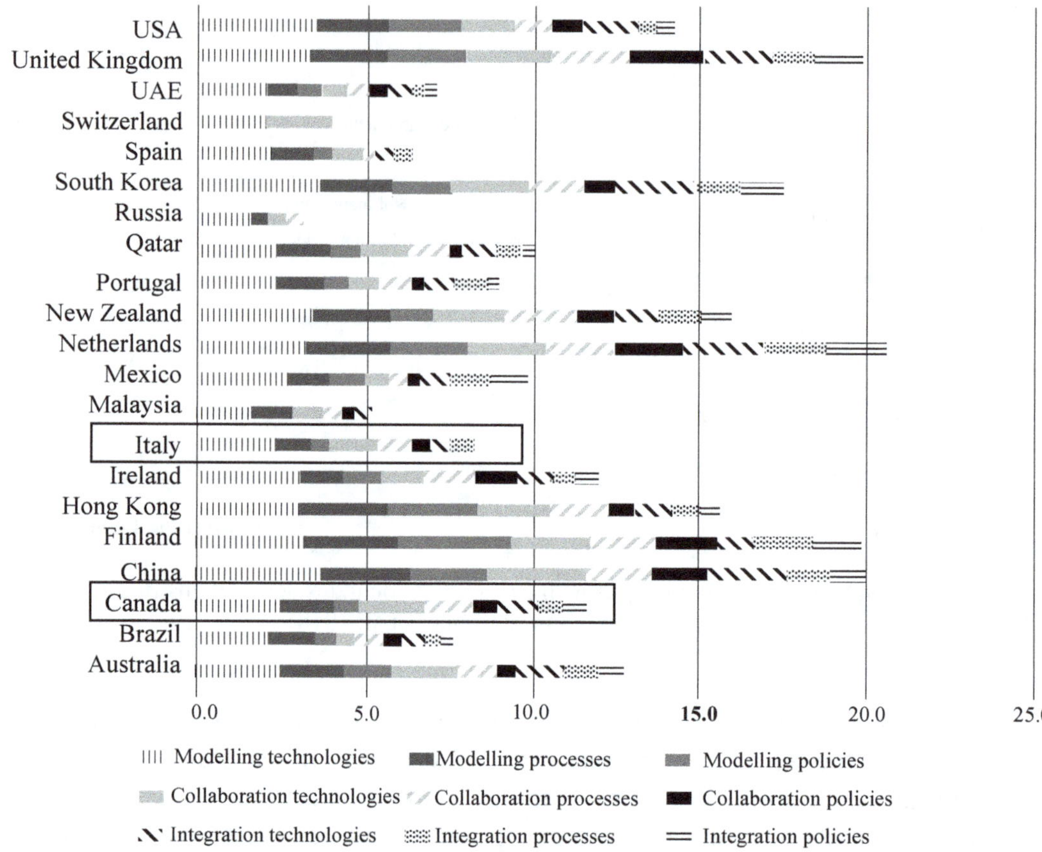

Fig. 2 Rates of the nine diffusion areas of BIM adoption across the 21 countries analyzed: highlighted in the red boxes Italy and Canada (Kassem and Succar 2017)

of the diffusion area 'integration policies' in Italy. Furthermore, Italy shows a significant difference between the technologies-related diffusion rates (colored in blue, purple, and light blue in Fig. 1) and the 'processes' and 'policies' ones. Canada shows a similar trend, but it is not missing any diffusion area. Kassem and Succar (2017) argue that a country with either an unbalanced distribution or missing diffusion area would possibly face different adoption challenges compared to a country with all the nine areas established and well distributed. Overall, this study shows the predominance of the 'modeling technologies' and 'modeling processes' diffusion areas, while minor rates are associated with 'collaboration/integration processes' and 'collaboration/integration policies'. These results point out the industry's tendency to focus more on the adoption of digital technologies and the related modeling processes, and less on the collaboration and integration processes and policies required to gain the full benefits of digitalization.

Besides the poor understanding of the collaboration and integration processes required to support digital innovation, also the organizational changes associated with digitalization are not understood fully yet. On the contrary, a recent report by the Boston Consulting Group (2016) argues the importance to couple organizational changes with process-oriented ones in order to leverage the full benefits of digitalization. When processes and technologies mature digitally, organizations should change accordingly and move toward networked organizational forms that are more conducive to collaboration and integration within and between firms (Picon 2016). Regarding the multidimensional nature of the changes brought on by digitalization, Poirier et al. (2015) proposes a conceptual framework, which is organized into three fields of change: *technology, process, and organization*. Author's belief is that the digital transformation requires a balanced distribution and management of changes in technology, process, and organizational setting. Furthermore, she underlines the importance of coupling these fields of change with a full understanding of the context of change, as well as the stage of implementation.

Starting from these considerations and aware of the significant effort already devoted from both researchers and practitioners on digital technologies and the related modeling processes, the overall objective of this research project has been rather to understand the process-oriented and organizational changes that the adoption of digital technologies bring on within architectural and engineering firms. To reach this overall objective, the project has focused on two distinct, but complementary levels (specific objectives):

1. The *process-oriented changes* that the adoption of digital technologies bring on within architectural and engineering firms and the new forms of process facilitating the achievement of the full benefits of digitalization;
2. The *organizational changes* that the adoption of digital technologies bring on within architectural and engineering firms and the new forms of organization facilitating the achievement of the full benefits of digitalization.

2 Research Approach

To achieve this, the research project has employed a mixed-method case-study approach. Case studies are a suitable methodology for explanatory research as they can help to answer the 'how' and 'why' questions related to a topic, especially when an in-depth analysis and observation is performed (Burawoy 1998; Eisenhardt 1989). The case-study selection for the purpose of this study included one architectural and engineering firm headquartered in Italy (<250 employees) and one in Canada (≥250 employees). The following reasons can justify the choice of an Italian and a Canadian firm:

- Both the Italian and the Canadian AEC industries cannot be considered first movers in digitalization. They shall rather be understood as evolving contexts still in transition;

- Both the Italian[4] and the Canadian[5] industries' workforce are mainly composed of small and medium firms (European Builders Confederation; Industry Canada 2014; Leung et al. 2008).

For clarity purposes, it must be underlined that this study has analyzed the two case studies distinctly and separately. Each of the case study focused on one of the two fields of change associated with digitalization: one case study on the process-oriented changes and the other one on the organizational ones. Specifically, in the analysis of the Italian firm, the multimethod approach adopted to perform the experimental research has meant investigating the process-oriented changes in practice through mixed-methods of data collection: observation, unstructured interviews, and process mapping. In the case of the Canadian firm, observation, surveys, and social network analysis (SNA) (Chinowsky et al. 2008, 2012; Zheng et al. 2016) have been rather used as methods to identify how digital transformation affects the organizational setting.

In both the case studies, the process of data collection, data analysis, and data representation has been performed by focusing on three different, but complementary, states of change:

1. *'As-is' state*. This step included data collection, analysis, and representation of the conventional design process and existing organizational structure;
2. *'Transition' state*. This phase included data collection, analysis, and representation of the design process and organizational structure in place at the time of the first adoption of digital technologies;
3. *'To-be' state*. This step included data collection, analysis, and representation of the envisioned digital design process and organizational structure to be achieved in order to gain the full benefits of digitalization.

One difference between the two case studies is that the analysis of the Italian firm has been performed at the project level. This has meant collecting, analyzing, and representing the changes of the design process implemented in the context of a project. The analysis of the Canadian firm has been rather performed at the company level. This has meant collecting, analyzing, and representing the changes of the organization itself and not in the context of a project.

[4]Italian small and medium enterprises (SMEs) account for 80% of the total share of nominal GDP in the construction industry. Small businesses only make up 70% of the Italian construction industry's workforce (European Builders Confederation).

[5]In 2012, small (between 5 and 99 employees) and micro (less than 5 employees) businesses made up 99.0% of the Canadian construction industry's workforce and accounted for 72.7% of the total share of nominal GDP in the Canadian construction industry (Leung et al. 2008).

2.1 Case-Study Analysis: Process-Oriented Changes in the Italian Firm

First, the 'as-is' state of the design process has been defined through an in-depth document analysis: the firm's quality management system (QMS) in particular has been used as the primary data source. Secondly, the 'transition' state of the design process has been studied through observation, interviews, document analysis, and process mapping of a BIM pilot project implemented by the Italian firm. Lastly, the 'to-be' state of the design process has been envisioned through the review of the literature and international standards about BIM and digital design process. To conclude, the three states of the design process ('as-is', 'transition', and 'to-be') have been cross-compared to identify the differences between one another. This last step has led to identify the process-oriented changes, which the adoption of digital technologies brought on within the firm, as well as the ones that still need to be implemented in order to facilitate the achievement of the full benefits of digitalization.

2.2 Case-Study Analysis: Organizational Changes in the Canadian Firm

The 'as-is' state of the firm's organizational structure has been defined, as first step, through an in-depth document analysis: the firm's human resources (HR) database, which included all the supervisor–employee couples, has been used as the primary data source. Secondly, the 'transition' state of the organizational structure has been defined by distributing a questionnaire to employees, which asked them to identify the peers they referred to for advice and help about digital technologies and processes. The data gathered have been then analyzed by using social network analysis. Lastly, the 'to-be state' of the organizational structure has been envisioned through a literature review about the new business models enhancing digital knowledge management, as well through interviews with the firm's top management.

3 Findings

3.1 Process-Oriented Changes

Data collection and analysis through the case study of the Italian firm has led to identification of the process-oriented changes that the adoption of digital technologies brought on within the firm (see Table 1), as well as the ones still to be implemented in order to gain the full benefits of digitalization (see Table 2). Because of the multidimensional nature of the changes associated with digitalization, also some

Table 1 Changes within the Italian firm brought on by digital technologies' adoption (Employer's information requirements; BIM execution plan)

Process	Organization	Technology
Improved *information exchange* between the architecture (A) and engineering (E) teams AND between the A and 'rendering' (R) teams	Enlarged *marketing and business* opportunities	New *hardware, software, and tools* implemented
Improved *collaboration* between A and E teams	New *roles and responsibilities*	
Integrated *decision-making* of the A and E teams	BIM *skills and capabilities* acquired by A and E teams	
Improved *consistency* of the A and E deliverables		
Reduced *coordination effort* (file exchange) between the A and R teams		
Reduced *file production* by the A and E teams		

Table 2 Changes still to be implemented to gain the full benefits of digitalization

Process	Organization	Technology
Preliminary agreement with the client about *BIM requirements* (EIR, BEP, etc.)	BIM *skills and capabilities* acquired by all the BU involved in projects	*Set-up of the BIM environment* and related procedures (graphic standard, BIM libraries, etc.)
Preliminary definition of project parties' *roles and responsibilities*		
BIM use *in all the phases* of the design process		
BIM adoption *by all the business units* (BU) involved in a project		
Preliminary evaluation of project parties' *BIM-skills and competencies*		

organizational and technological changes have been identified. These findings have been categorized according to the three primary dimensions of digitalization as proposed by Poirier et al. (2015). According to her model, in fact, key components of digitalization fall into three categories: process, organization, and technology.

3.2 *Organizational Changes*

Data collection and analysis through the case study of the Canadian firm has led to identification of the organizational changes that the adoption of digital technologies brought on within the firm, as well as the ones still to be implemented in order to facilitate the achievement of the full benefits of digitalization.

Specifically, data gathered through the questionnaire distributed to the firm's employees has led to identify a multidisciplinary network of professionals acknowledged by other peers as informal 'go-to' people for advice and help in digital technologies and processes. Although these employees were playing a significant role within the firm since they were helping other colleagues to get their work done, they were not formally acknowledged by the existing organizational structure ('as-is' state). Therefore, besides the formal roles and responsibilities as defined by the existing organizational structure, informal roles and relationships for advice and help in digital technologies and processes were in place too. This was the 'transition' state of the firm's organizational structure.

Because of the productivity inefficiencies associated with this 'transition' state, an organizational reconfiguration has been proposed to the firm's top-management defining a 'to-be' state of the organizational structure. This 'to-be' state meant the formal acknowledgment of this informal network of 'go-to' people through the creation of a support group for digital technologies and processes at the firm-level. New roles, responsibilities, and relationships have been therefore defined.

4 Discussion

To interpret and describe the significance of our findings, we framed the results of our study according to two out of the three BIM capability stages proposed by Kassem and Succar (2017) because of the minor distribution rate of the collaboration and integration processes in comparison to the modeling ones.

4.1 *Collaboration*

The findings of this study show that there is a need to enhance collaboration between the owner/client and the project parties starting from the early stages of the design process. In line with this view, it emerged the need for the owner to be more involved and active in defining and sharing with the involved parties the project requirements since the early phases of the process. Additionally, relationships, roles, and responsibilities within and between firms must be clearly defined in order to facilitate the set-up of a work environment that is conducive to digital innovation and that aim to enhance a dynamic network of teams.

4.2 *Integration*

Furthermore, there is a need to facilitate and strengthen integration between project participants in order to improve information exchange and performance simulation according to different parameters, like embodied energy, and so on.

Additionally, the preliminary assessment of the digital skills and capabilities of project participants is also a significant topic, which came out from the findings of this study, and that is recommended in order to avoid time and cost overruns. Lastly, it also emerged the need to define standards and procedures to be shared internally, that means within firms, and externally, alias across the supply chain.

5 Final Considerations and Future Directions

The digitalization of the AEC industry is rapidly increasing and shows potential to improve industry productivity and competitiveness, as well as process efficiency and products' quality. Researchers and practitioners are already devoting significant efforts to promote the adoption of digital technologies. However, a larger and three-fold focus (technology, process, and organization) is necessary to enable the achievement of the full benefits of digitalization. Interactions and effects of digital transformation on processes and organizational models used by most projects and firms need to be explored. This creates a demand for looking at new forms of process and organization, and the impact of these on the architecture of work, as well as the socio-technological context required for facilitating these changes.

To investigate this point, this research used a mixed-method case-study approach, which included an extended case-study analysis of two different A/E firms: one in Italy and one in Canada. Specifically, the case study of the Italian firm has meant understanding in practice the process-oriented changes and the new forms of process associated with digitalization. The case study of the Canadian firm has been rather analyzed to identify how the adoption of digital technologies affects the existing organizational setting and how this one can be reconfigured to facilitate the achievement of the full benefits of digitalization.

Regarding the process-oriented changes and the new forms of process, the results indicate that there is a need for implementing collaborative and integrated processes within and between architectural and engineering firms, in addition to the modeling ones associated with the adoption of digital technologies. On the other side, data collection and analysis about the organizational changes and the new forms of organization reveal that more dynamic and networked organizational models should be implemented in order to facilitate digital knowledge creation, diffusion, and utilization within firms and between firms.

Regarding the future directions of this research topic, both researchers and practitioners involved in the digital transformation of the AEC industry can keep to make use of analysis methods for change management and get a valid support from a set of applied investigations and research activities, such as:

- Collection and analysis of benchmarking data through the analysis of additional case studies;
- Interpretation of changes associated with the adoption of digital technologies by deepening investigations on industry best practices;
- Analysis and comparison of strategies for digital transformation in relation to different markets and diverse contexts (cultural, legal, etc.).

References

Articles

Barbosa F, Woetzel J, Mischke J, Ribeirinho MJ, Sridhar M, Parsons M, Bertram N, Brown S (2017) Reinventing construction: a route to higher productivity. McKinsey & Company

Burawoy M (1998) The extended case method. Sociol Theory 16(1):4–33

Chinowsky P, Diekmann J, Galotti V (2008) Social network model of construction. J Constr Eng Manag 134(10):804–812

Chinowsky P, Taylor J (2012) Networks in engineering: an emerging approach to project organization studies. Eng Project Organ J 2(1–2):15–26

Eisenhardt K (1989) Building theories from case study research. Acad Manag Rev 14(4):532–550

Gerbert P, Castagnino S, Rothballer C, Renz A, Filitz R (2016) Digital in engineering and construction. The transformative power of building information modelling. The Boston consulting group, pp 2–18

Industry Canada (2014) Establishments: construction (NAICS 23), Canadian Industry Statistics. Available at: http://www.ic.gc.ca/app/scr/sbms/sbb/cis/gdp.html?code=11-91andlang=eng (Accessed 19 March 2014)

Kassem M, Succar B (2017) Macro BIM adoption: comparative market analysis. Autom Constr 81:286–299

Leung D, Césaire M, Yaz T (2008) Productivity in Canada: does firm size matter?. Bank of Canada review, pp 5–14

Picon A (2016) From authorship to ownership. Archit Des 86(5):36–41

Poirier E, Staub-French S, Forgues D (2015) Embedded contexts of innovation. Constr Innov 15(1):42–65

Remes J, Mischke J, Krishnan M (2018) Solving the productivity puzzle: the role of demand and the promise of digitization. Int Product Monit 35:28

Zheng X, Le Y, Chan APC, Hu Y, Li Y (2016) Review of the application of social network analysis (SNA) in construction project management research. Int J Proj Manag 34(7):1214–1225

Websites

http://bimtopics.civil.ubc.ca/

http://www.ebc-construction.eu

https://www.ic.gc.ca/
http://www.ebc-construction.eu/

Execution Stage

Introduction

Bruno Daniotti, Marco Gianinetto

In the part III are reported research results useful for the optimisation of production and construction activities, where the supply chain is involved.

Some general studies are dedicated to define guidelines and platforms for BIM management by the construction companies. The research outlines a methodological and operational protocol for general contractors and the related supply chains and subcontractors, containing indications for the development of BIM-based workflows for the principal contractor's activities in building construction, operations and maintenance: off-site and on-site time schedules, site planning and building ergotechnics, accounting, project variances, non-compliance issues, test and commissioning, maintenance planning.

Other research activities are focused on innovative digital fabrication, 3D and 4D printing, data-driven Design for Manufacturing.

In particular the aim of this chapter is to shed light on the potentialities of smart material systems in architectural skins. The research on new materials has triggered new solutions, based on emergent technologies, for envelope design. Adaptive facades, exploiting stimulus-responsive materials and intelligent control systems, are able to change in response to environmental conditions, thus enhancing users' comfort and improving energy savings.

A specific focus is given as for the application of robotics for the development of a system for the production of disposable carbon fiber formworks. The desired three-dimensional object can be made by the robotic hot wire cutting of polystyrene, thus generating the shape around which the carbon fiber is deposited.

BIM Management Guidelines of the Construction Process for General Contractors

Salvatore Viscuso, Cinzia Talamo, Alessandra Zanelli and Ezio Arlati

Abstract Capability of construction companies to manage the digital interoperability between stakeholders and technicians is fundamental for the success of a work, both for new constructions and renovations. The research outlines a methodological and operational protocol for general contractors and the related supply chains and subcontractors. The Programme contains indications for the development of BIM-based workflows that implement the main activities in building construction, operations and maintenance: off-site and on-site time schedules, site planning and building ergotechnics, accounting, project variances, non-compliance issues, test and commissioning, maintenance planning.

Keywords Building information modeling · BIM protocol · Implementation plan · Pilot project · Building construction · Facility management

1 Introduction

Building Information Modeling (BIM) is a process focused on the development, use and transfer of a digital information model of a building project to improve the design, construction and operations of a project or portfolio of facilities. When properly implemented, BIM can provide many benefits to a project. Well-planned projects illustrate the value of BIM that yield: increased design quality through effective analysis cycles; increased innovation using digital design applications; greater prefabrication due to predictable field conditions; improved field efficiency by visualising the planned construction schedule. To improve the overall performance of the facility or a portfolio of facilities, operators in asset management, space planning and maintenance scheduling can use valuable information obtained at the end of the construction phase. Yet, there are also examples of projects where the company's technical operators—hereinafter referred to as the "team"—did not effectively plan

S. Viscuso (✉) · C. Talamo · A. Zanelli · E. Arlati
Architecture, Built Environment and Construction Engineering—ABC Department, Politecnico di Milano, Milan, Italy
e-mail: salvatore.viscuso@polimi.it

B. Daniotti et al. (eds.), *Digital Transformation of the Design, Construction and Management Processes of the Built Environment*, Research for Development,
https://doi.org/10.1007/978-3-030-33570-0_17

the implementation of BIM and incurred increased costs for the modeling services, schedule delays due to missing information, and little to no benefit. Implementing BIM requires detailed planning and fundamental process modifications for the project team members to achieve the value from the available model information (Eastman et al. 2011).

Currently the difficulties of a successful integration between BIM processes and production are directly related to the number of participants in the project. If, in architecture, the contemporary trend is to design building with technological complexity that distinguishes it as an identifying icon or landmark, the difficult task of managing a huge amount of information data and requirements in building construction and operations, often impossible to systematise if not through a predefined organizational approach, represents a challenge (Ciribini 2016).

Furthermore, in order to achieve a real synergy between the interoperable BIM design and the production fields, it is necessary to organise the work-sharing using cloud technologies; this necessary produces the application of standardised, secured rules to protect shared project milestones and work progresses.

To integrate BIM into the project delivery process, it is important to develop a detailed execution plan for companies involved in construction and asset management. Within the activities of a research contract with Pessina Costruzioni SpA, an Italian General Contractor specialised in construction and operations of healthcare facilities, the Department of Architecture, Built Environment and Construction Engineering (DABC) of Politecnico di Milano has developed a BIM Protocol that outlines the overall vision along with implementation details for construction and operations.

This Protocol provides a structured procedure for creating and implementing a corporate BIM Project Execution Plan that cover the main business activities of the company. The four steps within the procedure include:

1. the identification of high-value BIM uses during project planning, design, construction and operational phases;
2. the design of tailored BIM execution process by creating process maps;
3. the definition of BIM deliverables in the form of information exchanges;
4. the development of the infrastructure in the form of contracts, communication procedures, technology and quality control to support the implementation.

Specified sections of the BIM Protocol explain the details related to each step. Detailed templates have also been created to support each of these steps. These templates are included in the Appendices of the document.

The Protocol is compliant with the following standard: Industry Foundation Classes (IFC), defined by UNI EN ISO 16739:2016; UNI 11337:2017 (Building and civil engineering works—Digital management of the informative processes) that regulates interoperability between project objectives and digital modeling; BS EN ISO 19650:2018 (Organization and digitization of information about buildings and civil engineering works, including building information modelling—Information management using building information modelling).

2 Identify BIM Goals and Uses

The preparation of the process has as its first major step the definition of the project uses in relation to main Contractor's works. Their identification is fundamental, as it made possible defining the final objectives, correctly setting the models and identifying the parameters that allow the optimization of the scenario that best meets the purposes. An incorrect setting of the requirements can nullify all the modeling work done later (Deutsh 2011).

In accordance with a typically top-down decision-making process, a series of macro-categories define the criteria to evaluate projects, including the budget, the technical-construction feasibility, the internal distribution and the energy performances. The definition of these is carried out in close collaboration with the team and through some briefings involving suppliers and sub-contractors.

Once the team has defined measurable goals, both from a project perspective and company perspective, then the specific BIM uses on the project can be identified. The Protocol includes a list of common uses for BIM, which have been identified through analysis of project case studies, interviews with industry experts, and review of the literature. Each BIM use is a unique task or procedure on a project that can benefit from the integration of BIM into that process. The list is not comprehensive but provides a good representation of the potential uses of BIM within a construction company. It includes design authoring, 4D modeling, cost estimating, space management and record modeling. The Contractor's team should identify and prioritise the appropriate BIM uses that they have identified as beneficial to the project goal achievement (Holzer and Downing 2010).

To facilitate the BIM use review process, a BIM Worksheet template includes a list of the potential BIM uses, along with fields to review the value, responsible party, capabilities, additional notes, and the decision from the team on whether to implement the BIM use.

3 Design the BIM Execution Process

Once the team has identified the BIM uses, a process mapping procedure can clarify all the progressive steps to implement traditional workflows. Initially, a high-level map showing the sequencing and interaction between the primary BIM uses on the project allows all team members to understand clearly how their work processes interact with the processes performed by other team members.

After the high-level map is developed, then team members responsible for each detailed BIM use should select or customise more detailed process maps. For example, the high-level map will show how energy evaluation, cost estimating, 4D modeling and recording are sequenced and interrelated. A detailed map will show the detailed processes that the Contractor can perform by itself or, in some cases, with the consultancy of external organizations such as designers, sub-contractors or, as may be the case for energy task, LEED certifiers.

Maps representing for each process should clearly identify responsible parties involved in the workflow. For some processes, this may be an easy task, but for others it may not. It is important in all cases to consider which team member is best suited to complete the task successfully. Additionally some processes may have multiple responsible parties. The identified party will be responsible for clearly defining the information required to implement the process as well as the information produced by the process (Fig. 1).

Referring to graphical notation and information format for the processes within the BIM overview maps, each process should include a process name, project phase, and the responsible party. Each process should also include the detailed map title for the process. This detailed map notation is necessary since several processes may

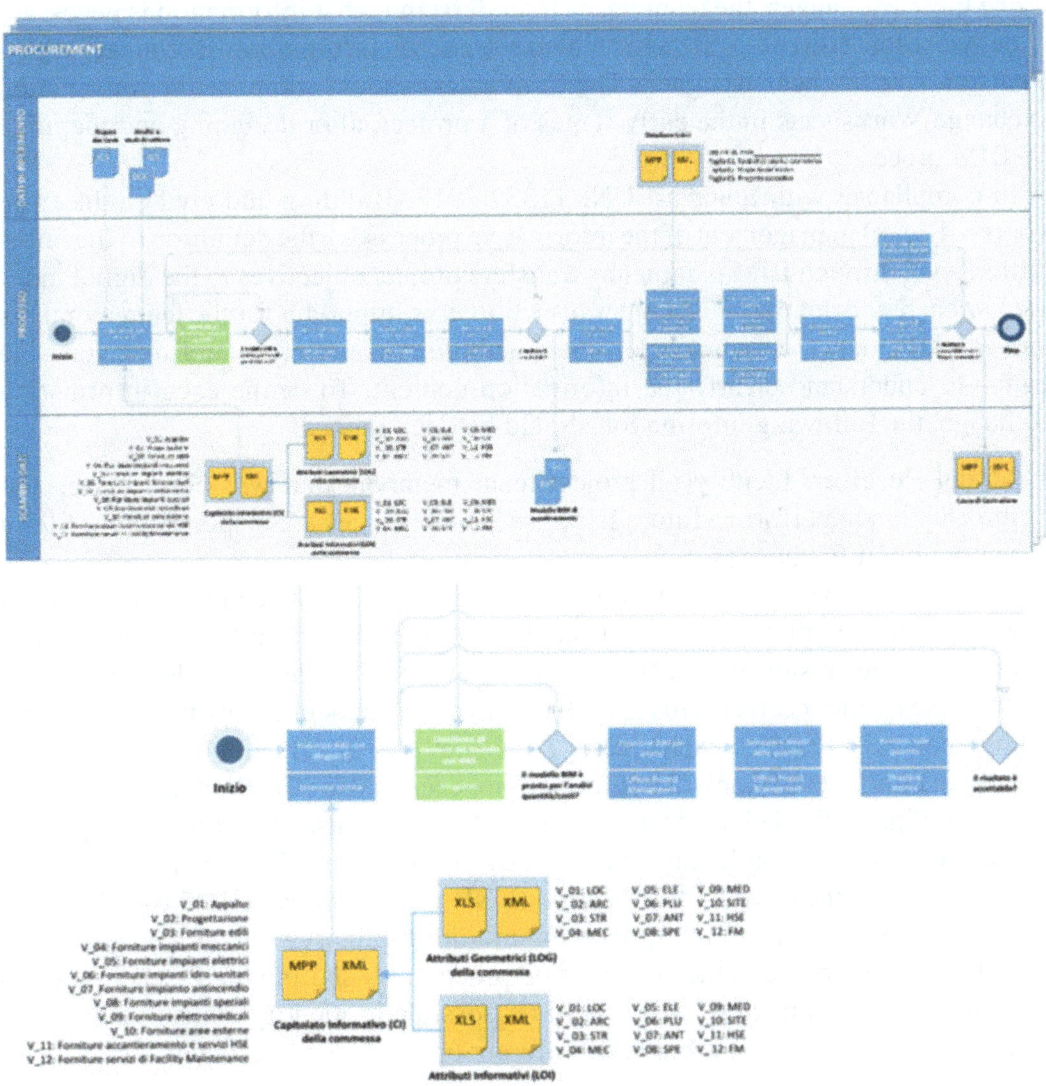

Fig. 1 Detailed process maps, organised in process, data exchange and reference data

share the same detailed map. For example, a construction management company may perform cost estimating from the building information provided from the designer. The Construction manager may perform this estimate during the schematic design, design development and construction document phase, but it may utilise the same detailed workflow—obviously with diverse detail levels of BIM models, reference documents and templates—to accomplish this task, which can be represented in a single detailed map.

4 Develop Information Exchanges

The goal of this task is to present a method for defining information exchanges between project processes that are crucial to successful BIM implementation. To define these exchanges, the team needs to understand what information is necessary to deliver each BIM use. To assist in this task, an Information Exchange Worksheet template has been designed. The Contractors should complete the Information Exchange Worksheets in the early stages of a project, after designing and mapping the BIM process.

In compliance with Standard UNI 11337:2017 (Building and civil engineering works—Digital management of the informative processes), the definition of the information level through BIM parameters transfers project objectives to the digital models. For each information exchange transaction, it is important for the team members and, in particular, the author—designers, sub-contractors etc.—and receiver—the team—to understand clearly the information content. To define each information exchange, the following information should be documented:

1 Model Receiver: Identify all project team members that will be receiving the information to perform a future BIM use. These parties are responsible for filling out the Input Exchanges.
2 Model File Type: define the Level of Geometry (LOG) for each BIM use that can be associated to the project. The task also lists the specific software applications, as well as the version and the file format used to manipulate the model during each BIM use by the receiver. This is pertinent in order to identify any interoperability that may exist between exchanges.
3 Information: identify only the information necessary for the BIM use implementation (Fig. 2). Predefined property sets (P-Sets) organise the information parameters representing building elements and components. Parameter list changes depending on the Level of Information (LOI) needed for each BIM use. A proper BIM maturity level can interpolate the multidisciplinary skills of designers and technicians involved in the construction process. By sharing the P-sets in a common data environment (CDE), it will be possible to identify and solve overlaps and process interferences between different operators and designers.

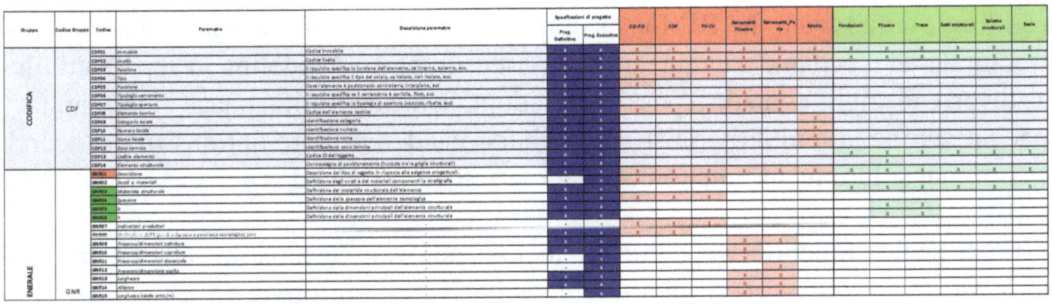

Fig. 2 Matrix parameters—building components, developed for schematic design, detailed design, construction design and as-built BIM models

5 Define Supporting Infrastructure

The final step of the BIM Protocol is to identify and define the project infrastructure required to implement effectively BIM as planned. Six specific categories support the BIM project execution process. After a deep evaluation of the literature review with the industry stakeholders and operators, the selected categories are Tender, Procurement, Construction, Recording, Asset Management and Quality Control.

Information for each category can vary significantly by project, therefore the goal of the description is to initiate discussion and address content areas and decisions that need to be made by the project team. Templates of Employer's Information Requirements (EIR) and BIM Execution Plan (Pre-contract and Post-contract) have been developed and shared with the team.

Quality control category is transversal to the different stages of the project. To ensure model quality in every phase and before information exchanges, procedures must be defined and implemented. Each BIM process activated during the lifecycle of the project must be pre-planned considering model content, level of detail, format and party responsible for updates; and distribution of the model and data to various parties. Each party contributing to the BIM model has a responsible person to coordinate the model that participates in all major BIM activities. He is responsible for addressing issues that might arise with keeping the model and data updated, accurate, and comprehensive.

6 Pilot-Projects

From January to December 2018 diverse pilot-projects involved the Contractor's team, in order to train and qualify the technicians that normally coordinate the company activities. The first project relates to the Public-Private Partnership proposal (PPP) for the construction of a novel Hospital Center in Lanciano (1st DEA level with 218 beds and net area of about 50,000 sm) in substitution of the obsoleted building. The proposal also included the Schematic design of the new hospital. The workflow started from the BIM uses definition, that involved in internal architects, structural and mechanical engineers and cost estimators. Modeling in separated files the diverse disciplines made it possible to adopt, in the definition phase of the sanitary space planning, a multicriteria strategy, that considered more options in the search for optimal solutions in function of predetermined weighted criteria (Viscuso et al. 2019).

Each health function was studied, in first analysis, evaluating all the interactions necessary within each department. This study therefore represented the basis for the development of a subsequent level of the individual departments' distribution. For defining different spatial and distribution scenarios and performing multicriteria analysis on them, indications provided by the facility managers guaranteed the compliance with the requirements needed for a possible Joint Commission Health Accreditation (Capolongo 2012). In order to evaluate the surfaces and volumes of this schematic design, the team interpolated the main data (surfaces, distances, air-light relationships) related to diverse design options and configurations: all medical and surgical specialties, diagnostic equipment and robotic equipment, laundry, sterilization, and catering. For a balanced pre-dimensioning, all human activities related to reception and waiting were also considered, such as commercial activities, bank, as well as all the babysitting services, gyms, etc. (Carpman and Grant 2016).

Discipline models were modeled separately at different Levels of Development. Initially three LOD 100 volumetric models of conceptual masses are created, useful for distributive analysis of three different design options, and for the verification of internal distribution specification (e.g. the separation between clean and dirty zones). Discipline models with LOD 200 were then realised, relating to the design option that was more in keeping with the health requirements (Fig. 3).

The second project consists in the development of the technical and economical documentation to be submitted for a construction bidding. The construction planning of the detail project of this tender—an 11-storey office building with net area of 27,000 sm—has needed the implementation of 4D modeling, in order to simulate and verify the following construction strategies (Fig. 4).

The company's team composed the Day-work Programme through a logical succession of construction stages (Shell & Core, Category A, Category B, Test & Commissioning), minimising their temporal overlap and carefully assessing the impact on the site in terms of duration, management and safety. The logical-temporal connection between the different activities allows a "continuum" of the work and a high level of prefabrication off-site (Das and Kanchanapiboon 2011).

The above-explained organization was modelled by the company's team for 4D BIM use, starting from IFC models included in the bidding documentation. The tri-dimensional visualization of Work Schedule also performed a better construction management per each stage: in fact, a correct distribution of workers in specified sectors reduces and minimises interferences between different activities, e.g., avoiding contemporary overlaps into the same sector or floor (Zhang et al. 2016).

The third implementation has involved the restoration and the complete renovation of a Historic building located within the "Pontificio Istituto Missioni Estere" (PIME) Complex, which is in Milan. The building was built at the beginning of the last

Fig. 3 ARC model of the new hospital center of Lanciano (LOD 200) and space planning optioneering (LOD 100)

century and is currently subjected to preservation by the Italian Superintendent of Archaeology, Fine Arts and Landscape. It consists of five floors for an overall net area of 5,857 sm. The detailed project–developed by General Planning Srl—designs commercial and exhibition functions at underground and ground levels, as well as offices and collective spaces and residences on the upper floors.

Within the pilot activities included in this research, the team have redacted a detailed Employer's Information Requirements (EIR) containing data, information and models to deliver at each construction progress report, together with the required Level of Definition. The purpose was to extract Shop drawings and Bills of quantity directly from BIM models per each Milestone, in order to provide to Procurement Office the exact information before authorising the purchasing of materials and services. Finally, the EIR contains the Level of Information to reach in the As-Built Model (LOD F) that will record the work as constructed and collect all the Building Log Book Documents (Fig. 5).

The last pilot project refers to the model check and the data population of BIM Models that are recently synchronised with the Facility Management software ARCHIBUS, in order to digitally drive operations and maintenance activities of the Ambulatory Care Centre "Navile" in Bologna. The set-up of web-based platform and the As-built modeling (Fig. 6) has been done by e-FM, company leader in the

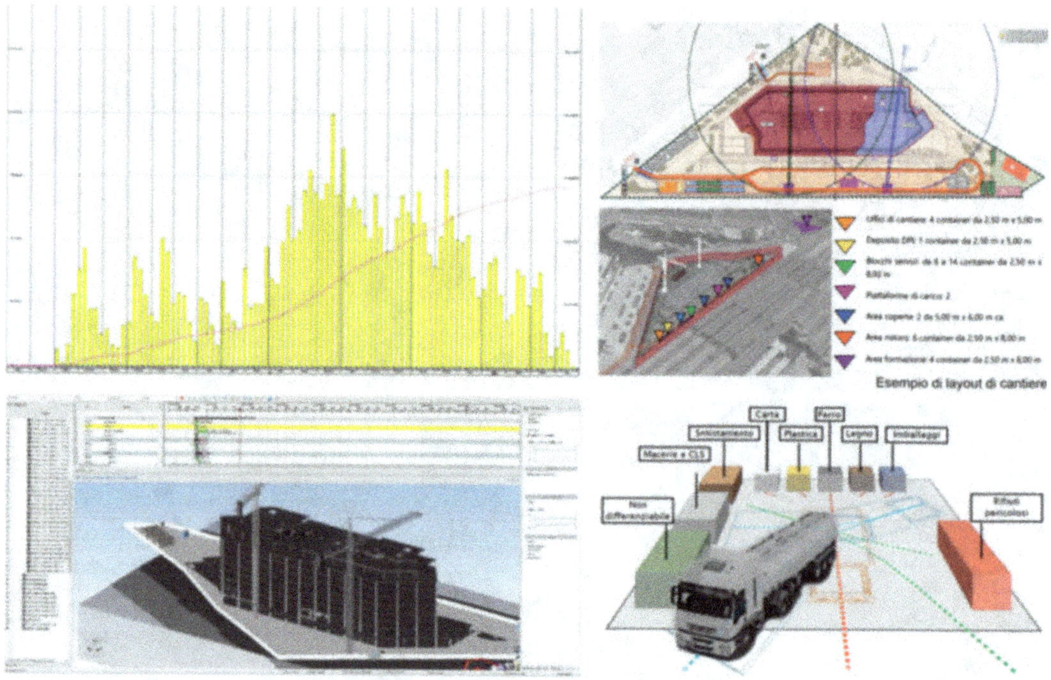

Fig. 4 Earned value analysis, 4D modeling and site layout are performed through the use of IFC models (LOD 350) included in the bidding documentation

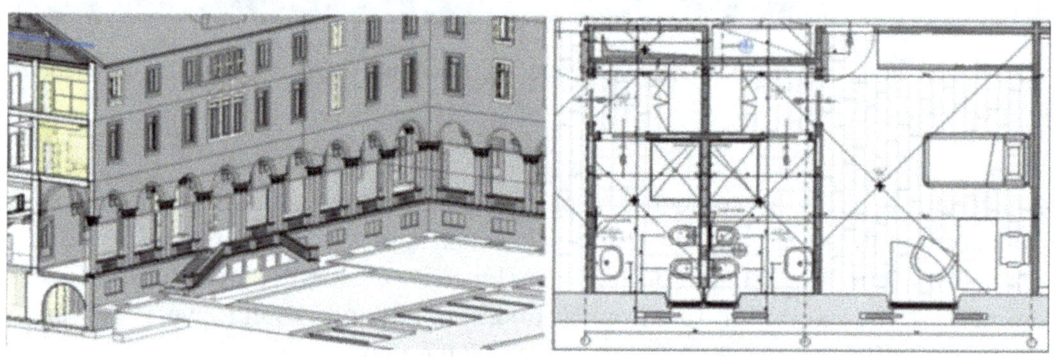

Fig. 5 Shop drawings—created with model authoring software—of PIME renovation

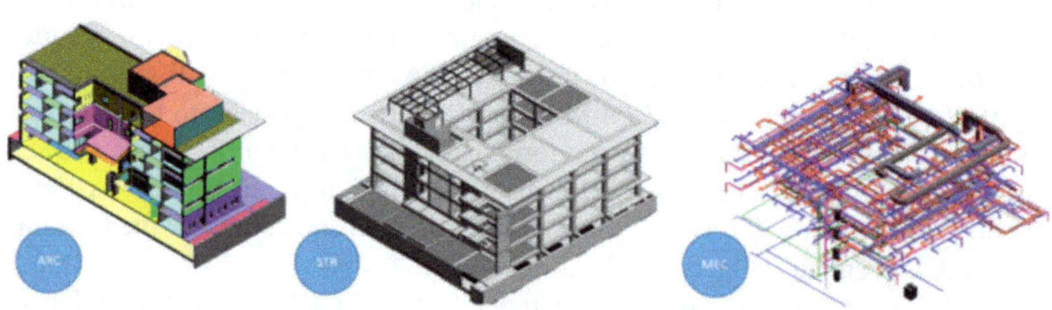

Fig. 6 As-built models of ambulatory care centre "Navile", organised per disciplines

Asset digital management. In the short-term, the aim is that empower departments seamlessly coordinate how they share, use and maintain assets. In the long-term, it will be possible to create a plan that aligns capitals and maintenance budget with company's business goals, thus optimising operations for the whole asset lifecycle (Tronchin and Manfren 2015).

7 Conclusion

The experience under examination shows how the development of the BIM Plan into a complex organisation, such as a construction company and relative sub-contractors, is necessarily a collaborative process. The planning of a shared work methodology is an essential condition for obtaining a synergic collaboration between the team components. An underestimation in the planning of the process, or in the minimum requirements needed for the advancement to a subsequent state of work, can nullify the effort that involves the construction of the BIM models for separate disciplines, which would not have a return of investment in geometric modeling alone (Ciribini 2016). BIM can be implemented at specified phases throughout a project, but the current technology, training, and costs of implementation relative to real benefits must always be considered when determining the appropriate areas and levels of detail needed in the information modeling processes. Teams should not focus on whether or not to use BIM in general, but instead they need to define the specific implementation areas and uses. A team should aim to implement BIM at the level needed to maximise value while minimising the cost and impact of the modeling implementation (Arlati and Viscuso 2018). This requires the team to selectively identify appropriate areas for BIM implementation and plan these implementation areas in detail.

References

Arlati E, Viscuso S (2018) La modellazione digitale nel processo di intervento edilizio. In: Appalti Pubblici Riserve, Varianti, e Strumenti di Precontenzioso - Tipologie e Contenuti, Collana Norme & Tributi, Il Sole 24 Ore, Milan, vol 9, pp 185–206

Capolongo S (2012) Architecture for flexibility in healthcare. Franco Angeli, Milano

Carpman JR, Grant MA (2016) Design that cares: planning health facilities for patients and visitors, 2nd edn. Jossey-Bass, San Francisco

Ciribini AL (2016) BIM e digitalizzazione dell'ambiente costruito. Grafill, Palermo

Das S, Kanchanapiboon A (2011) A multi-criteria model for evaluation design for manufacturability. Int J Prod Res 4(49):1197–1217

Deutsh R (2011) BIM and integrated design: strategies for architectural practice, 1st edi-tion. Wiley, New Jersey

Eastman C, Teicholz P, Sacks R, Liston K (2011) BIM handbook: a guide to building information modelling for owners, managers, designers, engineers and contractors, 2nd edn. Wiley, New Jersey

Holzer D, Downing S (2010) Optioneering: a new basis for engagement between architects and their collaborators. Archit Des 80(4):60–63

Tronchin L, Manfren M (2015) Multi-scale analysis and optimization of building energy performance—Lessons learned from case studies. Procedia Eng 118:563–572

Viscuso S, Dragoljevic M, Zanelli A (2019) Analisi multicriterio per la progettazione preliminare di una struttura ospedaliera. In: Proceedings of SITdA 2018 international conference "La Produzione del progetto", Reggio Calabria (ongoing publication)

Zhang C, Zayed T, Hijazi W, Alkass S (2016) Quantitative assessment of building constructability using BIM and 4D simulation. Open J Civ Eng (6), 442–461

BIM Methodology and Tools Implementation for Construction Companies (GreenBIM Project)

Claudio Mirarchi, Caterina Trebbi, Sonia Lupica Spagnolo, Bruno Daniotti, Alberto Pavan and Domenico Tripodi

Abstract Building Information Modelling (BIM) is recognized as one of the leading innovations in the construction sector. However, some studies highlight how BIM implementation is lagging behind in small and medium enterprises (SMEs). Moreover, the majority of the research is focused on the construction phase and there is a need for further studies able to demonstrate the possibilities associated with the introduction of BIM in construction companies. This paper presents the first set of results from the GreenBIM project focused on the development of a BIM implementation framework from SMEs in the construction sector. In this context, the paper focuses on the quantity estimation starting from building information models analyzed by the construction company.

Keywords Small and medium enterprises (SMEs) · Building information modelling (BIM) · Quantity estimation · Quantity take off (BIM) · Cost estimation

1 Introduction

Building Information Modelling (BIM) is nowadays recognized as one of the leading innovations in the Architecture, Engineering, Construction, Owner and Operator (AECOO) sector. BIM can be seen from different perspectives, interpreting its nature from the technological side as a tool or from the management side, identifying its introduction as a process shift (Jernigan 2007; Volk et al. 2014). The introduction of BIM requires an analysis of the context of application and the strategies adopted to optimize the implementation process can change according to a company's size, its geographical localization, business model, etc. Moreover, it is crucial to consider the societal impact related to the introduction of new technologies and processes

C. Mirarchi (✉) · C. Trebbi · S. Lupica Spagnolo · B. Daniotti · A. Pavan
Architecture, Built Environment and Construction Engineering—ABC Department, Politecnico di Milano, Milan, Italy
e-mail: claudio.mirarchi@polimi.it

D. Tripodi
GreenBIM, Reggio Calabria, Italy

B. Daniotti et al. (eds.), *Digital Transformation of the Design, Construction and Management Processes of the Built Environment*, Research for Development,
https://doi.org/10.1007/978-3-030-33570-0_18

(Mirarchi 2018). To provide a fertile environment to push the implementation of BIM in the construction industry, it is critical to involve the highest number of organizations in the construction chain. The market configuration in the construction sector (as well as in other ones (OECD 2000)) is characterized by a majority of small and medium enterprises (SMEs). This configuration is not related to a specific country, but it is common for the sector. For example, in Italy around 96% of the construction companies have less than nine employees (ANCE 2017), in the UK 93% have less than 13 employees (DTI UK 2006), and in France (all sectors combined) 98.8% of companies are SMEs (Tranchant et al. 2017).

Several studies in the literature explored the possible benefits related to the use of BIM in the construction sector. However, most of the existing studies are related to big and/or complex projects and focused on the design phase. Thus, several studies highlighted the lack of research for the SMEs and how SMEs are lagging behind in the adoption of BIM and in general in the implementation of new digital technologies along the lines of the industry 4.0 paradigm (Hosseini et al. 2016; Lam et al. 2017; Li et al. 2019).

The GreenBIM project (under development and in collaboration with the construction company *Berna Costruzioni*) is based on a practical approach to explore the implementation process of BIM in small and medium construction companies. The research is focused on two areas that still need to be broadly explored that are (a) the implementation process for SMEs and (b) the analysis of benefits and challenges in BIM implementation in the construction companies (i.e. in the construction and maintenance phase). With regards to the barriers in BIM implementation for SMEs explored in previous studies and presented in the next section, the study aims to provide a structured process to facilitate the diffusion of a shared implementation approach that can remove or at least reduce the impact of existing barriers demonstrating the effectiveness of BIM and how to work to achieve the required objectives. This paper presents the state of the art of the project exploring the results reached in the first stage of the study and provides the overview of future research actions.

The rest of the paper is organized as follows. The background section provides a brief summary of the existing studies in BIM implementation for SMEs. It also contains an introduction about the quantity of surveying activities which represent one of the main aims for the construction company in the integration of BIM. The implementation process section describes the activities and the organization structure proposed in the BIM implementation for construction companies focused on a specific objective, i.e., the quantity take off (QTO). Finally, the conclusion section summarizes the results of the research carried out and discusses future activities for the work.

2 Background

2.1 *BIM in Small and Medium Enterprises*

SMEs are usually defined according to their staff headcount, their turnover and/or their balance sheet total. In the European context a company is defined as SME if the staff headcount is under 250 and the turnover is under 50 million euros or the balance sheet total in under 43 million euros (European Union 2003). As already mentioned in the introduction, SMEs constitute most of the companies in the construction sector and their involvement in BIM implementation is crucial to guarantee communication between all parties involved in the construction process and thus define effective collaborative approaches.

However, the introduction of new technologies in SMEs is limited by several barriers. According to their limited investment capacity, SMEs are inclined to adopt consolidated and reliable methods that can guarantee a return on investment (ROI) (Poirier et al. 2015). The absence of practical evidence capable of demonstrating an effective ROI in BIM implementation is one of the main barriers in BIM adoption for SMEs (Hosseini et al. 2016). Li et al. (2019) identified six critical factors and challenges for BIM adoption in SMEs, namely: limited resources, collaboration challenges, lack of BIM awareness, legal disputes and uncertainties in policies, difficulties in meeting SMEs' needs and concerns about data and information. On the other hand, Hosseini et al. (2016) argued that the lack of awareness is no longer a barrier in SMEs in the Australian areas. Malacarne et al. (2018) identified in the lack of standardization a critical barrier which is also limiting the development of effective tools for the industry. In this picture, the proposed research will provide a possible path to follow in the BIM implementation and will demonstrate the benefits related to the use of BIM in construction companies limiting the above-mentioned issues. The results of the project will pave the way for BIM implementation in SMEs reducing the required investments (costs) thanks to a defined and tested roadmap and demonstrating the quantitative returns (ROI) in the investments on BIM adoption.

2.2 *Cost Estimating in BIM*

Cost estimating is a fundamental activity in construction projects. Focusing on construction companies, this activity is critical in the development of their business and it is directly related to the estimation of the resources required for the construction process (e.g., bricks, concrete, workforce, etc.). Some studies demonstrated the benefits of using BIM in the early phases of the construction process to improve the effectiveness of cost estimating activities (Valentini et al. 2017). On the other hand, BIM can support cost estimation activities in the other phases of the project as well providing precise quantity analysis and guaranteeing coherence between the design configuration and quantity estimation. Nevertheless, the information embedded in building information models are not sufficient to provide a cost estimate and the quantities obtained from models need to be combined with other information (Wu et al. 2014). On the one hand this activity requires the translation of the geometrical information contained in a model into geometrical information that is coherent with the estimation procedure (Ma et al. 2016). On the other hand, it requires the introduction of knowledge derived from the construction company that can translate the model quantity into quantities of materials, workforce, etc., and consequently into costs.

2.3 Implementation Process

BIM implementation in the construction sector companies, whether big or not, requires an articulated action considering among other elements the societal, market and technical context of application, the organizational structure of the company, the knowledge and skills of the personnel, the investment capacity and of course the objectives of the implementation. The main national standards and guidelines related to BIM (e.g., (American Institute of Architects 2013; BSI 2013; BIMForum 2015; UNI 2017)) are usually focused on the specific project and not on the organization. However, the structure of the documents required in the BIM process such as the employer information requirements (EIR) and the BIM execution plan (BEP) express the need to define a process at the organizational level that can be declined according to the specific needs of each project. This vision has been materialized in ISO 19650 part 1 (ISO 2018) where the concept of organizational information requirements (OIR) is introduced representing the information requirements defined according to organizational objectives. Even if not identified in the standard, it is clearly necessary to define not only the requirements but also the way to satisfy these requirements when the actions and the processes are developed directly in the organization. Hence, the BIM implementation process requires the clear identification of the objectives of the organization and the consequent planning of the actions to satisfy these requirements. In general, the objectives of construction companies are related to the need of increasing their productivity (reducing wastes, improving the quantity estimation activities, reducing reworks, etc.) and to the possibility of exploring new business models using innovative technologies, the development of new services, etc. While the GreenBIM project considers the overall picture, this paper presents a specific part of the research that is the definition of structured processes to use the building information models in the quantity estimation activities optimizing the integration between the construction companies' knowledge and the information contained in models.

BIM sees in the development of a shared collaboration environment commonly identified as the common data environment (CDE) (BSI 2007, 2013) one of its core principles. This environment is usually identified as a cloud where all the interested subjects can collaborate and share information throughout the construction process. However, even in the case of a perfect circular approach (which today is still not possible due to IT and contractual issues) there is always the need to check and guarantee the quality of the information used as input for the development of subsequent phases of the process. This leads to the need of defining specific requirements related to the development of information models and to check the quality of these models to guarantee the correct development of future activities. The quality of the information provided to the construction company has long been studied as one of the main areas of value loss in the construction process. Discrepancies between information in different documents, difficulties in the interpretation of the design, etc. are only a few examples. The use of BIM can limit these issues creating a model which, by simulating the actual building (or the one that will be constructed) and creating a CDE where all the information is coordinated and related, is able to avoid incoherence between project documents. However, on the one hand the use of BIM can reduce the proliferation of errors in the process. On the other hand, it introduces new needs in terms of data quality requiring a different approach and different competencies in the evaluation of the quality of the information generated and communicated during the process.

Hence, before starting any quantity estimation activity a quality check is required according to the organization requirements defined for the specific use of the information model. This checking activity can be synthesized in two main areas i.e. clash detection and information checking. The first activity must be structured according to the specific needs of the organization and can be organized according to a matrix where the elements that must be checked and those that do not need to be evaluated are highlighted (Table 1).

The translation between models and construction companies' resources is of crucial importance. The model developed by the designer is usually limited in terms of both detailed quantity information and construction site context. As a consequence, the quantity take-off application requires the identification of rules capable of translating the geometrical quantities provided by the models into real geometrical quantities that can reflect the needs of construction companies in their construction activities. Unfortunately, interoperability issues can hinder a fluent process requiring the precise definition of information flows where the output of one phase represents the input of the next one. Thus, starting from the information model defined during the design phase, the construction company needs to introduce its own expertise in the model to obtain the quantity take off model (Fig. 1).

The QTO activity can be developed according to different processes. On the one hand, if there is coherence between the modelling instrument used by the design team and those used by the construction company it is possible to define a direct connection through plug-ins or other direct relations between the geometrical information model and the QTO one. On the other hand, if the instruments are not compatible, it is possible to identify a specific translation of the design model using the IFC format. Nevertheless, the use of IFC requires a specific definition of the information to be exported and the specific analysis on the way in which the geometries are exported from the native model to the IFC one.

Table 1 Example of clash detection matrix

Models		Arc Model							Str Model			
	Classes	Ce	Dr	Sl	Ra	St	Wa	Wd	Sl	St	Pi	Be
Arc Model	Ce											
	Dr											
	Sl											
	Ra											
	St											
	Wa											
	Wd											
Str Model	Sl											
	St											
	Pi											
	Be											

Legend
Ce: Ceilings, Dr: Doors, Sl: Slabs, Rl: Railings, Sr: Stairs, Wa: Walls, Wd: Windows, Pi: Pillars, Be: Beams.
Arc: Architectural, Str: Structural.
Dark Grey cells: not relevant clash, light grey cells: duplicates

Once the QTO model is created, this model needs to be integrated with the history of the construction company which can be codified through an ERP (Enterprise Resource Planning). This stage, which is the integration process between the QTO model and the ERP, will be explored through future activities of the project.

3 Conclusion

This paper presented a first analysis of the information requirements relating to a specific use of building information models in construction companies, that is, the QTO operation required to pave the way for the definition of the resources for construction activities and consequently their costs. In this context the article proposed an analysis of the existing literature relating to the introduction of BIM in SMEs focusing on the need to provide effective means to help these companies in the digital transition. Starting from the results presented in this paper the GreenBIM project

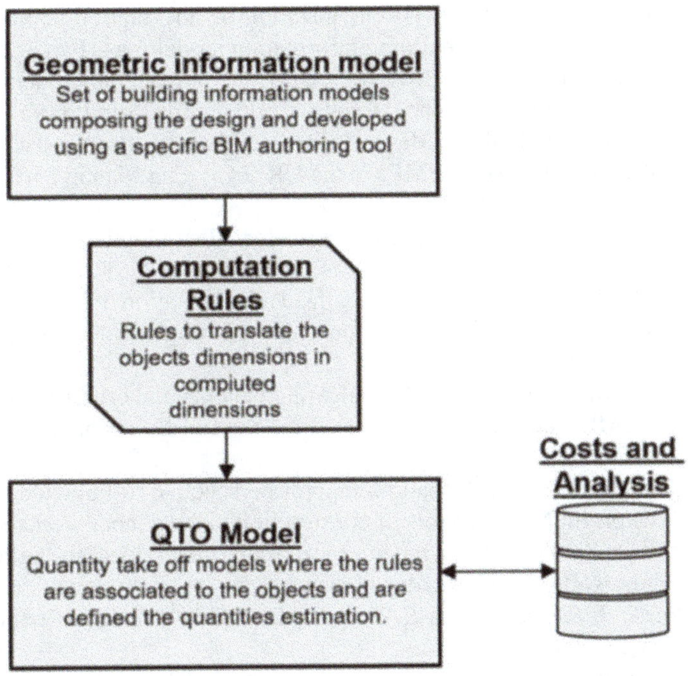

Fig. 1 Information flow from the (geometric) information model to the quantity and costs estimation models

saw the development of a structured process to integrate the QTO model into the construction company database (ERP) in order to match the expertise of the company with the information derived from the information models. Moreover, the project aims to explore the information requirements related to the management of the asset to promote the development of new services according to the maintenance needs and management in the use phases of the final products (buildings, infrastructures, etc.).

References

American Institute of Architects (2013) Digital practice documents - guide, instructions and commentary. AIA guidelines, pp 1–62

ANCE (2017) Osservatorio congiunturale sull'industria delle costruzioni

BIMForum (2015) Level of development specification 2015

BSI (2007) BS 1192:2007, collaborative production of architectural, engineering and construction information—Code of practice. The British Standards Institution, UK

BSI (2013) PAS 1192-2:2013, Specification for information management for the capital/delivery phase of construction projects using building information modelling. The British Standards Institution, UK

DTI UK (2006) Construction statistic annual, London

European Union (2003) Commission recommendation of 6 May 2003 concerning the definition of micro, small and medium-sized enterprises. Off J Eur Union. https://eur-lex.europa.eu/legal-content/EN/TXT/PDF/?uri=CELEX:32003H0361&from=EN

Hosseini MR et al (2016) BIM adoption within Australian small and medium-sized enterprises (SMEs): an innovation diffusion model. Constr Econ Build 16(3):71–86. https://doi.org/10.5130/ajceb.v16i3.5159

ISO (2018) ISO 19650-1—Organization of information about construction works—Information management using building information modelling—Part 1: concepts and principles

Jernigan F (2007) Big BIM, little BIM: the practical approach to building information modeling; integrated practice done the right way. 1st edn. 4Site Press

Lam TT, Mahdjoubi L, Mason J (2017) A framework to assist in the analysis of risks and rewards of adopting BIM for SMEs in the UK. J Civ Eng Manag 23(6):740–752. https://doi.org/10.3846/13923730.2017.1281840

Li P et al (2019) Critical challenges for BIM adoption in small and medium-sized enterprises: evidence from China. Adv Civ Eng 2019. https://doi.org/10.1155/2019/9482350

Ma Z, Liu Z, Wei Z (2016) Formalized representation of specifications for construction cost estimation by using ontology. Comput-Aided Civ Infrastruct Eng 31(1):4–17. https://doi.org/10.1111/mice.12175

Malacarne G et al (2018) Investigating benefits and criticisms of BIM for construction scheduling in SMEs: an Italian case study. Int J Sustain Dev Plan 13(1):139–150. https://doi.org/10.2495/SDP-V13-N1-139-150

Mirarchi C (2018) A spatio-temporal perspective to knowledge management in the construction sector. In: New frontiers of construction management workshop. Ravenna, Italy

OECD (2000) Small and medium-sized enterprises: local strength, global reach. Policy Brief. https://doi.org/10.1177/0022146511418950

Poirier E, Staub-French S, Forgues D (2015) Embedded contexts of innovation. Constr Innov 15(1):42–65

Tranchant A, Beladjine D, Beddiar K (2017) BIM in French smes: from innovation to necessity. WIT Trans Built Environ 169:135–142. https://doi.org/10.2495/BIM170131

UNI (2017) UNI 11337- 1—Building and civil engineering works—Digital management of the informative processes—Part 1: models, documents and informative objects for products and processes, Italy

Valentini V, Mirarchi C, Pavan A (2017) Comparison between traditional and digital preliminary cost-estimating approaches. Innov Infrastruct Solut. (Springer International Publishing), 2(1):1–8. https://doi.org/10.1007/s41062-017-0066-7

Volk R, Stengel J, Schultmann F (2014) Building information modeling (BIM) for existing buildings—Literature review and future needs. Automation in construction, vol 38. Elsevier B.V., pp 109–127. https://doi.org/10.1016/j.autcon.2013.10.023

Wu S et al (2014) A technical review of BIM based cost estimating in UK quantity surveying practice, standards and tools. J Inf Technol Constr 19(December):535–563

Adaptive Skins: Towards New Material Systems

Ofir Albag, Maria Anishchenko, Giulia Grassi and Ingrid Paoletti

Abstract The aim of this chapter is to shed light on the potentialities of smart material systems in architectural skins. The research on new materials has triggered new solutions, based on emergent technologies, for envelope design. Adaptive facades, exploiting stimulus-responsive materials and intelligent control systems, are able to change in response to environmental conditions, thus enhancing users' comfort and improving energy savings. Two different case studies are here illustrated: the first is a dynamic shading system based on shape memory alloys; the second example concerns a soft-robotics weather-responsive skin. Both projects are described following the same workflow structure in order to highlight similarities and differences. Indeed, the sections are outlining the steps for design, simulation and fabrication of the case studies, and are divided as follows: description of the project, methodology and design strategy, simulation and prototyping.

Keywords Adaptive facades · Shape memory alloys · Soft robotics · Computational design · Smart materials

1 The Role of Adaptivity

In the last decade, the concept of adaptive facade has arisen from the debate on energy efficiency in buildings. Indeed, the term "adaptive" refers to a system which constantly changes as a result of climate variations. A plethora of studies showed that envelopes are responsible in large part for buildings' energy consumption. Dynamic skins are able to optimally balance energy demand and occupants' comfort. Multi-performance skins were introduced as a paradigm for an "all season walls", providing protection in all weather conditions.

O. Albag
Milan, Italy

M. Anishchenko · G. Grassi (✉) · I. Paoletti
Architecture, Built Environment and Construction Engineering—ABC Department, Politecnico di Milano, Milan, Italy
e-mail: giulia.grassi@polimi.it

B. Daniotti et al. (eds.), *Digital Transformation of the Design, Construction and Management Processes of the Built Environment*, Research for Development,
https://doi.org/10.1007/978-3-030-33570-0_19

The research on facades is currently focusing on the design of mechanisms that allow adaptation, especially exploiting material properties or intelligent sensing/actuating devices. Thus, the advent of smart materials has led to a new conception of design and manufacturing, where the process can be simulated and optimized by embedding material properties and programming behavioural patterns.

This chapter is an attempt to identify fundamental morphological characteristics of kinetic adaptive facades by illustrating two case studies that has been developed within this research line.

2 Stimulus-Responsive Materials and Control Systems

Envelopes are considered as a demarcation line between external and internal conditions, a safety boundary that protects us, as well as a threshold between energy field transition (high-energy to low energy or different state of energy). However, studies on new materials and kinetic mechanisms are eliciting a new vision of facades, as the space where the transition happens, an interactive multifaceted space. Therefore, the role of the designer becomes designing, thus programming, the interaction.

Stimuli-responsive materials (SRMs) are able to enact intelligent behaviours in response to an external stimulus. According to Ritter (2007) SRMs can be classified on the basis of the effect produced by the stimulus as: property changing, energy exchanging and matter exchanging. Stimuli can be identified also with environmental external factors that influence human comfort, to name a few: solar radiation, outdoor temperature, wind, humidity, precipitation, noise, and pollution. The first two can be seen as the major elements impacting the design of the system from a performative perspective. Moreover, temperature and light are also the two most impacting triggers for activating SRMs. The effective control of environmental factors is a crucial point for the successful operation of adaptive facades. Indeed, without proper detection tools and control systems, the envelopes become senseless. According to Addington and Schodek (2005), control systems can be divided in two categories: extrinsic (active) and intrinsic (passive) (Table 1).

Extrinsic control systems rely on automated regulation and consist of sensors, processors, actuators and control logic. They are indeed capable of detecting feedback, by comparing the current configuration with the desired one and, if necessary,

Table 1 List of properties of extrinsic and intrinsic control systems

Extrinsic	Intrinsic
Gradual	Immediate response—direct actuation
Automated (sensors-processors-actuators)	Based on material properties
Users feedback	No user interaction
Energy intensive	Energy saving
Constant maintenance	No reaction to unexpected events

adjusting the behaviour of the building envelope. Notable applications of extrinsic systems are Al-Bahr Towers, One Ocean Pavilion and Institute du Monde Arab.

Intrinsic (or passive) control systems are characterized by the inherent adaptive capacity of materials, or kinetic components, constituting the facade. The environmental stimulus is the trigger for such self-adjusting behaviour. However, there is no filter: environmental inputs are directly transformed into actions. Therefore, it is of fundamental importance to carefully study the material response to the stimuli and simulating/ testing the whole functioning prototype. Some examples exploiting intrinsic control systems include: SmartWrap at Cooper-Hewitt National Design Museum and HygroSkin Meteorosensitive Pavilion.

The hybrid combination of the two control systems enables to take full advantage of smart materials and electronics. The following sections will illustrate two case studies of adaptive facades, shading light on design strategies for programmable materials.

3 Shape Memory Alloys for a Dynamic Shading System

The aim of this project is to design a kinetic system with low-maintenance and high degree of flexibility for its application. Hence, the main objective is to investigate an intrinsic technology in which environmental inputs could be directly transformed into actions, without external decision-making component. Stimuli-responsive materials constitute the perfect candidate, thanks to their ability of changing their shape as a response to an external trigger.

Shape memory alloys (SMA) are materials with the ability to recover their original shape after deformation, at the presence of proper conditions (Huang et al. 2010). SMA enables to create simple modules which can react to temperature change and modify their shape through an immediate response. This makes them universal materials that can work simultaneously as sensors and actuators, and react without the need of a control system. The shape memory effect is achieved thanks to the solid-state phase change of SMA. Most of the existing materials have three phases: solid, liquid and gas and react on a change of temperature, depending on characteristics of the material. A broadly similar process happens in a solid-state phase change. The material is a solid both before and after the transformation because the molecules remain very close together throughout. SMAs flip back and forth between two solid crystalline states called austenite and martensite.

3.1 Methodology and Design Strategy

The workflow is divided into three consecutive phases. It starts with the analysis of the daylight performance and understanding the need of the dynamic solar shading. The second step is to define the tools and methodologies. The final step of the research is the design of the shading module, the simulations of its effectiveness and the fabrication of a physical prototype.

The project is situated in southern Siberia, in the Russian city of Tomsk. The temperature there can fluctuate from −40 in winter to +30 in summer. According to the local construction regulations, fixed shading is prohibited due to the cold climate in winter. Nevertheless, shading in summer is needed. Evidently dynamic shading can play an important role and permit different solutions.

The objective of the work is to create a solar screening system which is light and easily deployable, but at the same time resistant to weather conditions. Therefore, the principle of origami, Miura-Ori pattern (Fig. 1a), coupled together with SMA joints, provide a dynamic shell mechanism. This pattern permits to create a light and thin structure, which is at the same time stiff enough to sustain external forces (Schenk and Guest 2014). The Shape Memory Alloy hinges can be actuated without any control system and be completely autonomous. In order to add manual control to the system, the alloys can respond to electrical stimuli as they heat up and subsequently change shape (Fig. 1b).

A large-scale application of such deployable mechanism requires high flexibility, stiffness and lightness. One material technology that well suits these requirements is "Wood-skin". It is created of digitally fabricated wooden tiles, sandwiching a nylon and polymer mesh in between (Mok 2016). The benefit of this material that the SMA foil can be inserted in-between the layers of wood, creating an invisible hinge.

The scheme of hinges allocation for the Miura-Ori pattern is following the logic of the origami folds. All the folds are divided in positive (+180°) and negative ones (−180°), the so-called "mountain and valley folds". The SMA hinge is attached to the folding-sheet with micro-nuts inside the layers of plywood.

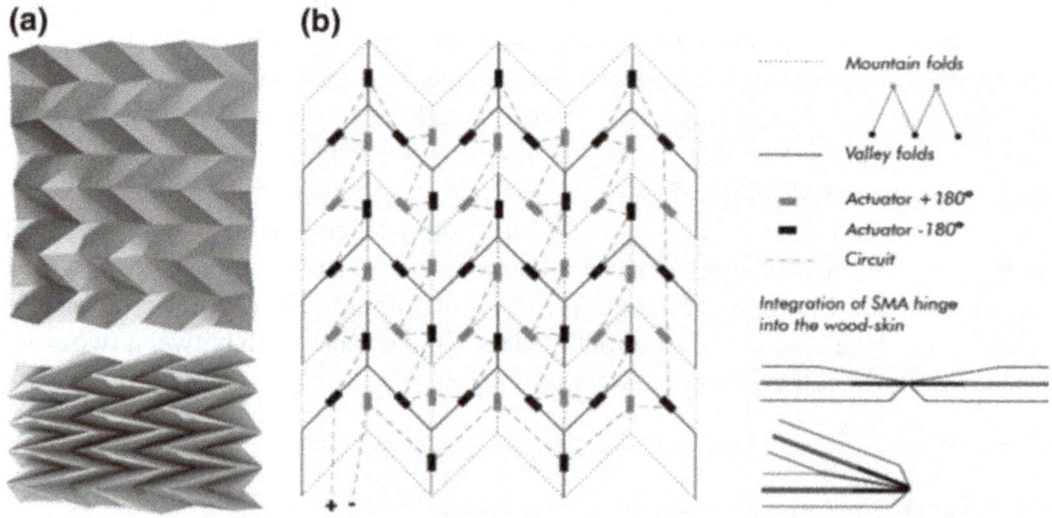

Fig. 1 **a** Miura-ori origami pattern. **b** Scheme of the allocation of the actuator

3.2 Simulation and Prototyping

The study of the Shape Memory Alloy is supported with a series of experiments with nitinol foil. The shape setting of the foil is done with the help of a high temperature furnace and high-voltage infra-red lamps for testing. The model is realized in a scale of 1:10. The wood-skin material is modelled with pieces of plywood of 1 mm thickness attached to the non-stretch fabrics.

To set the shape of the Nitinol it should be heated up to a temperature of 400–500 °C for 10–25 min, depending on the material's composition and thickness, then immediately cooled down to room temperature. Rectangular pieces of Nitinol with size 10–12 mm with thickness of 0.1 mm are folded to 180° and put on a ceramic base weighted by a 5 kg block of metal and left in the heated furnace on 400° for 15 min.

The pieces of preprogramed nitinol are then attached to the plywood with tape. The choice of tape is due to the small scale of the model. In the full-sized prototype, small bolts could be used instead to create a stiffer connection. Testing the small-scale prototype, the connections work well and react on the changes in temperature (Fig. 2a, b) (Anishchenko 2018).

The research results in the design of a shading device for a window (Fig. 3). A modular system can vary in size and be applied to new and existing buildings.

The technology's efficiency is simulated with Ladybug for Grasshopper. The simulation date is set to the July 1. The surface of the window glass is divided in smaller areas and the level of solar radiation is analysed in the centre of each area.

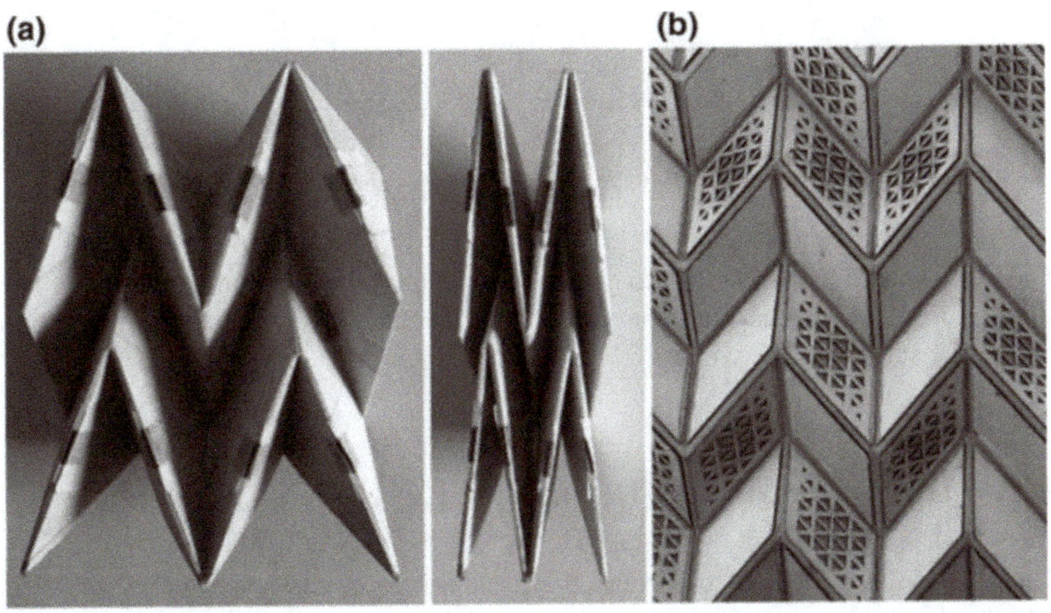

Fig. 2 **a** Physical prototype scale 1:10 with implemented SMA; **b** Physical prototype with cut out geometrical pattern

Fig. 3 Renders of exterior and interior views of the module

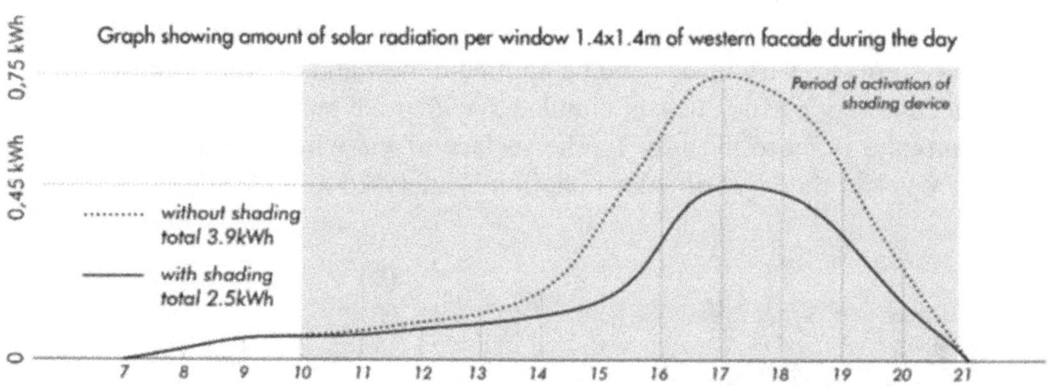

Fig. 4 Levels of solar radiation hitting the glass surface with and without the shading system

The results are then summed up to get the total amount of radiation per window. The simulation demonstrates that dynamic shading results in 35% reduction in radiation hitting the surface of the window (Fig. 4).

4 Soft-Robotic Weather-Responsive Envelope

The aim of the project is to harness the possibilities of the soft-robotic actuation into the development of a flexible and lightweight, adaptive building envelope. The innovative weather-responsive space for studying is equipped with a soft-robotic envelope system that transforms it radically from an open to enclosed thermally insulated space. Thus, it improves the climatic and energy performance. The design includes a bottom-up iterative process with use of simulations and optimization processes that focuses on the development of the soft robotic components.

Traditionally, all theories and techniques for robotic control, fabrication and sensing are based on a conventional definition of robots as a kinetic structure of rigid members. Recent advances in soft and smart materials, compliant mechanisms and nonlinear modeling have increasingly led to the popularization of the soft materials in robotics. This trend is driven by new scientific paradigms like biomimetics and morphological computation and by application requirements in biomedical engineering, service, rescue robots and other fields. The expectation is to make soft robots interact more easily and effectively with real-world environments (Mazzolai et al. 2012; Pfeifer et al. 2012).

The core concept of the soft robotics' is to fabricate a robot all made up of flexible and elastic components with the ability to change gaits easily and maneuver in very limited spaces, based on the use of shape changing materials and their composites (Yokoi et al. 1999). A variety of typologies of soft actuators were developed to this day, using different actuation mechanisms. Among these some of the more prominent are the PneuNets bending actuator, fiber-reinforced actuators, pneumatic artificial muscle, dielectric elastomer actuator and multi-module manipulator (De Falco et al. 2014).

4.1 Methodology and Design Strategy

The functional requirements for the design are derived from the need for temporary study and working space in institutions. Existing efforts usually suffer from compromised comfort. In this context, the requirements foresee the design of an outdoor study space designed to be easily assembled, disassembled and transported. It is to provide cover, sufficient illumination and ventilation and to be integrated with HVAC system and an insulating weather-responsive envelope to optimize energy efficiency.

With the soft robotics as a starting point, the design process takes a bottom-up strategy. It first focuses on the development of the soft envelope component and only then oversees in an iterative process the development of the overall design. The overall morphology of the envelope consists of eight identical faces of an octagonal dome configuration. Each of the envelope components requires two kind of motion: furl up to open and down to close (Fig. 5). This is possible by tweaking the inner structure with the pneumatic networks soft actuator (PneuNets).

The solution adapted to this project is a typology of PneuNets actuator that is designed specifically to create flat, compact soft robots that could squeeze through narrow spaces. These actuators use the PneuNets mechanism of work pressurizing a series of adjacent internal chambers. However, all chambers are embedded in a flat configuration inside the outlines of the body (Shepherd et al. 2011). Moreover, they use a variation in elastomer thickness to achieve their furling deformation. Attaching two embedded pneumatic networks back to back, we receive a compact bidirectional

Fig. 5 Exterior renders of the shelter in open and closed configurations

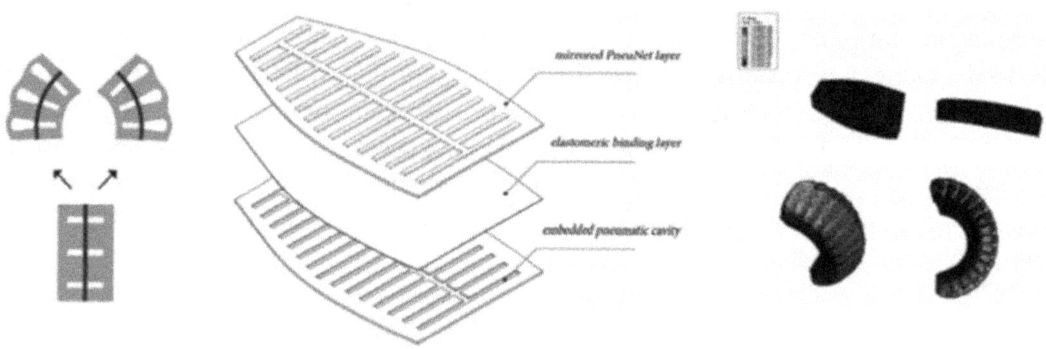

Fig. 6 Left to right: a diagram section of a flat bidirectional PneuNet actuator, when deflated and pressurized on each side; exploded axonometry of the soft-robotic skin component layers; FEM modelling and strain analysis for the soft component in full and sectioned view

actuator (Fig. 6). The result is a flat skin component that is able to adjust up or down, with a simple configuration, and only two air inlets to control the enclosure of the space.

The geometry is then optimized by design iterations, testing and adjusting through physical prototyping or digital simulations, using computational tools. This shows how the prototype will react under the different variables as pressure, gravity, wind and other loads; its deformation and trajectory. Once obtaining a satisfactory component design, the attention could be drawn to a broader scale of the skin system and its supporting structure.

4.2 *Simulation and Prototyping*

Designing a soft actuator, it is important to understand how the materials, morphology and geometry of the actuator do affect the behavior of the device. Dealing with elastic materials, it is difficult to accurately predict the effects of our design decisions due to their nonlinear behavior. Traditionally the design would be tested in a slow and costly process of physical prototyping. An increasingly more accessible approach is that of digital simulations using computational tools. These tools are used during the design phase in order to validate the design hypothesis and optimize it according to needs.

The first step with digital simulations is modeling the component in a 3d modeling software. Some widely accessible tools familiar to designers for simulating the behavior of simple tensile and pneumatic structures are parametric design plugins such as Grasshopper for Rhinoceros and Kangaroo. This real-time physical simulation tool suffers some serious drawbacks. It allows getting some fast and basic sense of geometry and behavior and to simulate simple inflation. However, it lacks some important features in order to successfully replace physical prototyping, such as the ability to assign properties of physically accurate materials or different thicknesses.

Less accessible but more accurate tool is the Finite Element Method (FEM) using Abaqus or similar. FEM analysis is more adequate for the design of elastomeric PneuNet actuators for its physical accuracy. The process generally starts with the 3D digital model of the actuator, simplified, divided and imported into the FEM software. Materials are then created inserting values of physical properties and assigned to the model. The parts are then positioned correctly relative to one another and merged. Physical loads are defined and applied on the relevant parts of the model, as well as boundary conditions (e.g., anchoring or self-interaction) in order to obtain a realistic behavior. The final step is meshing the geometry and running the simulation (Fig. 6). Although very accurate, FEM is limited due to its complexity of set-up and long processing time, rendering impossible to change and take informed decisions in real-time.

Unlike small-scale soft actuators, large-scale actuators require the deployment of different solutions and technologies, especially for the fabrication of big-scale elastomer casting molds. One option of a low-cost fabrication is to cut and assemble elastomer mold parts using a CNC instead of 3d-printing it in one piece. This and some other methods are limited to simple geometries and might not be adapted for obtaining complex internal structures.

As a part of this project, a small-scale prototype was fabricated in order to test and compare the design for any discrepancies. The prototype was casted using two custom 3d-printed molds in three steps. Finally it was connected with tubes to an air pump controlled by an Arduino board, programmed to activate it in response to a light sensor input as a proof of concept (Fig. 7).

5 Conclusion

The pursuit for adaptive facade has led to a quest for intelligent control systems that integrate together many diverse functions. Nevertheless, the incorporation of these advanced systems in buildings is still slow due to high costs of materials and technologies and to the necessity of a deep research and development phase.

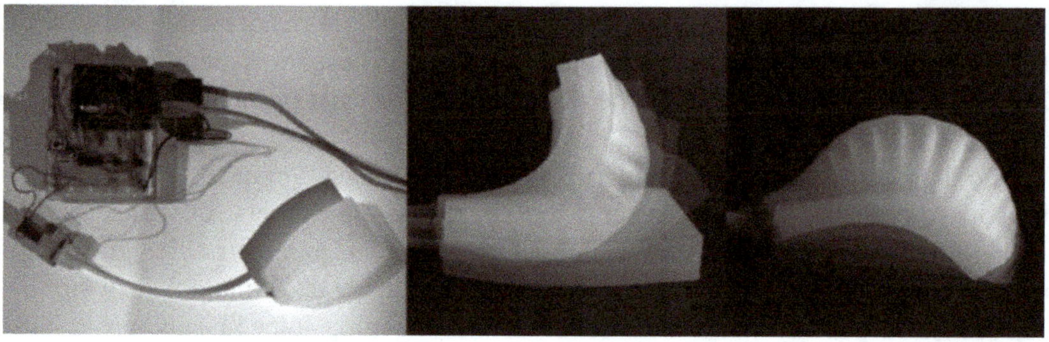

Fig. 7 Images of the scaled prototype with the Arduino control board configuration and actuation trajectory of the prototype in both directions

In fact, the two projects suggest that these systems need further investigations, accurate simulation software as well as a set of customized instructions for each application. On the other hand, they allow for a flexible design and endless possibilities of adaptation.

References

Addington M, Schodek DL (2005) Smart materials and new technologies: for the architecture and design professions. Architectural Press

Anishchenko M (2018) Smart memory materials in dynamic facade systems. Politecnico di Milano

De Falco I, Cianchetti M, Menciassi A (2014) A soft and controllable stiffness manipulator for minimally invasive surgery: preliminary characterization of the modular design. IEEE Eng Med Biol Soc

Huang WM, Ding Z, Wang CC et al (2010) Shape memory materials. Mater Today 13:54–61. https://doi.org/10.1016/S1369-7021(10)70128-0

Mazzolai B, Margheri L, Cianchetti M, Dario P, Laschi C (2012) Soft-robotic arm inspired by the octopus: from artificial requirements to innovative technological solutions, pp 338–339

Mok (2016) Wood-skin: composite material that's strong like wood, flexible like fabric. https://www.treehugger.com/sustainable-productdesign/%0Awood-skin-composite-wood-material-folds-like-fabrics.html. Accessed 20 May 2019

Pfeifer R, Lungarella M, Iida F (2012) The challenges ahead for bio-inspired soft robotics. Commun ACM 55(11):76–87

Ritter A (2007) Smart materials in architecture, interior architecture and design. Birkhäuser

Schenk M, Guest SD (2014) On zero stiffness. J Mech Eng Sci 228:1701–1714. https://doi.org/10.1177/0954406213511903

Shepherd R, Ilievski F, Choi F, Morin S, Stokes A, Mazzeo A, Chen X, Wang M, Whitesides G (2011) Multigait soft robot. Proc Natl Acad Sci USA 108(51):20400–20403

Yokoi H, Yu W, Hakura J (1999) Morpho-functional machine: design of an amoebae model based on the vibrating potential method. Robot Auton Syst 28:217–236

Development of a System for the Production of Disposable Carbon Fiber Formworks

Pierpaolo Ruttico and Emilio Pizzi

Abstract This article presents a new method for the production of disposable carbon fiber formworks for the casting of reinforced concrete columns. The desired three-dimensional object can be made by the robotic hot wire cutting of polystyrene, thus generating the shape around which the carbon fiber is deposited. The manufacturing technique is the so-called filament winding, where the polystyrene shape—obtained through hotwire cutting—is wrapped in fiber tape and then undergoes a curing process, which returns the carbon geometry in solid form. At the end of the process, the final piece of carbon is obtained by dissolving the positive polystyrene mold with a solvent. This gives rise to a production method capable of creating geometries that cannot be achieved by other means. The choice of creating disposable fiber formworks for concrete castings has considerable advantages: the possibility of creating structural elements with complex geometries that cannot be obtained by means of traditional formworks or other materials; saving of time in the casting phase and the advantage of not having the de-casting phase; facilitating the positioning of formworks and other structural elements thanks to the reduced weight of the material; the possibility of having always different geometries and a finish with a high aesthetic value and functional performance. The combination of concrete and carbon fibers offers above all considerable advantages from the structural point of view. The formwork not only has the function of giving shape to the finished element but also becomes a collaborator for static purposes. If used for pillars, the carbon fiber formwork has the function of a hoop, which makes it possible to obtain columns with the same static capacity but with lower sections and the elimination of the transversal reinforcement.

Keywords Carbon fiber · Robotic hotwire-cutting · Disposable formworks

---

P. Ruttico (✉) · E. Pizzi
Architecture, Built Environment and Construction Engineering—ABC Department, Politecnico di Milano, Milan, Italy
e-mail: pierpaolo.ruttico@polimi.it

B. Daniotti et al. (eds.), *Digital Transformation of the Design, Construction and Management Processes of the Built Environment*, Research for Development,
https://doi.org/10.1007/978-3-030-33570-0_20

1 Choice of Materials

Among the possible materials for the construction of the disposable formwork, the choice fell on the carbon fiber both for structural reasons related to the lightness of the material and its high mechanical resistance to traction and for the resistance to temperature changes and the effect of chemical agents. To achieve an effective rim, in fact, a high degree of rigidity is required and carbon has the highest elastic modulus among the fibers on the market. This material is also resistant to the alkaline environment typical of concrete. It was decided to combine carbon with a thermosetting matrix, particularly epoxy; compared to a thermoplastic matrix, the former guarantees better mechanical characteristics, together with a more effective method of applying the fibers. The most suitable and economical production process for the deposition of the fiber on molds is that of filament winding. To make molds for molds that are always different through digital manufacturing methods, it is necessary to use an inexpensive material that is easy to work with, sufficiently resistant to the compression applied during the winding phase of the fibers and inert against resins and carbon fiber. Remaining among the most popular and well-known materials in the building industry, extruded polystyrene was chosen. This material can also be dissolved, thus offering the possibility of making molds for non-extractable geometries. This allows us to overcome one of the biggest limits of filament winding, that is the possibility to realize only shapes that can be removed from the mold at the end of the production (Figs. 1 and 2).

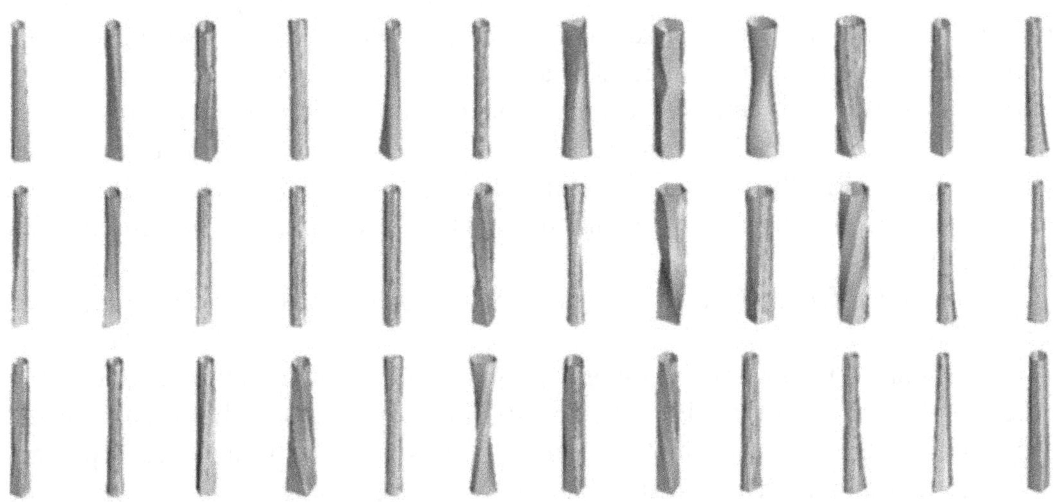

Fig. 1 Algorithmic design of the formwork—Indexlab diagrams

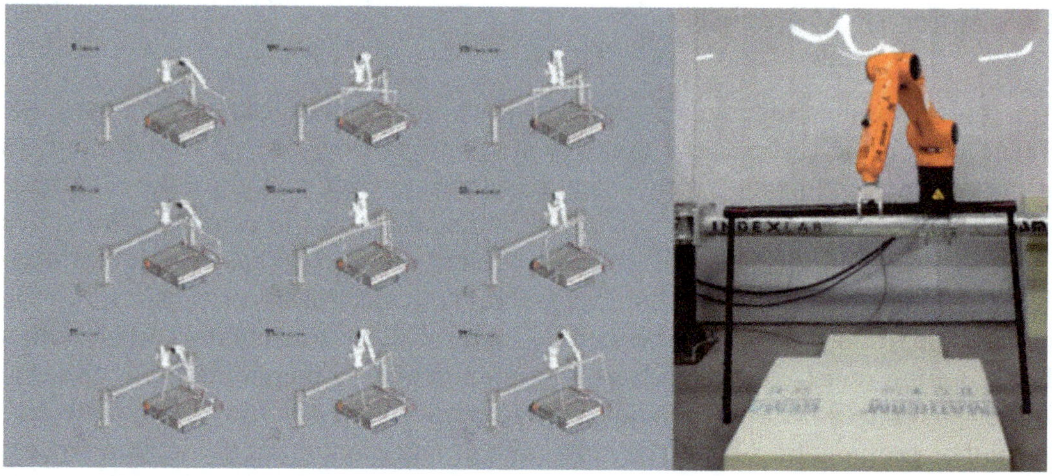

Fig. 2 Algorithmic programming of the robot

2 Innovation in Design, Process, and Product

Designing a disposable formwork means designing the final geometry of the pillar. The pillar is a three-dimensional structural element with a much larger size than the other two; its role is to support the structure above, withstand transverse loads and transmit the forces to the underlying structures or foundations. The section of a traditional pillar is generally square, rectangular or circular, depending on functional, physical, and production reasons. The use of these forms in the past also had other reasons: in the absence of automatic calculators, simple geometries with a constant section facilitated the determination of all the data necessary for the design, such as volume, weight, barycentric axes, moments of inertia and modules of resistance. These problems can now be overcome thanks to the use of three-dimensional modeling programs, together with finite element analysis software, which allows you to analyze in detail any type of geometry. It is therefore easy to design complex geometries.

Robotic manufacturing also makes it economically viable to produce such geometries.

In this context, the idea was born of exploiting robotic manufacturing for the construction of formworks that allow the actual implementation of what was designed. The process used is that of robotic cutting of polystyrene with hot wire; a process that is up to a hundred times faster than numerical control milling. The molds are machined using grooved surfaces and are composed of two halves to facilitate fixing on the rotating axis of the lathe during the filament winding phase (Fig. 3).

The winding of the mold takes place by applying the first layer of epoxy resin in the form of adhesive tape and the subsequent deposition of the pre-impregnated carbon fiber tape with moderate and constant inclination, so as to cover the surface uniformly without creating wrinkles in the material. The molds are covered by two

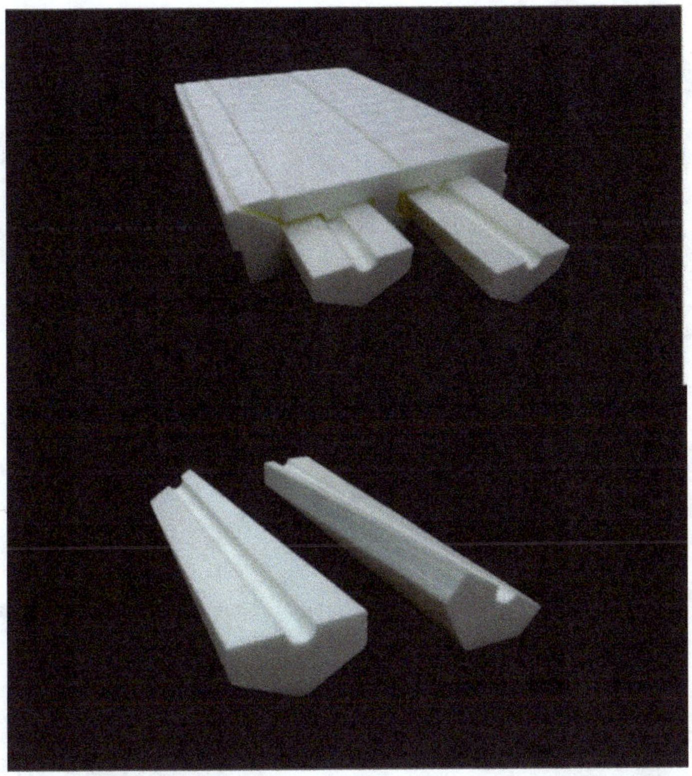

Fig. 3 Molds cut by means of a hot wire, mounted on a robot as end effector

windings, each one continuous for the whole length of the piece, made with the same inclination but starting from the two opposite ends of the block, thus creating an opposite winding (Fig. 4).

Once the winding phase is completed, the mold is removed from the axis of the lathe and put into the autoclave. The temperature, pressure, and care time depend on the mix of materials used, the size and number of coils made. The low resistance of the extruded polystyrene to high temperatures requires low-temperature cycles in the autoclave, with a consequent lengthening of the treatment time. To obtain the disposable formwork from the semi-finished product obtained, it is necessary to eliminate the polystyrene mold. Working with complex geometries, it is not possible to remove

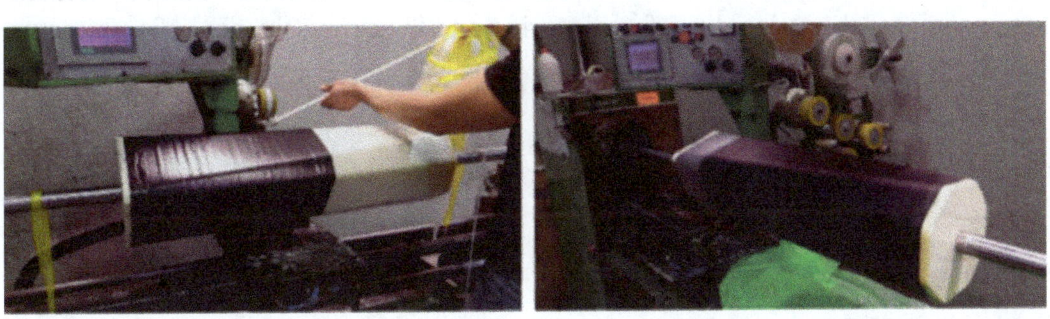

Fig. 4 Filament winding of the mold with carbon fibers

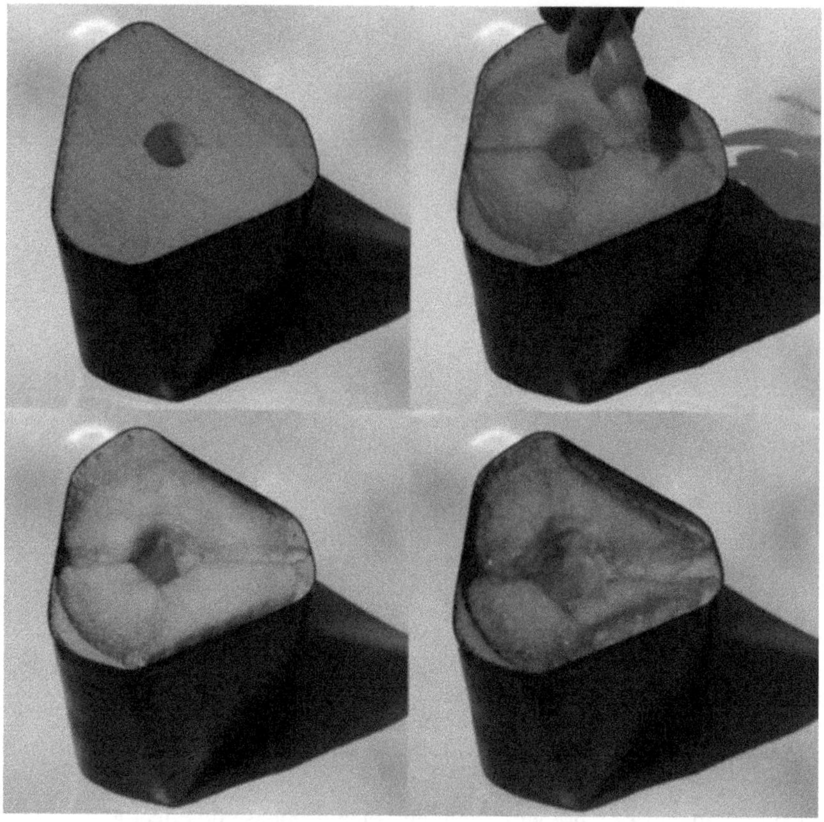

Fig. 5 Return of the formwork, ready for the concrete casting

the mold from the carbon body. The acetone's ability to dissolve polystyrene is therefore exploited. In a few minutes, the reduction in the volume of the mold—due to the contact between acetone and polystyrene—is sufficient to allow the extraction of the mold and thus obtain the hollow shape of the disposable fiber formwork. The formwork is thus ready for casting (Fig. 5).

Laboratory tests show that, in terms of cross-section, a wheeled concrete specimen with a disposable formwork made with this system offers an ultimate compressive strength almost three times greater than an unwheeled specimen (97 tons compared to 36 tons).

In short, the proposed method expresses the desire to create unique elements that encompass the rationality of engineering and the expressiveness of architecture (Fig. 6).

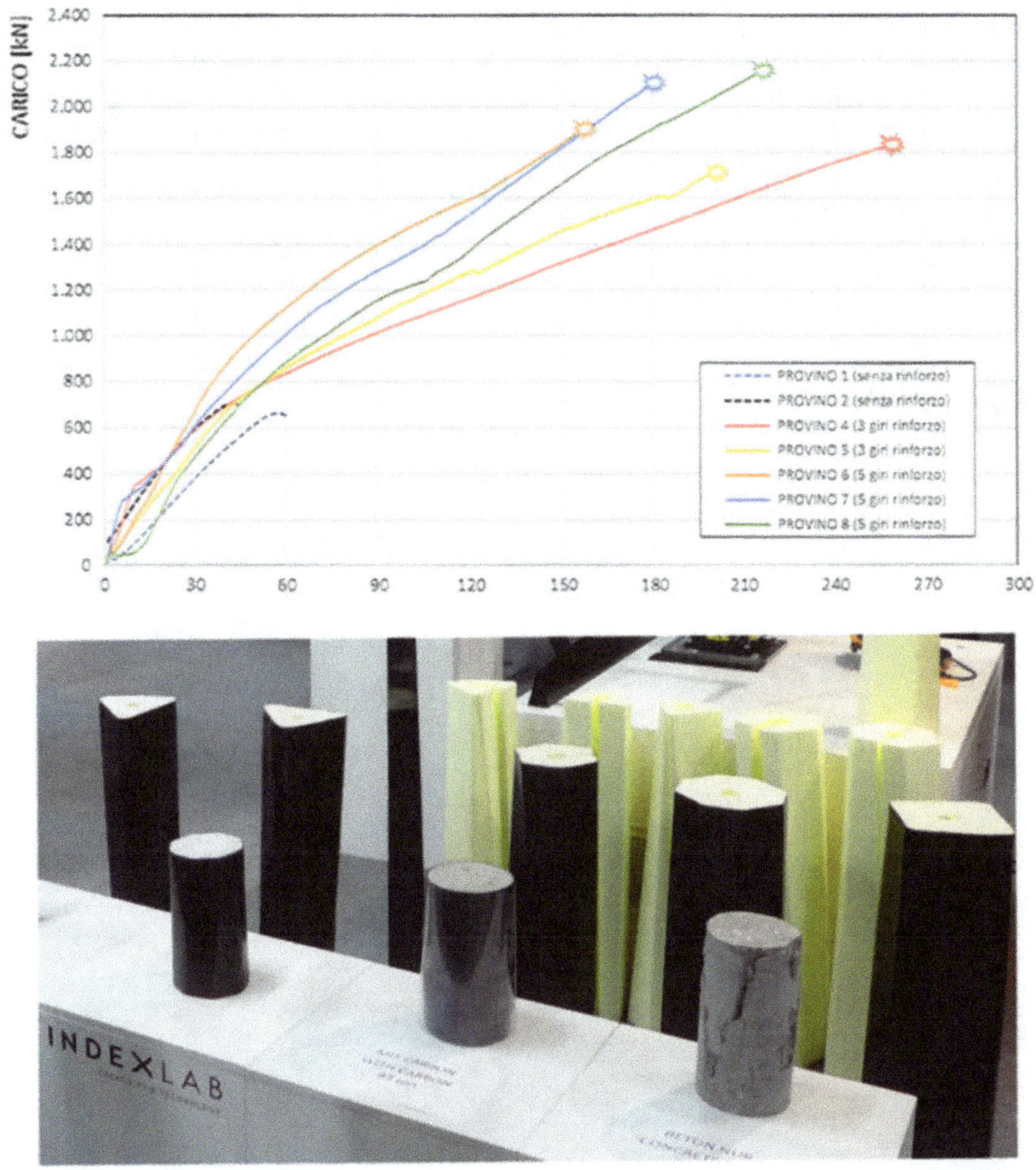

Fig. 6 Results of tests of ultimate resistance to uniaxial compression

Acknowledgements This research was conducted by Pierpaolo Ruttico and Massimo Rota (Indexlab - Politecnico di Milano), in collaboration with Federico Carmona (Carmon@carbon) and with the support of Francesco Braghin (Polimi - Mecc).

Management Stage

Introduction

Bruno Daniotti, Marco Gianinetto

The part IV is aimed to present the most relevant research results about the digital transformation about the management of existing buildings, in general and for historical heritage.

The Facility Management (FM) sector is undergoing a profound transformation of practices, processes, tools and references due to the adoption of novel ICT (Information and Communication Technology) solutions which nowadays promise to improve the traditionally-conceived FM streamlining processes, making new knowledge bases available to support data-driven decision-making processes and embracing a network approach to stakeholder management.

A specific focus is at first dedicated to researches about Built Heritage Information Modelling/Management. The topic has a strategic value, as the relevant new tools enable the kind of Knowledge Management required by the Planned Conservation vision. The key point is the step from implementing BIM authoring software to build a single parametric model, to the definition of several domain-specific parametric models fit for each of the many activities involved in built heritage conservation process.

Then a set of research activities is reported about digital asset management to improve facility and property management.Some studies are dedicated to analyse the impact of digital transformation, IoT and Big Data in the Facility Management sector.

Digital Asset Management is presented as a methodological framework for asset management business processes reengineering. Through the application of the proposed set methods and procedures, it is possible to leverage innovative Information and Communication technologies (ICTs) for the development of improved information manage-ment practices in digital built environment management. The case studies developed demonstrated the possibility to effectively implement innovative Digital Asset Management processes and address different core areas of the discipline.

An information management framework for the use phase of assets has been developed in a perspective ofwith the aim to streamlined real estate management processes. The framework allows users to access and update at any time technical, administrative and maintenance data both at the building and at the district level. This allows for the integrated management of physical assets and enhances the risk prevention strategy.

Other research activities are focused on real time sensing and digital 3D control: It's possible to enhance building responsive operation by gathering granular data on passive systems' operation and space occupation by setting a building component, that houses a sensor network.

Specific attention is then dedicated to integrating BIM for the asset and facility management and BIM digital platform to support emergency interventions: This text illustrates the development of a horizontal organizational model for the intelligent smart and dynamic control of complex buildings through the creation of an innovative digital Web Based platform capable of integrating Building Information Modeling (BIM) technology with a "Facility Management platform".

Built Heritage Information Modelling/Management. Research Perspectives

Stefano Della Torre and Alessandra Pili

Abstract The paper deals with the research issues concerning Historic Building Information Modelling, developing the outcomes of the national project BHIMM, carried out in 2017. The topic has a strategic value, as the relevant new tools enable the kind of Knowledge Management required by the Planned Conservation vision. The key point is the step from implementing BIM authoring software to build a single parametric model, to the definition of several domain-specific parametric models fit for each of the many activities involved in built heritage conservation process. As designed for specific aims, the models can differ both in accuracy and in identification of the objects. This vision moves the research agenda towards exchange platforms and semantic interoperability, pointing out the development of specific ontologies as a crucial task to be carried out in the next years.

Keywords Historic BIM · Parametric models · Conservation process · Ontologies

1 Introduction

The implementation of digital informative modeling on historic buildings are a quite recent research topic, whose developments started more or less ten years ago, producing growing scientific literature. Undoubtedly the process can be described as a technological transfer process that is the transfer of BIM from the construction industry field to the cultural heritage world, where innovation often happens through the reuse of practices. This entails theoretical and practical issues, as the new tools go into a frame of accustomed attitudes, often without changing mind and underestimating, or wasting, the making of the new procedures and techniques in order to achieve targets unsought before. Both academics and the industry are still looking for

S. Della Torre (✉) · A. Pili
Architecture, Built Environment and Construction Engineering—ABC Department, Politecnico di Milano, Milan, Italy
e-mail: stefano.dellatorre@polimi.it

B. Daniotti et al. (eds.), *Digital Transformation of the Design, Construction and Management Processes of the Built Environment*, Research for Development,
https://doi.org/10.1007/978-3-030-33570-0_21

the most effective roadmap to a real implementation of digital techniques on this very peculiar market. Nevertheless, two important premises deserve to be highlighted, in order to show the vision, the making and some biases of these researches.

The first premise of the development of Historic BIM has been the availability of digital survey techniques, ending up in 3D models characterized by better and better accuracy. The evolution of acquisition tools in Geomatics field has been the precondition for the capability of 3D models to support the management of existing buildings. Architects and Conservators got accustomed with Point-clouds, while the costs of automatic survey became competitive with the costs of direct measurements. Then, a large deal of the literature on HBIM deals with documentation and the step from surveying to modelling.

The second premise concerns the step from restoration to preventive conservation, which implied not only a focus on prevention and maintenance, but also a growing interest for knowledge management and databases supporting long-term care activities. Therefore, HBIM answers a twofold expectation on digital innovation, on one side looking for more accurate survey and documentation, on the other side looking for a more effective storage and retrieval of information. HBIM should exactly bridge these two opportunities, giving the way to opportunities not yet put into practice. If this is the research goal, an integrated multidisciplinary approach is a basic condition: exactly what the BHIMM project aimed to set up.

BHIMM (Built Heritage Information Modeling Management) project was launched in 2011, when this kind of researches were still not so popular worldwide, as a PRIN granted by the Italian Ministry of University and Research (Della Torre 2017). It involved six Research Units: besides Politecnico di Milano, the Universities of Rome, Genoa and Brescia, Politecnico di Torino and the Bari based unit of CNR. The research groups involved academics from different fields (architectural conservation, construction technology, building physics, structural analysis, survey and documentation…). BIM techniques as implemented in the heritage domain have been tested on several outstanding historic buildings: for instance, the basilica of S. Maria di Collemaggio in L'Aquila, damaged by earthquake in 2009 and under reconstruction (Oreni et al. 2014; Trani et al. 2015; Brumana et al. 2018); Milan Cathedral (Fassi et al. 2015); the Albergo dei Poveri in Genua (Musso and Franco 2014; Napoleone 2017; Vecchiattini and Babbetto 2017); Masegra Castle in Sondrio (Barazzetti et al. 2015). Different modelling solutions were tested, and in many cases not only the most popular BIM authoring software, but other modelling software was used, in order to overcome the "rigidity" of the BIM. Almost all the activities have been explored, which follow each other along the cyclic management process of an historic building, with the perspective of making those activities as much as possible digital by means of interoperable techniques. The Research Unit from Rome developed a deeper investigation on ontologies for cultural heritage, laying the bases for important further researches still developing (see below).

The outcomes of BHIMM project contributed to a renovated approach and opened to further research in the direction of more performing technological transfer of BIM potentialities to the domain of historic buildings.

2 Built Cultural Heritage Process

The many pilot researches carried out in the frame of BHIMM project on several Italian historic buildings demonstrated that the transition from the traditional representation by 2D views associated with a restoration project to a 3D parametric model cannot be just a matter of tools and procedures to be customized: instead, it is necessary to understand the making of parametric tools to renovate the whole heritage conservation and valorization process.

According to the most established literature, parametric objects should: contain geometric information and associated data and rules; have non-redundant geometry, which allows for no inconsistencies; have parametric rules, which automatically modify associated geometries when inserted into a building model, or when changes are made to associated objects; have the ability to be defined at different levels of aggregation; have the ability to receive, or to export, or to link sets of attributes (such as structural materials, acoustic data, energy data, and cost) to other applications and models. (Eastman et al. 2011: 626). Parametric modeling refers to a virtual construction with fully defined objects that know where they belong, how they relate to other objects and what they consist of (Barazzetti 2016).

The transition from drawings to parametric modeling is parallel with the transition from restoration as an event to conservation as a process (we use to speak of planned conservation). Therefore, it is a transition from a scheme that used to see restoration as a construction process (on existing premises rather than on free land), to the vision of a long-term process, in which many different activities (management, maintenance, monitoring, conservation, heating, restoration, etc.) are carried out by different actors, who need overcome asymmetries and cognitive gaps as they exchange a lot of information.

The conservation process of built cultural heritage is radically different from the construction process, and therefore substantial challenges arise in implementing BIM tools in this context. The topic of Built Heritage Information Modeling, or Historic BIM, requires to be approached through a deep understanding of the processes, well before working on the questions concerning the development of the tools. Researches of this kind have been carried out as the foundation of proposals aimed at innovating the practices in the preservation field (Della Torre 2018). The point is the continuity of the process, which requires an important investment in knowledge management and digitalization of the procedures. Conservation works on protected buildings entail continuity in knowledge management, so that they require at any step a high level of information and description. In the best case, the scientific report produced by a previous intervention should be available, implemented by following maintenance and management activities (Della Torre et al. 2018).

3 Survey and LoDs

As the available BIM tools have been developed first to design and assemble serial building components, they can hardly deal with the irreducible variety of ancient buildings, which are still in use although well beyond the limits of their service life. Once a building had been recognized as heritage, its conservation has to comply with the dimensions of authenticity.

The basic problems emerging when dealing with the step from point-clouds to parametric models for architectural conservation have been largely dealt with in the literature:

- collected data cannot be limited to external surfaces of the objects, which have to be investigated and someway represented also by their thickness, layers and internal features, characterizing elements as three-dimensional not only under the geometric point of view, but also for materials and techniques;
- as the point-cloud has to be converted in models made of discrete and significant objects, corresponding to the needed operations for design, computation or documentation purposes: given the irregular shapes of historical buildings, NURBS (non-uniform rational B-splines) curves and surfaces have supplied a satisfactory solution, which enables to keep the level of accuracy obtained thanks to the advanced surveying techniques;
- the typified components downloaded from libraries, provided by commercial parametric software, can be enriched and customized (Dore and Murphy 2013; Oreni et al. 2013), but any historic building is the outcome of construction processes, besides changes and layering in progress of time, which produce the variety that is just the core of the recognition of authenticity, and therefore of value. The question of harmonizing the oneness character with typology approach is surely not new, as this was one of the most discussed topics in Conservation theory. The solution of producing individual objects embedded in the model instead in the software library has been successfully explored, but it entails giving up a series of possible automations.

The available readymade families of elements typical of historic architecture, classified by periods, often coming from the plates of architectural treatises, can surely help to make better-looking and more realistic 3D models, but if these models should serve to enhance data sharing, there is the risk of disseminating a level of detail, which is not exhaustive for the management of the whole process.

The trend, from both the research and industry sides, has been to bridge the gap between survey and parametric modeling, empowering the most popular software tools. Nevertheless, also the sense of this step has to be clarified, by means of some reflections on operational needs, that is on the required levels of description and information. In the field the acronym LoD is used, meaning “Level of Development” or “Level of Detail”.

Being definitely different from the design of a new building, the work carried out on an existing building, especially when it is dealt with as a heritage object to be conserved and not just retrofitted, is not progressive: it is a matter of reverse engineering or downloading from data banks, and "required functionality determines the LoD and the resulting cost and effort associated with BIM creation" (Volk et al. 2014).

For the Italian norm UNI 11337 Stefano Della Torre suggested a definitely peculiar approach to LODs, overturning the parallelism with the progressive LODs described for the design of new constructions: BIM practices as derived from the new construction lead to deal with details and accuracy in developing terms, from rough to detailed, as the design and construction process goes on. The Italian norm proposes to think in terms of different parallel models derived from one accurate survey, and the use of the terms "level", or "grade" does not mean any progression from one level to the next one, but just differences in terms of the detail or accuracy consistent with the purposes and the characteristics of the parametric model required by each domain. This enables to embody the historic complexity since the starting phase (Brumana et al. 2019) to better support the assessment and decision making, to limit the unexpected expensive interruption of the construction site, adopting heritage monitoring as a strategy for planned conservation across the time (Fregonese et al. 2018).

4 Model, Platform, Common Data Environment

The conceptual distinction between a (unique) interoperable model and the concept of platform, which enables data exchange among several "domain specific" parametric models, or BIMs, is a basic point established in BIM research. Rothenberg said that "a model represents reality for the given purpose, the model is an abstraction of reality in the sense that it cannot represent all aspects of reality" (Rothenberg 1989). In other words, a single model is not enough to represent the reality.

Domain specific models (i.e., BIM for Architectural design, Structural analysis, Energy analysis, construction and site design, cost analysis, management, maintenance, etc.) will undoubtedly perform better than the huge single model, which should also work as the exchange platform. Some experience can already be shown, e.g. focused on energy retrofitting (Gholami et al. 2015) and maintenance (Kiviniemi and Codinhoto 2014).

For each model, a different individual identification of BIM objects can be carried out; each specific model should download from the general data base just the useful data, and upload its outputs in order to make them available for other domains through the "platform", or to say better the Common Data Environment (CDE).

The high level of detail in HBIM models may not be needed in some technical domains even if related to cultural heritage. For example, single domains such as structural analyses, energy simulations, behavioral studies, cost computation, etc. could refer to simplified versions of the model. However, sometimes even single technical activities require an advanced level of detail, as well as different definitions of the parametric objects. For example, sometimes a structural or energy analysis may need to take into account an entire element, such as a wall or a column, and sometimes just its layers, parts, stones, etc. At other times, the entire level of detail of the model would be required for other kinds of analyses (Della Torre et al. 2017).

An example may clarify this point. As dealing with a wooden beam floor in an historic building, laser scanning produces a point cloud, which shows a non-regular geometry. Usually, beams are seen as inflected in the points-cloud, due to an elastic reaction to actual loads, and/or to permanent inelastic deformations, and/or to their original shapes. As a result, each beam is different. Furthermore, the soffit is often decorated by paintings. The question is which features should be reflected in the parametric model. But the problem could be put otherwise, if we imagine not one comprehensive model, but special parametric models fit for each activity. Then, the question concerns the purpose of the models, and which features should be present in the parametric environments.

So in this example, we can imagine to split the problem in at least three branches: the architectonic perspective, dealing with the values, the uses and the performances of the spaces; the structural perspective, dealing with the load-bearing capacity of the beams and their efficiency; the conservator-restorer's perspective, dealing with the conservation issues of the surfaces and the materials. The architectural project needs a complete representation of all the features concerning spatial and performance issues and cultural values. But for these aims in the deformation, or irregular shape of the beams has no relevance. This means that a survey can be fit for such a model even if the accuracy is in the order of some centimeters, as it happens with regular laser scanning. The structural analyses take into account the quality of the wooden beams, but as these deformations have less than second-order effects on the stresses, there is no sense in overloading the parametric models with these details, although considered in the interpretation of the results. Again, the accuracy of some centimeters is enough. In the perspective of conservation-restoration issues, however, the shapes and the thin layers of the surfaces are definitely important: the geometry is needed as fine as possible, so that a better surveying technique may be required, such as digital photogrammetry, in order to reach an accuracy in the order of millimeters. This geometry may be usefully turned into a parametric model, because undoubtedly the work to do benefits from dealing with fully defined objects, knowing their position, consistency and relationships.

So, it is possible to imagine different BIMs, each one related to the same CDE, with elements described with different accuracy in one or more point-clouds. This can be a direction for future research, focusing on exchange procedures and dynamic shift from CDE to single BIMs, and related costs and required skills.

The vision of several BIMs referring to a common data environment has been argued as the way to ensure the highest interoperability (Laakso and Kiviniemi 2012). It is also the way to carry on a very important change in heritage field, because the required anticipation of detailed investigation, as well as the availability of tools that enable a continuous control, will enhance the quality of the whole process.

5 Towards Semantic Interoperability

Interoperability is the ability of a system to understand and to be understood when it interfaces with other systems. If two systems have syntactic interoperability, they can communicate with each other and exchange data. Instead, semantic interoperability allows the system to automatically interpret information. In this case, the content of the information is unambiguously defined: what has been transmitted is the same as what is understood. The aim of further HBIM research is to define open standards for semantic interoperability among the various models concerning Cultural Heritage activities. The mandatory premise in order to enable interoperability among diverse information systems is the definition of specific ontologies (Doerr 2009; Noardo 2015; Bruseker et al. 2017).

Ontologies, as computer systems, work in a hierarchical way where the classes are descripted from general to detail (Kupcík et al. 2012; Khan and Safyan 2014) according to father-son logic. In an ontologies structure to transfer the property from "father" to "son", i.e., from class to subclasses, is an important characteristic to describe correctly the classes. Properties (functional, transitive, symmetric, or asymmetric, reflexive or irreflexive) are used to define the domain and the range of classes, as well as to describe and define the classes.

Ontology notion derives from classical philosophy and means "study of the Being". Categories are used to describe the properties and relationships of the "Being". Ontology is an explicit specification of a conceptualization (Gruber 1992), so according to the sentence that "what exists can be represented" ontology concept has been transferred in other disciplines such as Physics and Computer Science. In the latter field, an ontology defines a common vocabulary for sharing information in a domain (Noy and McGuinness 2001). Therefore, the term "ontology", transferred from the language of logic to the language of semantic web, expresses that formalization, which enables to describe in the most complete and faithful way the characteristic concepts and relationships of a given knowledge domain. As conservation and valorization of built cultural heritage are complex activities, in which diverse subjects operate using different languages, a well-defined conceptualization is a mandatory condition for avoiding misunderstanding and waste. This may entail all the risks embodied in schematization, but it is also necessary for enabling any information exchange.

BuildingSMART International encourages several working groups in the definition of more and more efficient IFC standards, in order to extend interoperability to infrastructure, but also to landscape, etc., always as open standards. Currently, the specific needs of heritage buildings seem to be out of the scope, leaving the option to give up, or to continue in the development of a Historic BIM, which proposes a careful modelling, but in the reality reduces the historic building to the same logic of a new one. The scientific gap to be filled can be identified in the immaturity and limitedness of the specific ontologies and IFC standards and in their inadequacy to represent in an exhaustive and effective way the activities on built cultural heritage.

The topic of ontologies has been developed at different levels in the various implementation fields. The use of ontologies for heritage, which requires a clear understanding of the cultural values and of the reasons of decay, has been definitely established in the field of museums (CIDOC-CRM) (Le Boeuf et al. 2018), while it is still object of research in the field of archaeology (CIDOC archaeo e CIDOC ba [Nicolucci 2015]).

The elaboration of a dedicated ontology adds the opportunity to fill typical conservation issues into HBIM models (Cacciotti et al. 2015; Zalamea et al. 2018; Beltramo et al. 2019) and to make the data embodied in Spatial Geographic Systems available to other kind of software, beginning from HBIM Systems (Acierno et al. 2017; Fiorani 2019; Acierno and Fiorani 2019).

The relationship between GIS and BIM platforms has been widely studied and further developed. Nevertheless, the challenge of sharing data and exchanging information through the two domains is still open. In fact, the accustomed syntactic approaches targeted to such integration do not allow a complete exchange of semantic and geometric information from BIM to GIS field and vice versa. It is worthy to point out the recent study on the 14 c. bridge across the Adda River at Lecco, which besides the elaboration of the HBIM themes of 3D survey and parametric modelling through NURBS (Barazzetti et al. 2016), has developed the topic of linking BIM model and geospatial data useful in the infrastructure domain (Barazzetti and Banfi 2017). The relationship ifcOWL-CityGML is being further explored in several ongoing researches (Matrone et al. 2019).

The new target is to formulate a systematic proposal of IFC standards compatible with the most used ontologies for the management of geospatial data, fundamental in the cultural heritage sector. An implementation of such an interoperable knowledge management can be developed with focus on prevention, a key activity, whose lack often causes dramatic losses. Making prevention, that is managing risks, requires exactly the possibility to cross geospatial data on hazard and exposure, with vulnerability data, which are usually referred to single properties and produced through correct maintenance activities; on the other hand, an operating prevention can be carried on both through interventions on the environment and strengthening works on single buildings: therefore, through actions requiring the cooperation of diverse actors. Thus, the needed step goes towards an effective data sharing, in order to support the decision-making process and to share the program of the interventions. Interoperability concerns the management of data as well as the repercussion on the processes.

References

Acierno M, Fiorani D (2019) Innovative tools for managing historical buildings: the use of geographic information system and ontologies for historical centers. Int Arch Photogramm Remote Sens Spatial Inf Sci XLII-2/W11:21–27. https://doi.org/10.5194/isprs-archives-XLII-2-W11-21-2019

Acierno M, Cursi S, Simeone D, Fiorani D (2017) Architectural heritage knowledge modelling: an ontology-based framework for conservation process. J Cultural Heritage 24:124–133

Barazzetti L (2016) Parametric as-built model generation of complex shapes from point clouds. Adv Eng Inform 30:298–311

Barazzetti L, Banfi F (2017) BIM and GIS: when parametric modeling meets geospatial data. In: ISPRS annals of the photogrammetry, remote sensing and spatial information sciences, Geospace 2017, vol IV-5/W1, 4–6 Dec 2017, Kyiv, Ukraine, pp 1–8. https://doi.org/10.5194/isprs-annals-IV-5-W1-1-2017

Barazzetti L, Banfi F, Brumana R, Gusmeroli G, Oreni D, Previtali M, Roncoroni F, Schiantarelli G (2015) BIM from laser clouds and finite element analysis: combining structural analysis and geometric complexity. In: The international archives of the photogrammetry, remote sensing and spatial information sciences, vol XL-5/W4, 2015 3D virtual reconstruction and visualization of complex architectures, Feb 2015, pp 25–27, Avila, Spain

Barazzetti L, Banfi F, Brumana R, Previtali M, Roncoroni F (2016) BIM from laser scans… not just for buildings: NURBS-based parametric modeling of a medieval bridge. ISPRS Ann Photogramm Remote Sens Spatial Inf Sci III-5:51–56. https://doi.org/10.5194/isprs-annals-iii-5-51-2016

Beltramo S, Diara F, Rinaudo F (2019) Evaluation of an integrative approach between HBIM and architecture history. Int Arch Photogramm Remote Sens Spatial Inf Sci XLII-2/W11:225–229. https://doi.org/10.5194/isprs-archives-XLII-2-W11-225-2019

Brumana R, Della Torre S, Previtali M, Barazzetti L, Cantini L, Oreni D, Banfi F (2018) Generative HBIM—modeling to embody complexity: surveying, preservation, site intervention. The Basilica di Collemaggio (L'Aquila). In: Applied geomatics, 1 SI: Geomatics and restoration, Springer, pp 545–567

Brumana R, Banfi F, Cantini L, Previtali M, Della Torre S (2019) HBIM level of detail—geometry—accuracy and survey analysis for architectural preservation. Int Arch Photogramm Remote Sens Spatial Inf Sci XLII-2/W11:293–299, https://doi.org/10.5194/isprs-archives-XLII-2-W11-293-2019

Bruseker G, Guillelm A, Carboni N (2017) Cultural heritage data management: the role of formal ontology and CIDOC—CRM. In: Vincent ML et al (eds) Heritage and archeology in digital age. Quantitative methods in the humanities and social sciences. Springer, Berlin

Cacciotti R, Blasko M, Valach J (2015) A diagnostic ontological model for damages to historical constructions. J Cult Herit 16:40–48

Della Torre S (2017) Un bilancio del progetto BHIMM. In: Della Torre S (ed) Modellazione e gestione delle informazioni per il patrimonio edilizio esistente. INGENIO-WEB, pp 1–6

Della Torre S (2018) The management process for built cultural heritage: preventive systems and decision making. In: Van Balen K, Vandesande A (eds) Innovative built heritage models—reflections on cultural heritage theories and practices, CRC Press—Taylor and Francis Group, London, pp 13–20. ISBN 9781138498611

Della Torre S, Moioli R, Pili A (2018) Digital tools supporting conservation and management of built cultural heritage. In: Van Balen K, Vandesande A (eds) Innovative built heritage models—reflections on cultural heritage theories and practices. CRC Press—Taylor and Francis Group, London, pp 101–106. ISBN 9781138498611

Della Torre S, Mirarchi C, Pavan A (2017) Il BIM per la conservazione, Rappresentare e gestire la conoscenza. ANANKE 82:108–115

Doerr M (2009) Ontologies for cultural heritage. In: Staab S, Studer R (eds) Handbook on ontologies. Springer, Berlin, pp 463–486

Dore C, Murphy M (2013) Semi-automatic modelling of building façades with shape grammars using historic building information modelling. In: International archives of the photogrammetry, remote sensing and spatial information sciences, 3D-ARCH 2013—3D virtual reconstruction and visualization of complex architectures, 25–26 Feb 2013, vol XL-5/W1, Trento, Italy

Eastman C, Teicholz P, Sacks R, Liston K (2011) BIM handbook: a guide to building information modeling for owners, managers, designers, engineers and contractors, 2nd edn. Wiley, Hoboken

Fassi F, Achille C, Mandelli A, Rechichi F, Parri S (2015) A new idea of BIM system for visualization, web sharing and using huge complex 3D models for facility management. Int Arch Photogramm Remote Sens Spatial Inf Sci XL-5/W4:359–366. https://doi.org/10.5194/isprsarchives-xl-5-w4-359-2015

Fiorani D (2019) Il futuro dei centri storici. Digitalizzazione e strategia conservativa. Roma. Edizioni Quasar. ISBN 978-88-7140-925-2

Fregonese L, Rosina E, Adami A, Bottacchi MC, Romoli E, Lattanzi D (2018) Monitoring as strategy for planned conservation: the case of Sant'Andrea in Mantova (Mantua). Appl Geom 10(4):441–451

Gholami E, Kiviniemi A, Sharples S (2015) Implementing building information modelling (BIM) in energy efficient domestic retrofit: quality checking of BIM model. In: Proceedings of the 32nd CIB W78 conference 2015, 27th–29th 2015, Eindhoven, The Netherlands

Gruber TR (1992) A translation approach to portable ontology specification. Knowledge system laboratory, Technical Report KSL, pp 92–71

Khan S, Safyan M (2014) Semantic matching in hierarchical ontologies. J King Saud Univ Comput Inf Sci 26:247–257

Kiviniemi A, Codinhoto R (2014) Challenges in the implementation of BIM for FM—case Manchester Town Hall complex. In: Computing in civil and building engineering

Kupcík M, Šír M, Bradác Z (2012) Interoperability through ontologies. In: Programmable devices and embedded systems, pp 196–200

Laakso M, Kiviniemi A (2012) The IFC standard—a review of history, development, and standardization. J Inf Technol Constr (ITcon)" 17:134–161. http://www.itcon.org/2012/9

Le Boeuf P, Doerr M, Ore CE, Stead S (2018) CIDOC-conceptual reference model, definition. Version 6.2.3

Matrone F, Colucci E, De Ruvo V, Lingua A, Spanò A (2019) HBIM in a semantic 3D GIS database. Int Arch Photogramm Remote Sens Spatial Inf Sci XLII-2/W11:857–865. https://doi.org/10.5194/isprs-archives-XLII-2-W11-857-2019, 2019

Musso SF, Franco G (2014) The "Albergo dei Poveri" in Genova: conserving and using in the incertainty and in the provisional. In: Della Torre S (ed) ICT per il miglioramento del processo conservativo. Firenze, Nardini, pp 41–50

Napoleone L (2017) Ricerca storica e sistema informativo: l'Albergo dei Poveri di Genova. In: Della Torre S (ed) Modellazione e gestione delle informazioni per il patrimonio edilizio esistente, INGENIO-WEB, pp 508–518

Nicolucci F (2015) Un'infrastruttura di ricerca per l'archeologia: il progetto Ariadne. Archeologia e calcolatori, suppl. 7:44

Noardo F (2015) Ontologie e modelli di dati per l'informazione spaziale dei Beni Architettonici. ASITA 2015:893–900

Noy NF, McGuinness DL (2001) Ontology development 101: a guide to creating your first ontology. In: Knowledge systems laboratory, pp 32

Oreni D, Brumana R, Cuca B, Georgopoulos A (2013) HBIM for conservation and management of built heritage: towards a library of vaults and wooden beam floors. In: CIPA 2013 XXV international symposium, ISPRS Annals, vol 164, pp 1–6

Oreni D, Brumana R, Della Torre S, Banfi F, Barazzetti L, Previtali M (2014) Survey turned into HBIM: the restoration and the work involved concerning the Basilica di Collemaggio after the earthquake (L'Aquila). In: ISPRS annals of the photogrammetry, remote sensing and spatial information sciences, technical commission v symposium, vol II, 23–25 giugno, Riva del Garda, pp 267–273

Rothenberg J (1989) The nature of modeling. In: AI, simulation & modeling, pp 75–92

Trani M, Cassano M, Della Torre S, Bossi B (2015) Construction site information modelling and operational planning. In: Heritage and technology mind knowledge experience—Le Vie dei Mercanti _XIII Forum Internazionale di Studi, pp 1383–1392

Vecchiattini R, Babbetto R (2017) I tiranti metallici preindustriali nella modellazione BIM di edifici in muratura. Dalla conoscenza alla rappresentazione parametrica. In: Della Torre S (ed) Modellazione e gestione delle informazioni per il patrimonio edilizio esistente, INGENIO-WEB, pp 355–369

Volk R, Stengel J, Schultmann F (2014) Building Information Modeling (BIM) for existing buildings—literature review and future needs. Autom Constr 38:109–127. https://doi.org/10.1016/j.autcon.2013.10.023

Zalamea O, Van Orshoven J, Steenberghen T (2018) Knowledge-based representations applied to built cultural heritage. In: Van Balen K, Vandesande A (eds) Innovative built heritage models. CRC Press and Balkema, Leiden, pp 93–100

Digital Asset Management

Fulvio Re Cecconi, Mario Claudio Dejaco, Nicola Moretti, Antonino Mannino and Juan Diego Blanco Cadena

Abstract Digital Asset Management is a key discipline enabling a sustainable and high-quality built environment. The physical asset is nowadays more and more integrated within the digital environment, therefore it produces a great amount of information during its life cycle. This information should be used to improve process management during the use phase of the asset, according to a servitised and cross-disciplinary approach. Accordingly, a methodological framework for asset management business processes reengineering is here presented. Through the application of the proposed set methods and procedures, it is possible to leverage innovative Information and Communication technologies (ICTs) for the development of improved information management practices in digital built environment management. The case studies developed demonstrated the possibility to effectively implement innovative Digital Asset Management processes and address different core areas of the discipline.

1 Shifting Towards a New Real Estate Market

The real estate market is shifting its scope form the traditional meaning of the asset, conceived as tangible good exploited as a means for achieving a primary objective, to the integration of physical assets into the set of services delivered to the client (Moretti et al. 2017). This new way of thinking about the real estate, brings tangible and intangible goods and services together, as a whole new product to be sold on the market. The output of the real estate development industry becomes a system of both project components and added services (Baines et al. 2008). Moreover, management of the built environment has been characterised over the last years by an increase in the complexity of physical assets, as well as the high number of stakeholders and

F. Re Cecconi (✉) · M. C. Dejaco · N. Moretti · A. Mannino · J. D. Blanco Cadena
Architecture, Built Environment and Construction Engineering—ABC Department, Politecnico di Milano, Milan, Italy
e-mail: fulvio.rececconi@polimi.it

B. Daniotti et al. (eds.), *Digital Transformation of the Design, Construction and Management Processes of the Built Environment*, Research for Development,
https://doi.org/10.1007/978-3-030-33570-0_22

pervasive use of Information Communication Technologies (ICTs) (Centre for Digital Built Britain 2018). Physical assets can be considered complex systems featuring tangible and intangible performances. The shift from the traditional paradigm (the building as a product) to the contemporary one (the building as a service) is enabled by the digitisation leading to a new complexity, to be managed through approaches enabling modelling and management of information, for achieving a more sustainable built environment. These dynamics raise the issue of how digital-based processes can be encompassed within AM and which are the most suitable tools and practices to be employed to catch the new complexity of the built environment. Therefore, the research triggers process innovation for Asset Management, exploiting existing tools and practices, combined and reshaped to achieve enhanced performances of the built environment.

2 State of the Art

The built environment is currently more and more integrated with the digital environment, enabling new processes in the asset management. The Internet of Things implementation in Architecture, Engineering, Construction and Operation (AECO) is gaining momentum, allowing the sector to reach advanced building performances (Wong et al. 2018). Higher level of effectiveness and automation for comfort control and adjustment (Fan and Xia 2015), continuous commissioning and maintenance management of physical assets (Dixit et al. 2019) are only a couple of the features that can be achieved through the integration of the physical asset with a dense network of sensors and actuators and with automated systems. The building, in this context, must be considered as an entity, able to recognise external and internal changes and adapt its behaviour to provide users with defined performance levels. These information exchanges between the user and the building are typical of the cognitive building, namely an asset which is able to adapt itself according to the behaviour of the users who, conversely, receive information from the asset and can be involved in the control loop in a bi-directional way (Rinaldi et al. 2016). The great number of sensors, actuators and advanced automated systems that characterise the contemporary built environment raises new issues concerning the management of the wide information flow (Boton et al. 2015). Accordingly, IT tools should be exploited to tackle cross-sector problems, since traditional approaches are no longer suitable with the increasing complexity of the built environment. Therefore, methodologies for analysis and interpretation of Big Data could be exploited for handling the unstructured, real-time data flows, to be interpreted and processed in order to achieve a higher knowledge and effectiveness in asset management.

Structured information can be integrated in the Building Information Modelling approach which, according to ISO 19650-2:2018 (2018), can be intended as a digital object-oriented process for design, construction and management of buildings or infrastructures. BIM enables professionals to make informed decisions thanks to reliable and updated information, during the entire life cycle of the asset (Sanchez et al. 2016). Accordingly, BIM cannot only be considered as a 3D-oriented design methodology, but as a digitally based methodology, exploiting different Information and Communication Technology (ICT) tools and techniques, which allow to effectively manage the physical assets in a sustainable way.

3 Aim of the Research

The ongoing research at the Politecnico di Milano—ABC Dept. aims at enhancing Asset Management business processes, through new information management approaches enabled by availability of ICT tools. The employment of this innovative approach allows for the optimisation of existing Asset Management processes and the creation of new ones. Therefore, a methodological framework to improve or develop new Digital Asset Management processes has been developed. The methodological framework is articulated in three phases, bringing from a strong knowledge and understanding of the traditional Asset Management processes, to the optimisation or development of new Digital Asset Management ones.

4 Methods and Tools

The methodological approach is organised in three steps: the AM process mapping, the process modelling and the process reengineering. The first concerns the identification and categorisation of the AM core functions, the second implies a standardised methodology for business process modelling, the third concerns the optimisation and creation of new digital-based and servitised AM processes.

AM encompasses a wide array of processes and sub-processes to be implemented in order to achieve acceptable levels of performance and maintain their value during their entire life cycle. Engineering Asset Management can be divided into 14 areas, classified by typology (strategic, tactical, operational) and according to the lifecycle phases in which they take place. Processes in Fig. 1 are the most frequently implemented in asset management. The business process mapping is the first step to achieve a strong knowledge of the AM discipline and to define the boundaries of Digital Asset Management: it allows us to identify processes to be managed in order to make informed decisions on assets.

Each area is composed by a sequence of input, core processes, sub-processes and output. The Business Process Modelling (BPM) technique has been adopted to model the AM business processes, identifying the information flows among processes. Therefore, for each process identified in the business process mapping phase, inputs, main processes, sub-processes and outputs are modelled, to classify the main flows (relationships) and transformations (activities). Modelling of the information flows of core AM functions enables the digital-based servitised reengineering which takes places in the third phase.

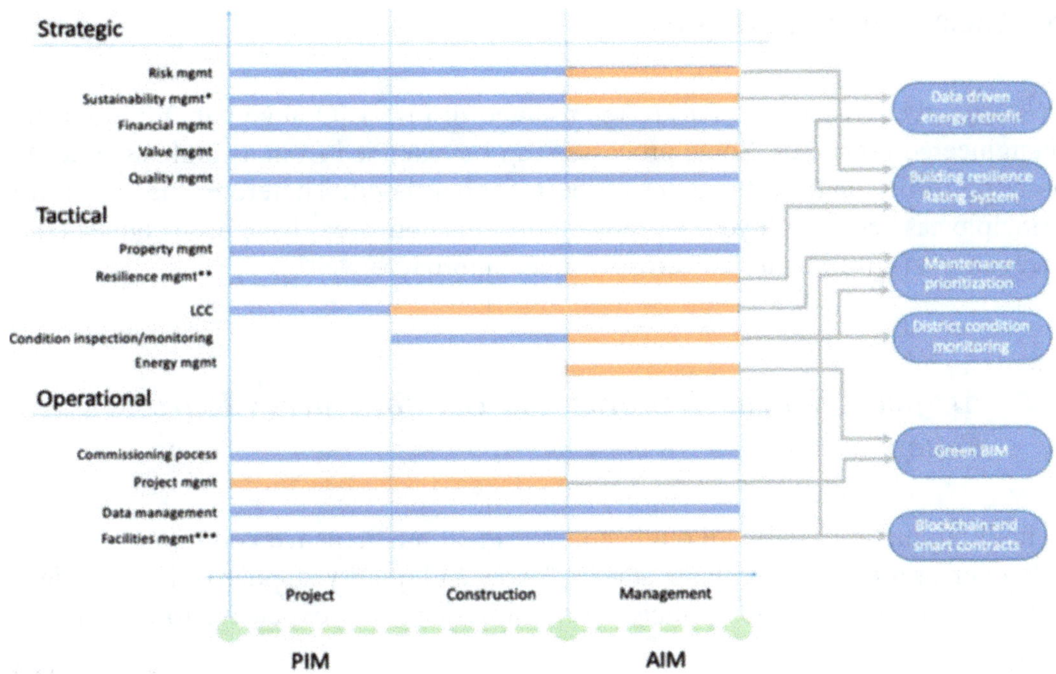

Fig. 1 Digital Asset Management core areas and reengineering outcomes (case study)

The third phase concerns the Business Process Reengineering. Reengineering means the optimisation or creation of new business processes, in order to achieve a digitally based and servitised outcome.

This operation may lead to the complete redevelopment of the processes previously employed by the organisation. In this phase, the identification of a number of additional key features of the processes is crucial. The key features can be summarised as: the typology of the organisation; the operating context; the organisation's constraints (financial and regulatory); the needs of the organisation and the involved stakeholders. These five features allow to properly shape the process reengineering operations and outcomes, without losing the standardisation. To the five features a sixth must be added: the organisation's data availability and accessibility. This is a primary driver for the business process reengineering in Digital Asset Management and for achieving the benefits deriving from it (e.g. more efficiency and effectiveness, reduction in uncertainty, increased reliability, etc.). This assessment also informs on which are the most suitable processes to be reengineered. Once the definition of the organisation's context and constraints has been defined, processes can be optimised through the use of enabling ICTs. The reengineering process takes places according to the principles of modularity, scalability and cross-domain. Currently, nine core AM functions have been reengineered.

5 Case Studies and Tools

Processes reengineered so far are represented in Fig. 1. Orange bars represent the reengineered processes. Some processes can be extended from a single asset to the asset portfolio, or from the asset to the neighbourhood. Therefore, the scalability principle has been employed. Moreover, different core areas have been connected to meet an inter-disciplinary and cross-domain approach.

5.1 Artificial Neural Networks and GIS for Energy Retrofit Policy

School buildings in Italy are outdated, in critical maintenance conditions and they often perform below acceptable service levels and quality standards. Nevertheless, data supporting renovation policies is missing or is very expensive to be obtained (Fig. 2).

This case study proposes a method for evaluating buildings' energy savings potential, using the Building Energy Certification (Certificazione Energetica degli Edifici—CENED) open database (2018). The aim of the study concerns the development of a data-driven set of methods, based on the use of open data, machine learning (ML) and Geographic Information Systems (GIS) to support regional energy retrofit policies for school buildings. The main advantage concerns the possibility to predict the post-retrofit energy savings, avoiding the expensive on-site Condition Assessment (CA) phase. Data has been first clustered to identify the most common thermo-physical properties of the envelope, then three retrofit scenarios have been defined, to allow for the retrofit of homogeneous types of buildings. The energy saving potentials have been evaluated through the implementation of eight Artificial

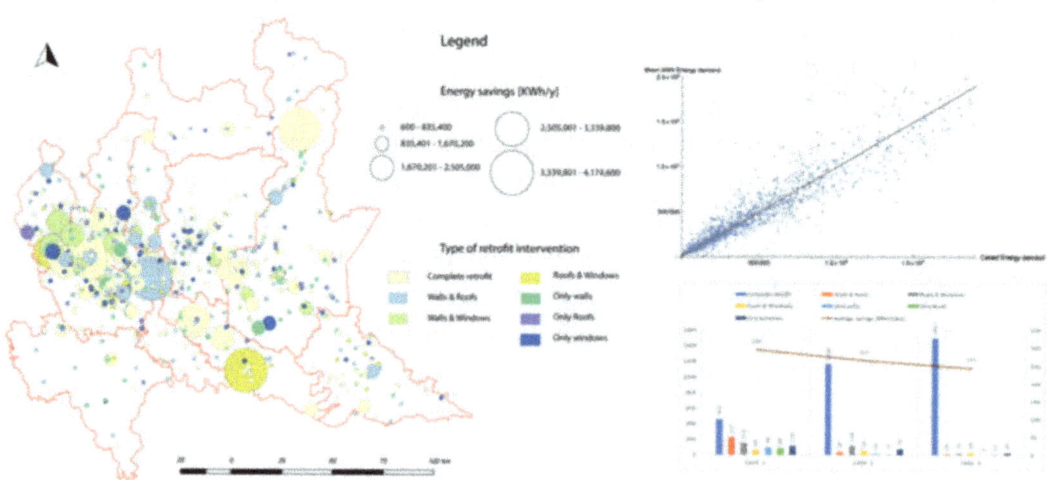

Fig. 2 School portfolio refurbishment map and scenario analysis

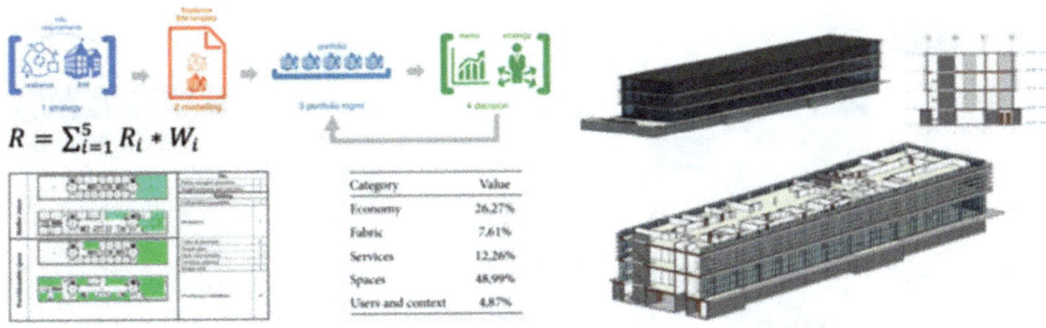

Fig. 3 Research diagram, resilience rating equation, and 3D model of the case study building

Neural Networks (ANN). Finally, data has been geolocated and further processed to support the definition of the energy retrofit policies for the most critical regional areas. The Lombardy region has been chosen as case study to test the robustness of the proposed methods. The results of the case study proved that school buildings energy retrofit policies can be supported and defined using available open data, ML and GIS. The future developments of the research concern the further integration of GIS for retrofit cost assessment and scenario analysis (Re Cecconi et al. 2019a).

5.2 *BIM-Based Building Resilience Rating System*

Measuring and rating resilience of assets is a key enabler for asset and portfolio management. Through this case study a resilience rating system for buildings has been developed by utilising a Building Information Modelling (BIM) approach. The assessment is carried out through a calculation following the Analytical Hierarchy Process (AHP). This methodology can be applied to different types of buildings, without a loss of precision or reliability. This resilience rating forms an integral part of a more comprehensive array of Key Performance Indicator (KPI) frameworks for asset and portfolio management, and therefore can significantly influence strategic investment choices for designers, engineers, and building owners (Re Cecconi et al. 2018) (Fig. 3).

5.3 *Building Maintenance Budget Allocation*

Available data on asset condition and performances can be conveyed into different Key Performance Indicators (KPIs). Many KPIs measuring technical, functional and economic/financial asset performances can be found in the literature. Nevertheless, they are often strictly related to a specific scope, thus they provide an incomplete depiction of the whole assets' performances. The objective of this case study is to provide facility managers and asset owners with an easy instrument to prioritise

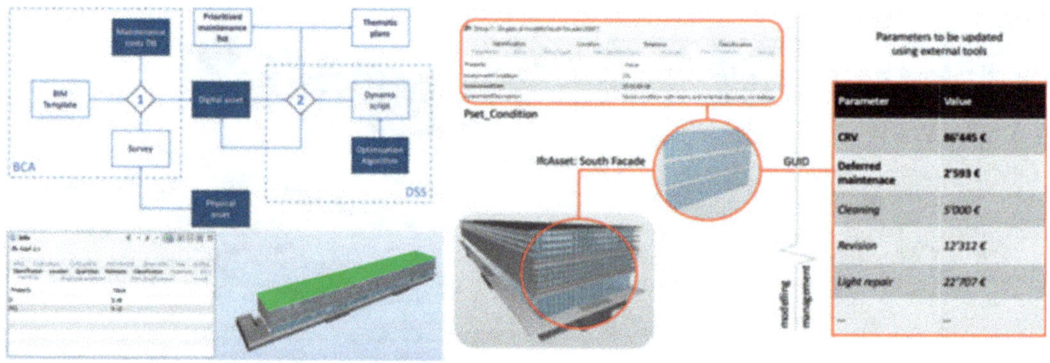

Fig. 4 Maintenance budget allocation research schema and interface

maintenance. In order to reduce costs related to its use, the instrument, developed in the form of a Decision Support System (DSS), is based on existing and reliable performance metrics and leverages new technologies like Building Information Modelling (BIM). Accordingly, the Facility Condition Index (FCI) is combined with the D index, a KPI related to the age of building components, developed by the authors. The joint use of the FCI and the D index, allows facility managers to make more conscious decisions. The proposed DSS helps in the definition of the best maintenance plan, providing a ranking of building components which require more urgent maintenance interventions. Although the DSS should be tested measuring its ability to preserve buildings and their performances in the long term, the first results are positive, as confirmed by the application to a case study on an office building in Italy. Moreover, the usability of the instrument has been appreciated by the users in a medium size Italian company (Re Cecconi et al. 2019b) (Fig. 4).

5.4 *Green BIM*

Rating systems are assumed as instruments to endorse architectural quality, reliability, energy efficiency, economic convenience and, finally, assign a sustainability label. Moreover, these tools can be tied to a BIM model. The aim of this case study is test a lean methodology to fulfil Common European Sustainable Built Environment Assessment (CESBA) requirements through Construction to Operations Building information exchange (COBie) in projects on existing buildings adopting the BIM Bronze approach from the UK Ministry of Justice (MoJ). This will allow to develop a semantic model and to extract sustainability reports in a post-construction phase, thereby minimising the cost of gathering information on existing buildings. An illustrated example regarding energy criteria of CESBA protocol has been developed to further the proposed approach (Maltese et al. 2017) (Fig. 5).

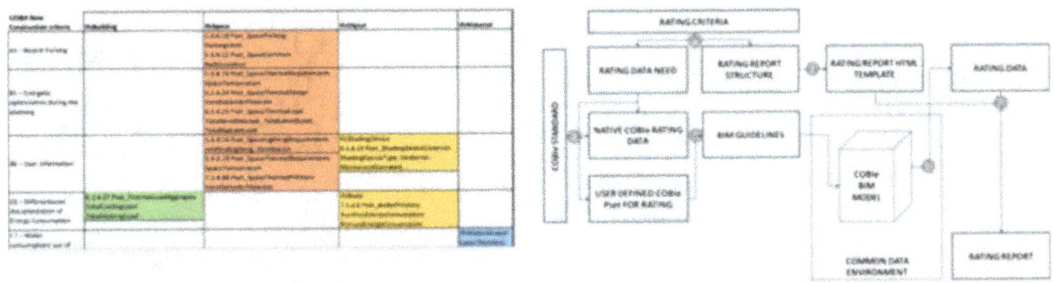

Fig. 5 Green BIM research schema and COBie mapping template

6 Smart Contract for Optimised Facility Management of Cognitive Buildings

Building operations require fine Facility Management (FM) when it comes to achieving satisfactory performance levels of spaces within buildings. It is known that occupancy monitoring fosters energy efficiency in buildings through dynamic adaptation of indoor conditions according to the variation in occupancy. A similar approach can enhance maintenance contract management too, especially if coupled with a contract management system such as blockchain. Here a methodology is presented for the optimisation of maintenance contracts, through occupancy monitoring with an arranged sensor network. For this, two ultrasonic sensor modules have been employed for tracking the use intensity of a corridor within an office building in Milan. Tests were performed for determining the proper configuration for their installation, to accurately read the use intensity flow. Data can be processed and stored within a digital asset model, associated with the maintenance plan. Once cleaning requirements reaches a predefined threshold, which is agreed and defined in the maintenance plan using as an indicator the occupants flow, the system triggers a maintenance alert to the contractor, who can then activate the cleaning intervention (Fig. 6).

The cleaning need threshold stored in the BIM model can be considered as the oracle of the maintenance smart contract which is activated once the cleaning operation has started. The maintenance smart contract allows users to automatically validate the transaction, once the maintenance intervention is performed and verified. Users' privacy issues are surpassed since the sensors measure the distance from the nearest object for determining the flow of people, without registering any personal information.

The proposed approach enables an enhancement for the automation of maintenance management operations in a cost-effective manner. However, further validation and trials are required with respect to the flexibility of its application (different space types), especially when dealing with abrupt occupants' displacement (Blanco Cadena et al. 2019).

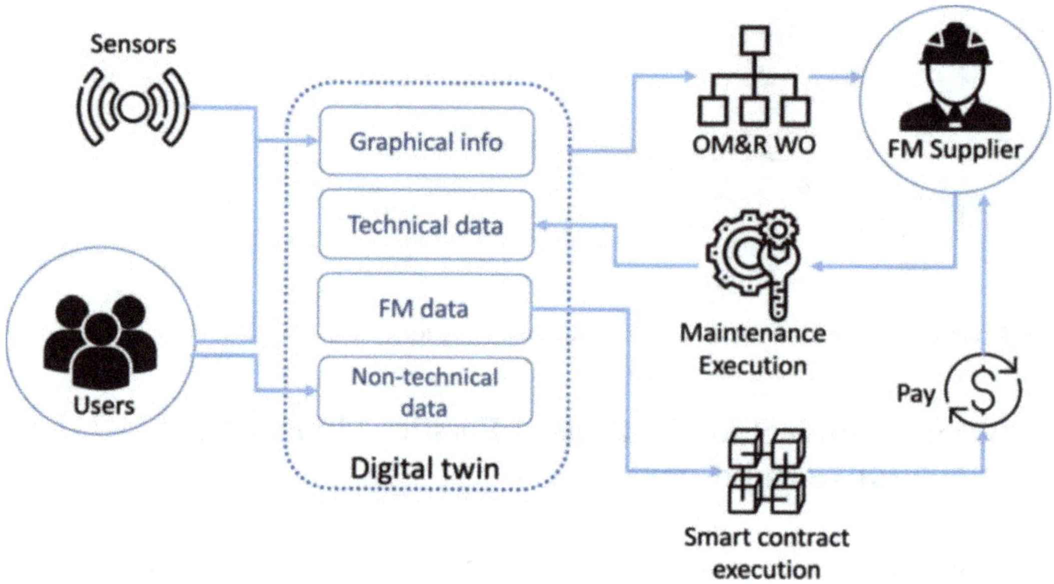

Fig. 6 Smart contracts for improved facility management research diagram

7 Conclusions and Next Development

The methodological approach has been defined and organised in three steps which start from the deep knowledge of AM processes leading to their reengineering, through the information flow modelling enabled by ICT technologies and tools. The digitisation potential of the processes is defined through the analysis of availability of information and takes places thanks to the modularity, scalability and cross-domain principles. This can be considered a streamlined assessment for the initialisation of a reengineering process. The proposed three-tiered methodology is the framework within which the reengineering of business processes takes place.

Through the proposed case studies, it has been demonstrated how Digital built Asset Management processes can be developed according to a data-driven and servitised approach. The case studies, which can be considered as models with an immediate business application (ready-to-use tools), demonstrate the possibility of achieving positive results in the implementation of the DAM processes. In conclusion, further applications and validation of the proposed methods could strengthen the research and business outcomes and foster new partnerships with the industry.

Acknowledgements The authors would like to acknowledge Rigamonti Francesco e Figli S.p.A. and Tekser S.r.l. for funding these studies. Moreover, the authors thank Dr. Sebastiano Maltese for supporting and contributing to this study.

References

Baines TS, Lightfoot HW, Benedettini O, Kay JM (2008) The servitization of manufacturing: a review of literature and reflection on future challenges. J Manuf Technol Manag 20(5):547–567

Blanco Cadena JD, Moretti N, Poli T, Re Cecconi F (2019) Low-cost sensor network in cognitive buildings for maintenance optimisation. Tema: Technol Eng Mater Archit 5(1):93–102. ISSN 2421-4574, https://doi.org/10.17410/tema.v5i1.218

Boton C, Halin G, Kubicki S, Forgues D (2015) Challenges of big data in the age of building information modeling: a high-level conceptual pipeline. Lecture Notes in Computer Science (including subseries Lecture Notes in Artificial Intelligence and Lecture Notes in Bioinformatics), vol 9320, pp 48–56

Centre for Digital Built Britain (2018) Year one report. Towards a digital built Britain

Dixit MK, Venkatraj V, Ostadalimakhmalbaf M, Pariafsai F, Lavy S (2019) Integration of facility management and building information modeling (BIM). Facilities [Internet]. 37(7/8):455–483. https://www.emeraldinsight.com/doi/10.1108/F-03-2018-0043

Fan Y, Xia X (2015) A multi-objective optimization model for building envelope retrofit planning. In: Yan J, Shamim T, Chou SK, Li H (eds) Energy Procedia [Internet]. Elsevier Ltd., Amsterdam, pp 1299–1304. https://www.scopus.com/inward/record.uri?eid=2-s2.0-84947063926&doi=10.1016%2Fj.egypro.2015.07.193&partnerID=40&md5=3abdbf62a628cbe604e3f42eb80317dd

ISO. BS EN ISO 19650-2:2018 (2018) Organization and digitization of information about buildings and civil engineering works, including building information modelling (BIM)—Information management using building information modelling. Part 2: delivery phase of the assets. https://www.iso.org/committee/49180.html?fbclid=IwAR14dRLyXIH_C3jrlfzD0Uke_HimIXeR_w5_CJoFwC2gsJAL4hrOfRLFJ20

Maltese S, Moretti N, Re Cecconi F, Ciribini ALC, Kamara JM (2017) A lean approach to enable sustainability in the built environment through BIM. Techne-J Technol Archit Environ 13:278–286

Moretti N, Dejaco MC, Maltese S, Re Cecconi F (2017) The maintenance paradox. ISTeA 2017—Re-shaping. Constr Ind 234–242

Re Cecconi F, Moretti N, Maltese S, Dejaco MC, Kamara JM, Heidrich O (2018) A rating system for building resiliencel Un rating system per la resilienza degli edifici. Techne 15:358–365

Re Cecconi F, Moretti N, Tagliabue LC (2019a) Application of artificial neutral network and geographic information system to evaluate retrofit potential in public school buildings. Renew Sustain Energy Rev [Internet] 110:266–277. https://www.sciencedirect.com/science/article/pii/S1364032119302941?dgcid=author. Accessed 12 May 2019

Re Cecconi F, Moretti N, Maltese S, Tagliabue LC (2019b) A BIM-based decision support system for building maintenance. In: Advances in informatics and computing in civil and construction engineering [Internet]. Springer, Cham, pp 371–378. http://link.springer.com/10.1007/978-3-030-00220-6_44. Accessed 5 Nov 2018

Regione Lombardia (2018) Database CENED +2—Certificazione ENergetica degli EDificilOpen Data Regione Lombardia [Internet]. https://www.dati.lombardia.it/Energia/Database-CENED-2-Certificazione-ENergetica-degli-E/bbky-sde5. Accessed 15 Nov 2018

Rinaldi S, Bittenbinder F, Liu C, Bellagente P, Tagliabue LC, Ciribini ALC (2016) Bi-directional interactions between users and cognitive buildings by means of smartphone app. In: 2016 IEEE international smart cities conference (ISC2) [Internet]. IEEE, pp 1–6. http://ieeexplore.ieee.org/document/7580819/. Accessed 28 Feb 2018

Sanchez AX, Hampson KD, Vaux S (2016) Delivering value with BIM: a whole-of-life approach [Internet]. Delivering value with BIM: a whole-of-life approach. Taylor and Francis Inc., London, pp 1–344. https://www.scopus.com/inward/record.uri?eid=2-s2.0-84979718392&doi=10.4324%2F9781315652474&partnerID=40&md5=d49efd9f72d1714050550ef876bbf331

Wong JKW, Ge J, He SX (2018) Digitisation in facilities management: a literature review and future research directions. Autom Constr [Internet] 92:312–326. https://www.sciencedirect.com/science/article/pii/S0926580517309020. Accessed 2 Nov 2018

Building and District Data Organization to Improve Facility and Property Management

Mario Claudio Dejaco, Fulvio Re Cecconi, Nicola Moretti, Antonino Mannino and Sebastiano Maltese

Abstract Effective Facility Management services are a key issue for a high-quality built environment, both in the public and private sectors. Within this context, an information management framework for the use phase of assets has been developed with the aim to streamline real estate management processes. The framework allows users to access and update at any time technical, administrative and maintenance data both at the building and at the district level. This allows for the integrated management of physical assets and enhances the risk prevention strategy.

Keywords Facility management (FM) · Urban facility management (UFM) · Building/district information model (BIM/DIM)

1 Introduction

Production, collection and management of information relating to the built environment involve different players, belonging to the public or private sector (Volk et al. 2014). The public player assumes a role that is twofold: on one hand it has a legislative role making decisions also on the types of information which must be collected, stored and provided (when requested) to validate interventions on buildings. On the other hand, it is itself one of the players which must produce, collect and preserve that kind of mandatory information.

Moreover, it plays the role of controller; since it is responsible for monitoring the reliability and accuracy of the documentation which the private player must deliver, during the execution of the building process (design, execution, use and management

M. C. Dejaco (✉) · F. Re Cecconi · N. Moretti · A. Mannino
Architecture, Built Environment and Construction Engineering—ABC Department, Politecnico di Milano, Milan, Italy
e-mail: mario.dejaco@polimi.it

S. Maltese
Institute for Applied Sustainability to the Built Environment, University of Applied Sciences and Arts of Southern Switzerland—SUPSI, Canobbio, Switzerland

B. Daniotti et al. (eds.), *Digital Transformation of the Design, Construction and Management Processes of the Built Environment*, Research for Development,
https://doi.org/10.1007/978-3-030-33570-0_23

of buildings), both in the case of a public or a private investments. According to this, it can be stated that public administration plays both a proactive and coercive role.

The private player is obliged to produce at least the documentation required by law, relating to its real estate properties. Moreover, it is responsible for updating and managing this information. The private player collects and produces data, since it is obliged by the public player. Nevertheless, in the case of a transaction or modification of existing assets, the owner can refer to that documentation, in order to set a better business process and ease the due diligence process.

In this context, tools for data collection and management assume a key role at the building and district level, especially in the context of facility and property management, in those places and areas where the competence and responsibility between the players is not clear (Lee et al. 2018; Mignard and Nicolle 2014). Therefore, the Building and the District Logbook could be considered risk prevention tools, both for public and private operators.

2 Information Management for Built Environments

One of the most critical issues to consider in built environment management concerns the collection, management and use of information, as different skills, needs and operators coexist, and where competencies among players are not always clear. Another problem concerns management areas where, for instance, connections of systems from the public to the private network are not well defined, when private areas used as public spaces and occupation of public property are not clear. In this situation rules and competencies are not always defined properly, with regards to both private and public responsibilities.

Usually, not considering peculiar cases that need to be analysed carefully, the city environment can be divided into open space and buildings (Gabellini 2012). Tangible, but also intangible goods that can be found in this context must be managed, in order to achieve a higher level of quality of spaces and life. Typically, the discipline the management of these matters is Facility Management (FM). Recently, FM has been declined at the urban level as Urban Facility Management (UFM), in order to achieve a more comprehensive strategy for quality improvement of the built and open environment (Sharifi and Murayama 2014). Within this context, the UNI 11447: 2012 (2012) provides a breakdown structure, useful for the determination of UFM services.

According to this classification, services can be applied to buildings, neighbourhoods and the city as a whole: the scale of complexity, in terms of stakeholders, elements, systems, entities to be managed, methods and procedures increase remarkably (Lotfi et al. 2008).

At the building and urban level, the ownership and stakeholder structure, in most cases, can be defined precisely, though there are certain situations where this subdivision cannot be performed easily; these contexts might cause disputes between owners and/or public administration and private bodies.

The tools proposed in this paper, the Building and District "Logbooks", aim to streamline data management relating to building and urban spaces in order to facilitate the interaction among stakeholders and different players involved in management of the built environment. The proposed tools are not only intended to manage technical issues, but also to support a legal and economic dialogue.

The Building and the District Logbooks are both based on a sound structure for information management and they can be implemented in different moments of the assets' life cycle. Information can be exploited for different purposes: knowledge of building/urban areas; collection and management of technical/legislative issues; safety; conservation of economic value; law/regulation compliance of the building and the urban district (Dejaco et al. 2017a).

According to the phases of the physical assets' life cycle considered, the level of information can vary, though a minimum amount of data must always be present. Furthermore, the District Logbook can be considered a higher level of information repository to be employed for specific issues and characteristics relating to the single building, and vice versa (Dejaco et al. 2017b).

3 Building Logbook

The Building Logbook can be considered the repository of the documentation relating to a building. Its structure varies according to the characteristics of the building under analysis, according to the function, complexity and typology. In Table 1 its basic contents are identified.

Table 1 Contents of a Building Logbook (Dejaco et al. 2017b)

Section	Information to be collected
Building registry info	Concerning the urban registry information and the updated internal subdivision into sub-units
Technical information on building elements	– building breakdown – description of technical, typological, functional characteristic of components (actual conditions)
Information on the property, management and tenancy	– updated documentation on ownership – updated documentation on tenancy and related contracts and agreements – documentation concerning the management of the building (concerning leasing contracts and technical management contractors)
Operative information for management and maintenance	– documentation relating to the technical, administrative, economic management of the building – safety and certification

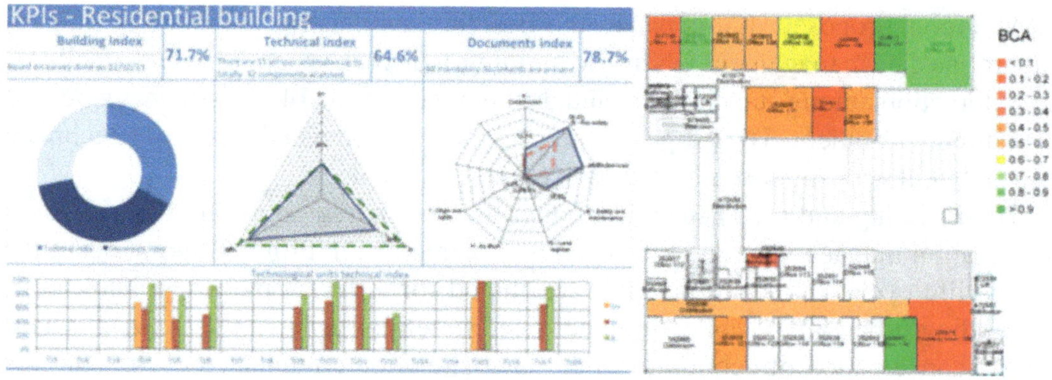

Fig. 1 Final synthetic report of a building assessment (Dejaco et al. 2017c; Re Cecconi et al. 2017)

This data is useful for all the stakeholders and for defining the conditions of a building during its life cycle from a technical, legislative and economic point of view. In the case of a new building, all the data come from the design and execution documentation; for an existing building it would be necessary to develop a specific building assessment and a documentation check. The result of this is represented in Fig. 1. Starting from a visual assessment of the building, it is possible to define a "technical index", representative of the state of its technical components; at the same time, the verification of the available documentation produces a similar "document index"; finally, a synthetic "building index" is computed. All of this data can be depicted in a graphic representation, highlighting technical elements, and/or those that referred to specific areas of the building. Detailed data (technical, documental, economic, …) is collected into a databases, available for following analysis on behalf of different stakeholders.

Data entered through the use of digitized forms that can be filled out in the office during the design phase, or on site with a mobile device, during the construction and the use phases of the building (Figs. 2 and 3).

The data of this type, summarized and detailed, are useful in order to:

- define/check building maintenance and budget allocation;
- prioritize maintenance/refurbishment works and investments;
- compare conditions of the building with market value;
- program medium/long-term real estate management;
- support technical, economic and legal disputes between private or public operators.

4 District Logbook

The district logbook is a similar tool developed for information management at the urban level. The contents and the structure of this tool have been defined through an analysis of needs of the heterogeneous stakeholders involved in the management of

FORM DATA		COMPONENT DATA		
			CODE	NAME
FORM NUMBER		COMPONENT		
CODE		TECHNOLOGICAL ELEMENTS CLASS	C.V.01	Opaque envelope
	C.V.01.01.03.01–S	TECHNOLOGICAL ELEMENT	C.V.01.01	Vertical opaque envelope
NAME		TYPOLOGY	C.V.01.01.03	External finishing
	FORM -	MATERIAL	C.V.01.01.03.01	Plaster on masonry
		ASL (Actual Service Life)		

SERVICE LIFE INDEX	ASL≤RSL	$D^+=$	
	ASL>RSL	$D^-=$	
DEGRADATION INDEX		$A_C=$	1.000

ANOMALIES					
TYPOLOGY	NAME	DESCRIPTION	PRESENCE [Y/N]	INTENSITY	EXTENSION
LOW anomalies that compromise plaster visual performances	Colour changing	Variation of one or more parameters that define the color (hue, clarity, saturation), discoloration of the finish, oxidation and tarnishing of surfaces, rust spots and permanent stains on plaster and cement		Visibility of the degradation, contrast level and residual brightness of finishing	
	Surface deposits	Accumulation of urban atmospheric dust or other foreign material, of variable thickness, inconsistent and not adhering to the surface of the coating		Nature, texture and thickness of the deposits	
	...	...		...	

Fig. 2 Example of an assessment schedule for a technical element (Re Cecconi et al. 2014)

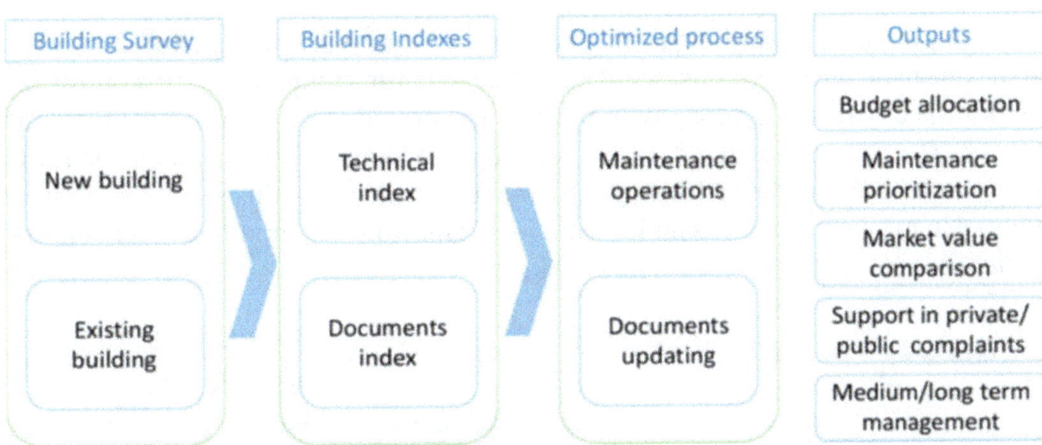

Fig. 3 Building Information Logbook process scheme

the city environment context. The main structure of the logbook has been defined as described in Table 2.

The contents can vary according to the life cycle phase in which the logbook is developed and to the specific case. In the management phase the completeness of the information is a remarkable issue, however, in this case, information can be collected gradually, starting from compulsory documentation concerning the city environment and its components. Also, for this tool the digitized forms can be filled out during different phases and by different operators, using a computer or through an on-site evaluation with a mobile device, developed by technical operators or involving the

Tab. 2 Contents of a District Logbook (Dejaco et al. 2017b)

Section	Information to be collected
Urban and Building registry information	– general information about district – quantitative data and references to infrastructures, urban facilities and buildings – urban planning information and forecasts
Technical information on urban goods and elements	– location – ownership, responsibility, manager – geometric info
Population registry information, property management and tenancy	– data on the population of the district – data on city users – data on management entities
Operative information for management and maintenance	– technical condition – instructions/procedures/guidelines for maintenance and management – safety and certification

community. In this case, adequate procedures for collection and filtering of data should be defined Fig. 4.

It is important to underline that this information can be intended in different ways. On the one hand, it can be considered the data useful for a municipality to control and monitor the city districts, along with the services provided, to optimize the planning process. The data integrated in a proper database is useful in the setup phase of public tenders, and in the following implementation phase. In this sense, the District Logbook can be conceived as an information framework for monitoring tenders and public works.

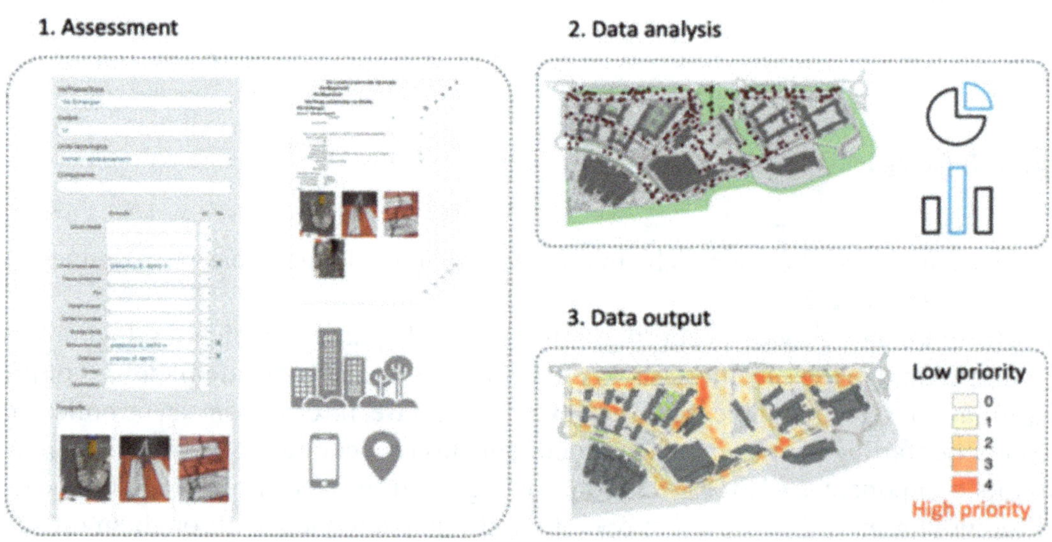

Fig. 4 Survey process, with the different phases of assessment, data analysis and data output (Moretti et al. 2018)

On the other hand, the contents of the Logbook can be seen in terms of compliance with urban laws and codes, urban parameters and taxation issues, licenses and permissions, (a connection between the public and private domains). Therefore, in this case it can be seen as a tool supporting the coercive function of the public administration, in granting the safety and validity of the process, use of spaces and property.

5 Facility and Property Process Support

In a continuous improvement process, the information collected and managed with the Building and District Logbook application should be exploited to enhance the effectiveness of processes in the Public Private Partnership (PPP), leading to a shared approach to urban transformation and management. Data collected and updated during the life cycle (use and operational phases) of the building and/or of the district, can be employed for the condition assessment of the elements of the physical asset (building or urban area). Moreover, this data is a valuable support for making informed decisions Fig. 5.

The Building and District Logbook can be considered as two complementary tools, all the information managed can be intended as a support for the bidirectional sharing of data between public and/or private stakeholders, from the urban to the building level and vice versa.

The tools are able to handle data in different formats, including:

- graphical documentation (design, execution, use and operational phases);
- quantitative historical data concerning consistencies and costs;
- documentation concerning the use of the building (contracts, certifications, …);
- ultimately the kind of data related to the specific use and typology of the property under investigation.

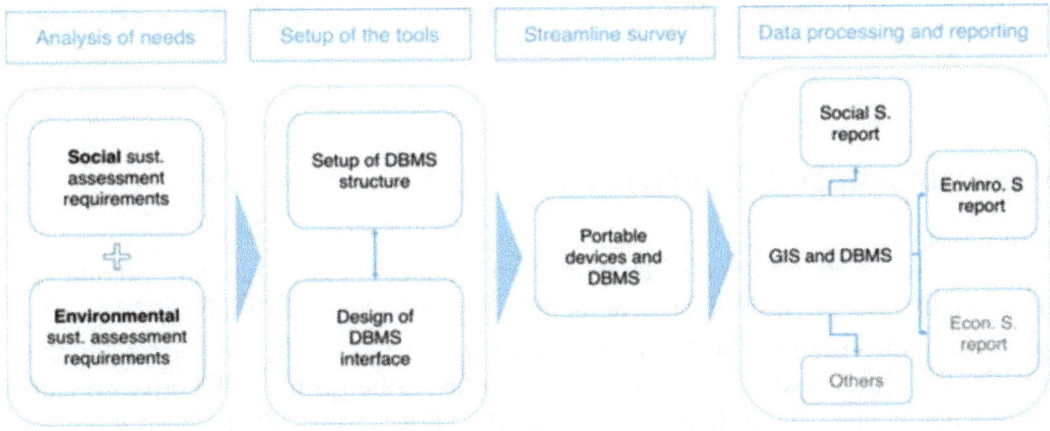

Fig. 5 District Information Model (DIM) as a consequence of the use of district logbooks (Moretti et al. 2017)

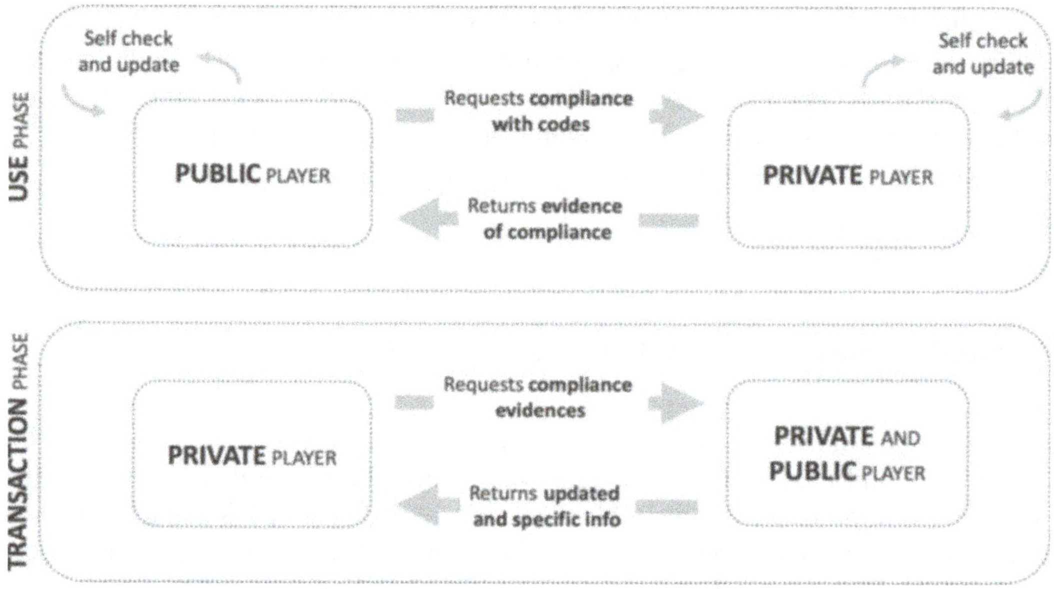

Fig. 6 A model/hypothesis of data flows in use and transaction phase (Dejaco et al. 2017b)

This data is a key for a variety of stages of the lifecycle of an asset, e.g., during the transaction and management phases (Fig. 6). Specifically, these tools should be used by the private players in terms of code and compliance checking and to have access to updated compulsory documentation concerning the asset. The public administration, on the other hand, assumes a double role of legislator and controlled entity, since it is simultaneously the issuer and the controller of codes and standards. Moreover, the public and the private players are in a mutual relationship: the public player is responsible for checking compliance of the codes and standards, applied by the private player. At the same time, the latter in the case of discrepancies between decisions made by the public player and the actual facts, can refer directly to the District and the Building Logbook, resolving the dispute in a simple and effective manner. The flows are not related to the whole data compiled in the Logbooks, it is possible and advisable, both for private and public stakeholders, to verify only certain specific issues.

6 Discussion and Conclusions

The tools presented can be considered useful for data collection, update and management with regards to the built environment, at different levels of analysis. In one case considering the building, the control of private properties can be considered the main aim. On the other hand, considering the urban level, the primary objective concerns the management of the public spaces. The Building and the District Logbook can be considered as a "support tool" to prevent disputes among stakeholders, caused by a potentially scarce knowledge and/or availability of the district's status and of the building's condition.

Often, when information is requested, the lack of data due to the scarce standardization and attention during storage and management phases, causes the loss of a great amount of time and crucial resources. Moreover, when the competences of private and public actors are not well defined, or there is an overlapping between management of technical, economic and operational issues, the presence of a suitable data collection like the logbook allows users to overcome such conflicts.

The tools, in a wider contest, can be inserted in the data engineering process, this is an approach that should be adopted, when dealing with management of the digital built environment. This could lead to the development of an Asset Information Model (AIM), namely a modular information system able to support informed decision-making and asset management processes. The AIM is featured as a digital model of the asset. It can be composed of a 3D graphical model of the building, information concerning technical characteristics of elements and a collection of documentation relating to safety, permissions, and responsibility with regards to the asset and its components. Thus, concerning the documental part, it can be considered as a translation of the Logbooks into digital terms.

A hypothesis of application of what has been developed concerns the connection between data relating to the energy certifications of buildings and their positioning on the territory. In this context, information was considered to be included in databases for the registration of energy certificates of buildings (CENED), with information that is typical of the Building Logbook, both from the point of view of the description of technological aspects and the condition assessment of the buildings.

The building is located in specific areas/municipalities (whose data can be collected in the District Logbook), which have specific rules of use. Through the proposed tools it is immediately possible to create thematic maps through which to represent the "quality" of the building from the point of view of energy consumption. The use of this procedure could potentially have multiple effects, for example:

- the private investor could verify the presence of low energy consumption areas in the territory as well as individual buildings with the best performance, and direct their investments accordingly;
- linking the thematic map of the energy survey with the real estate values of the different municipal areas, it is possible to verify the influence that energy certification can have on the trend of the real estate values themselves;
- the municipality can verify on the territory which and how many buildings are compliant with current laws in terms of energy consumption, discriminating between new and old buildings, checking, in case of the request for incentives, that the works are carried out properly;
- by locating the presence of non-certified public buildings, their intended use (e.g., schools) and their "influence" on the surrounding area (e.g., distribution of students by age group), a municipality could plan the renovation and/or divestment and request/predispose the necessary budgetary resources appropriately.

It is clear that the set of necessary information is not only linked to the proposed tools but is part of a broader model for managing a variety of data, to be analyzed and managed according to different levels of detail.

A future challenge to be addressed concerns the precise definition of processes for data collection and management relating to the definition of the operators involved and intended use.

References

Dejaco MC, Maltese S, Re Cecconi F (2017a) Il fascicolo del fabbricato. Maggioli Editore, Santarcangelo di Romagna, p 204

Dejaco MC, Moretti N, Re Cecconi F (2017b) Streamlined management of the built environment: the district and the building logbook as risk prevention tools. ISTeA 2017—re-shaping. Constr Ind [Internet]. 185–194. https://re.public.polimi.it/handle/11311/1032396#.WnHPYpM-euU. [cited 2018 Jan 31]

Dejaco MC, Re Cecconi F, Maltese S (2017c) Key performance indicators for building condition assessment. J Build Eng [Internet] 9:17–28. http://linkinghub.elsevier.com/retrieve/pii/S2352710216302893

Gabellini P (2012) Tecniche urbanistiche. Carocci Editore, Roma, p 516

Lee P-C, Wang Y, Lo T-P, Long D (2018) An integrated system framework of building information modelling and geographical information system for utility tunnel maintenance management. Tunn Undergr Space Technol 79:263–273

Lotfi S, Habibi K, Meshkini A, Koohsari MJ (2008) New multicritera approach for urban facility management by applying GIS. World Appl Sci J [Internet]

Mignard C, Nicolle C (2014) Merging BIM and GIS using ontologies application to urban facility management in ACTIVe3D. Comput Ind [Internet] 65(9):1276–1290. http://dx.doi.org/10.1016/j.compind.2014.07.008

Moretti N, Maltese S, Dejaco MC, Re Cecconi F (2017) GIS supporting surveys for urban sustainability assessment. In: LC3 2017 volume I—proceedings of the joint conference on computing in construction (JC3), 4–7 July 2017, Heraklion, Greece. Volume I (July), pp 905–912

Moretti N, Dejaco MCC, Maltese S, Re Cecconi F (2018) An information management framework for optimised urban facility management. In: 35th international symposium on automation and robotics in construction (ISARC 2018), pp 381–387

Re Cecconi F, Dejaco MC, Maltese S (2014) Efficiency indexes for building condition assessment. Int J Hous Sci Appl [Internet] 38(4):271–279. http://www.housingscience.org/html/publications/pdf/38-4-7.pdf. [cited 2019 Jun 24]

Re Cecconi F, Maltese S, Dejaco MC (2017) Leveraging BIM for digital built environment asset management. Innov Infrastruct Solut [Internet] 2(1):14. http://link.springer.com/10.1007/s41062-017-0061-z

Sharifi A, Murayama A (2014) Neighborhood sustainability assessment in action: cross-evaluation of three assessment systems and their cases from the US, the UK, and Japan. Build Environ 72:243–258

UNI (2012) UNI 11447:2012. Urban facility management services. Guidelines to set and program contracts

Volk R, Stengel J, Schultmann F (2014) Building Information Modeling (BIM) for existing buildings—literature review and future needs. Autom Constr [Internet] 38:109–127. http://dx.doi.org/10.1016/j.autcon.2013.10.023

Digital Transformation in Facility Management (FM). IoT and Big Data for Service Innovation

Nazly Atta and Cinzia Talamo

Abstract The latest innovations in Information and Communication Technologies (ICTs) along with the establishment of the new paradigms of Internet of Things (IoT) and Big Data Management are opening up innovative scenarios with respect to cognitive and decision making processes related to the management of the built environment. The novel availability of information offered by these digital technologies can lead to the definition of strategies aimed at significantly reducing management costs and improving building performance and service quality. Although this potential is now widely recognized by the various operators in the sector, experimentations have not yet led to a harmonization and standardization of procedures, processes and enabling technologies applications. The paper proposes strategies and tools for supporting the various FM operators in the choice and implementation of IoT technologies as well as in the management of Big Data and their sources, aiming at optimizing and innovating the current FM processes, models and services.

Keywords Facility management (FM) · Building management · Internet of things (IoT) · Big data · Information and communication technology (ICT) · Information management · FM digitalization

1 Information and Communication Technologies as Driver for Innovation of FM Sector

The Facility Management (FM) sector is undergoing a profound transformation of practices, processes, tools and references due to the adoption of novel ICT (Information and Communication Technology) solutions which nowadays promise to improve the traditionally conceived FM processes, making new knowledge bases available to support data-driven decision-making processes and embracing a network approach to stakeholder management. The digital transformation today is affecting the FM

N. Atta · C. Talamo (✉)
Architecture, Built Environment and Construction Engineering—ABC Department, Politecnico di Milano, Milan, Italy
e-mail: cinzia.talamo@polimi.it

B. Daniotti et al. (eds.), *Digital Transformation of the Design, Construction and Management Processes of the Built Environment*, Research for Development,
https://doi.org/10.1007/978-3-030-33570-0_24

sector and its operators that are trying to keep up with it, even if still in a convulsive way since there is a clear lack of consolidated and shared references and tools (Konanahalli et al. 2018). Indeed, the experimentations on ICT-FM integration are still at an infantile stage. However the operators of the FM sector already see the potential of technologies not only as mere tools but as generators of communication and collaboration opportunities able to increase their business value (Ahmed et al. 2017). In fact, when speaking about digital transformation, it is important to understand that it is not just a technological shift at the operational level, but it involves also the strategic level of the top management since it allows to create reliable knowledge bases concerning the several different aspects of the built environment allowing also new possibilities for a widespread continuous monitoring of relevant parameters (Talamo et al. 2016). Therefore, the recent technological innovations in the field of ICT should not be meant just as mere operational tools, but they have to be recognized as valuable sources of data and information, on which it is possible to improve management activities and to establish novel collaboration between company departments—deriving mutual benefits—and also to develop new business opportunities. In other words, the goal is no longer just to increase the operational efficiency—which however remains an important objective for facility managers—but also to re-engineer processes, procedures and activities, thanks to the adoption of digital technologies. The role of the facility managers—who manage the so-called "no core business"—has always been to support the needs of the Client by ensuring a high quality of building services, the efficiency of the infrastructures and the promptness of interventions. However, in the last years we have witnessed a paradigm shift in the Facility Management sector: Clients increasingly require service strategies aimed at predicting events instead of responding to them. The market is asking companies for an increasingly pressing level of flexibility and innovation, forcing them to migrate from the widespread traditional culture of doing to new strategies aimed at predicting future trends (Konanahalli et al. 2018). This shift marks the transition from corrective or planned strategies to preventive and predictive strategies. These new demands, which today can be met and satisfied thanks to the integration of the new ICT solutions, lead facility managers to face profound changes in their traditional practices. Indeed, this new scenario requires a drastic rethinking of FM models, processes and services both at the strategic and operational levels, in order to meet the expectations of Clients, ensuring their competitiveness and increasing their added value. In such a frenetic and convulsive but also so challenging and exciting context, which marks this transitional period of digital transformation of the sector towards more advanced high-performance scenarios, the facility manager:

- can no longer be considered a mere maintenance or service manager, but must be able to rely on the tools of digitization—by gaining skills, know-how and experiences also related to the ICT field—in order to contribute to the development of the digital innovation of the FM sector;
- has important tasks and owns responsibilities related to several interconnected topics (as energy, security, quality of workplaces, data flow management, etc.) that need a constant and deep collaboration with all the company functions. Facility

managers should ensure the horizontal collaboration among the different departments in charge of the different topics of interest by following a network approach rather than a siloed one, also relying on the new digital IoT platforms for communication and information sharing for identifying and exploiting possible synergies, sharing knowledge bases and deriving mutual benefits;
- must ride the wave of this digital transformation staying competitive and able to offer innovative cutting-edge FM services. Facility managers can reach this goal by experimenting and using the new technological tools enhancing their own capabilities of shaping innovative ideas, designing new FM solutions and fully implementing them anticipating Clients needs and expectations.

The common thread that links these aspects—today of vital importance for facility managers to promote and boost innovation—is the capacity of exploiting ICTs which represent the innovation factor that enables and supports this development and evolution of Facility Management. In particular, main important technological phenomena such as the Internet of Things (IoT), Big Data Management and Data Analytics are nowadays supporting the change of processes and functions of the facility management business (Fukada et al. 2018; Wong et al. 2018).

2 FM-Related Information: The Role of IoT and Big Data Management

ISO/IEC 20924:2018 Internet of Things (IoT)—Vocabulary[1] defines Internet of Things (IoT) as an "infrastructure of interconnected entities, people, systems and information resources together with services which processes and reacts to information from the physical world and virtual world" (ISO/IEC 20924:2018). Therefore, the IoT can be meant as a network of connected devices (e.g. fixed, mobile and wearable sensors) that have communication functions (Lee and Kim 2018). The IoT is rapidly becoming one of the core technologies of the digital transformation of the FM sector because of its capability of connecting building users, building components and services merging the physical and virtual worlds and letting them communicating through intelligent digital interfaces. Moreover, the IoT increases the capabilities of facility managers to create updated knowledge bases—thanks to real-time sensor data—store them in the cloud and process data when needed through intelligent systems as the Information Platforms.

[1]The International Standard ISO/IEC 20924 *Internet of Things (IoT)—Vocabulary* has been prepared by the *ISO/IEC Joint Technical Committee 1 (JTC1)—Information Technology* and in particular by its *sub-committee 41—Internet of Things and related technologies*. The aim of this standard is to provide a definition of Internet of Things (IoT) along with the definitions of IoT-related terms and concepts, ensuring in this way a common and shared IoT terminology.

In the IoT-based FM vision, sensors[2] and IoT devices[3] play a decisive role.

According to IoT Analytics (2018), the global IoT Market will reach $1,567B by 2025.[4] Statista (2019) forecasts that the amount of devices installed and connected to the Internet will arrive at 75.44 billion worldwide by 2025.[5] These forecasts confirm that the IoT, enabled by the already ubiquitous Internet technology, is a reality today. The technological revolution and digital transformation are already underway in the main social and business sectors with the promise of making the world a connected place. Sensors and IoT devices (e.g. wearables, smartphones, Radio-Frequency IDentification—RFID, smart meters, etc.) collect data—concerning the different aspect of the built environment—which are then analyzed and stored. In particular, sensors are embedded in/installed on physical objects (Things) and they are able to give to the physical objects communication and information exchange capabilities. Therefore, sensing technology is used to give to the objects virtual identities in order to acquire from them a broad range of data (e.g. relevant parameters of interest such as position, motion, etc.) (Konanahalli et al. 2018; Lee and Kim 2018) in the form of Big Data. According to the International Standard *ISO/IEC 20546:2019 Information technology—Big data—Overview and vocabulary,*[6] the term Big Data[7] implies datasets that are extensive in volume, velocity, variety and/or variability (Table 1).

Nowadays sensors and IoT devices—as well as traditional meters connected to a smart network which in this way gain communication capabilities—are able to collect FM-related parameters, in the form of Big Data, regarding—among others—for example, environmental conditions such as temperature, humidity, heat, atmosphere

---

[2]ISO/IEC 20924:2018 defines a sensor as an "IoT device that measures one or more properties of one or more physical entities and outputs digital data that can be transmitted over a network" (ISO/IEC 20924:2018).

[3]ISO/IEC 20924:2018 defines an IoT device as an "*entity of an IoT system that interacts and communicates with the physical world through sensing or actuating*" (ISO/IEC 20924:2018).

[4]Source: IoT Analytics 2018—available at: https://iot-analytics.com/state-of-the-iot-update-q1-q2-2018-number-of-iot-devices-now-7b/ (Accessed in April 2019).

[5]Source: Statista 2019—https://www.statista.com/statistics/471264/iot-number-of-connected-devices-worldwide/ (Accessed in April 2019).

[6]The International Standard *ISO/IEC 20546:2019 Information technology—Big data—Overview and vocabulary* has been prepared by the *Sub-committee SC 42—Artificial intelligence* of the *Joint Technical Committee ISO/IEC JTC 1—Information technology*. The standard aims at providing a taxonomy, articulated in a set of terms and definitions, needed to promote the common understanding and the improved communication regarding the Big Data topic. Moreover, in order to reach a better understanding of the topic, the standard provides a conceptual overview of the Big Data field and of its relationship with other main technical areas and fields.

[7]ISO/IEC 20546:2019 defines Big Data as "*extensive datasets—primarily in the data characteristics of volume, velocity, variety, and/or variability—that require a scalable technology for efficient storage, manipulation, management, and analysis*" (ISO/IEC 20546:2019).

Table 1 Big Data features. Adapted from Talamo and Atta (2018)

"V"	Features	Description
Volume	Data size and amounts	Extensive amounts of Data (Petabyte, Exabyte, etc.) available for performing analysis to extract valuable information
Velocity	Data in motion	Fast data streams transmitted—through communication networks—from one source to one or more destinations
	Data collection and storage	Data streams instantly generated and, then, collected and stored at high speed in extremely fast times
	Data lifetime	Period of time in which data remain valid, significant and reliable
	Real-time data analysis	Data can be extracted, aggregated, processed and analyzed in real-time
Variety	Heterogeneity of sources	Possible sources: RFID tags, sensors, databases, storage systems, logs or accesses to public web, business applications, social media, etc.
	Typologies and shapes of data	Possible data typologies and shapes: images, video, audio, live streams, etc.; information from databases and data storage systems (e.g. SQL, NoSQL, doc repository, etc.); digital documents (e.g. txt, PDF, Excel, HTML, XML, etc.)
	Diversity of data formats	Possible data formats: structured data (numeric, strings, alphanumeric); semi-structured data (HTML and XML files); unstructured data (free text, videos, voice messages, images)
Variability	Data semantic	Change of data meaning according to the reference context
	Data format	Variable data structure/format/shape
	Data quality	Variable data interpretation according to different users

composition, light, sound, etc. Being able to transmit data to a central monitoring system, connected[8] sensors and devices can be used to remotely control the main building systems, such as air conditioning, heating, lighting, etc. Moreover, by combining external data coming from the surrounding urban environment—such as external temperature, external humidity, weather conditions, etc.—with FM-related operational data it is possible to increase the capability of facility managers to better understand the impact of these surrounding external conditions on the building

[8]Networks (e.g. WPAN, Bluetooth, Wi-Fi, 3G, 4G, LTE, broadband network, Ethernet, etc.) connect the physical and the virtual world, enabling the possibility for people, things, and services to become connected, to communicate and to exchange data and information.

service delivery. In this way, extending the detection to the urban scale (Table 2), it is possible to exploit interesting synergies and to increase the knowledge of external on-going phenomena that can influence—positively or negatively—the building and services performance (Barkham et al. 2018; Paganin et al. 2018). Indeed, for instance, the gained insights can be used to proactively plan resource requirements, increase response and organization skills and prevent possible issues according to the real-time external conditions (e.g. weather) and, above all, to the short-/long-term external conditions forecasts.

Table 2 Example of Urban Big Data. Adapted from Paganin et al. (2018)

Urban Big Data	Sensors-generated data	Users-generated data
Possible sources	Connected systems of sensors (e.g. WSN) and other IoT networks in urban areas	Participatory sensing systems
	Public utilities sensor systems	Social media, blogging and Web2.0
	Building management systems (BMS)	Accesses and log-ins on web applications
	Smart grids	Global positioning system (GPS)
	Surveillance system	Global navigation satellite systems (GNSS)
	Geographic information systems (GIS)	On-line social networks
	Satellite earth observation service	Mobile applications
Examples of types of Urban Big Data	Environmental data (T, H, p, etc.)	Users position
	Seismic, hydrological, and geological data	Users preferences
	Data on mobility (delays in public transport, real-time traffic data, etc.)	Users online activities
	Data on public utilities (distribution of energy, electricity water, gas)	Socially-generated or shared data (posts, links, etc.)
	Data from monitoring of use and consumption (heating, lighting, etc.)	Replies to on-line surveys

3 IoT Platforms for Advanced FM Service Management

Sensors and devices—essential for collecting and extracting data from objects—are connected to terminal devices for data collection. It is necessary that these sensors are connected to the Internet which enables the data exchange communication functions. To exploit the potential of sensor data, it is also necessary to guarantee interoperability of Internet communication, an application system and an embedded system capable of providing user interfaces. These functions are performed by IoT platforms. The International Standard *ISO/IEC 30141:2018 Internet of Things (loT)—Reference Architecture*[9] provides a standardized IoT Reference Architecture highlighting six domains and related entities which communicate and exchange data by means of the network, as shown in Fig. 1.

In particular, the six domains of the IoT Platform identified by ISO/IEC 30141:2018 are (ISO/IEC 30141:2018):

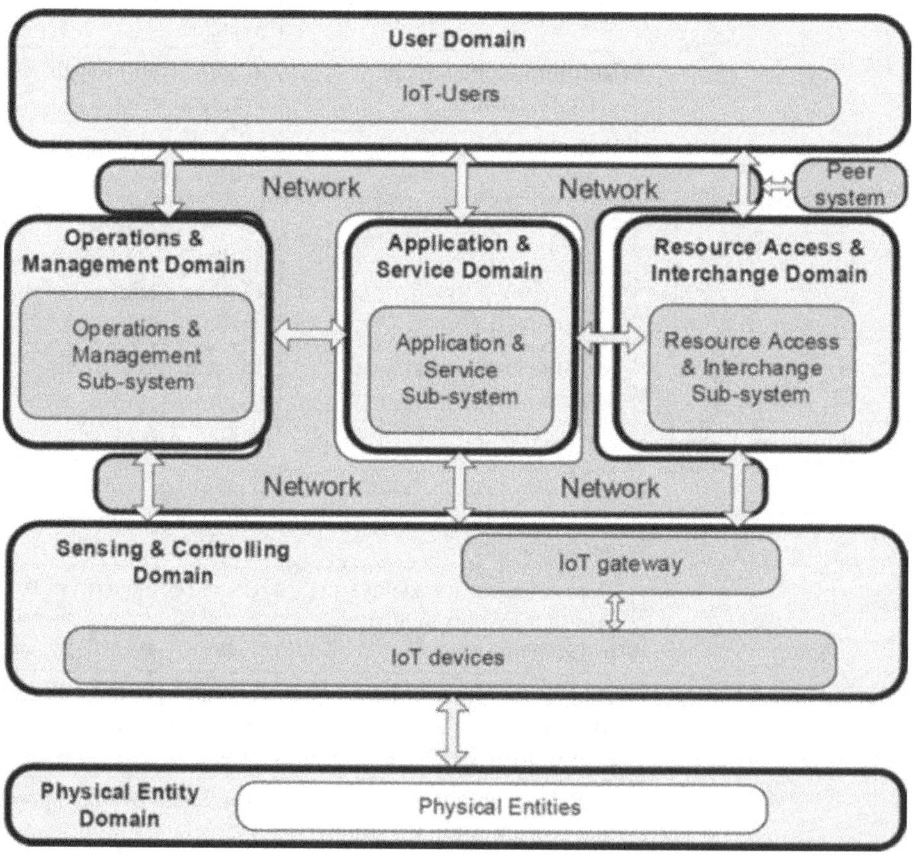

Fig. 1 Architecture of an IoT platform. *Source* ISO/IEC 30141:2018

[9]The International Standard *ISO/IEC 30141 Internet of Things (loT)—Reference Architecture* has been prepared by the *sub-committee 41: Internet of Things and related technologies* of the *ISO/IEC Joint Technical Committee 1 (JTC1): Information technology*. The standard proposes a general IoT Reference Architecture, including the definition of the system characteristics and its conceptual model.

- User Domain (UD). In this domain the actors are human users who can interact with the offered digital services through user devices and applications, e.g. smartphones, tablets, computer desktops, dashboards, control panels, etc.
- Physical Entity Domain (PED). This is the domain of the physical entities (e.g. building elements, plants, equipments, etc.). It is the receiver of actions taken by facility managers based on the results of the sensing, monitoring, and controlling.
- Sensing and Controlling Domain (SCD). This is the domain of sensors, actuators, smart meters, etc. These IoT devices are able to collect information allowing a continuous monitoring of different aspects of the PED. This is a fundamental domain for facility managers (that can be in charge of this domain) who can collect, process, analyze and store data, exploiting data value for taking informed decisions and actions that can influence the physical entities of the PED.
- Operations and Management Domain (OMD). This is the domain of system operators and managers who have to maintain overtime the overall health of the systems of the whole IoT Platform.
- Resource Access and Interchange Domain (RAID). This domain gives to users the access to the capabilities of the IoT system, offered through smart service interfaces. The access can be different according to the different users (e.g. access to different functionalities, different authorities, etc.).
- Application and Service Domain (ASD). The main actors of this domain are the service providers, who offer applications and services to the IoT-User in the User Domain (UD). Moreover, the applications and services can also interact with the entities in the sensor, devices and actuators of the SCD in order to collect data and/or drive actions in the PED (ISO/IEC 30141:2018).

The IoT platform, therefore, enables the creation of new digital services, based on the possible interaction—thanks to smart interface applications—between people, things and services. Thus, the new services created and offered will be based on the convergence of data and services, using data collected through sensors and other heterogeneous data sources (Lee and Kim 2018). In particular, the platform allows facility managers to offer value-added services thanks to the new availability of real-time data, their processing, analysis, extraction and archiving as well as the contextualization, interpretation of on-going and future data, information and events, and the creation and analysis of alternative scenarios.

4 Definition of FM IoT-Based Strategies

With the aim of favoring data integration but also, above all, of allowing the integration and collaboration among FM stakeholders, nowadays the concept of IoT Platform is gaining more and more interest, being able to deliver a unique and integrated environment for the storage, analysis and retrieving of data and for an advanced management of services which embrace the concept of centralized data exploitation. In particular, IoT Platforms enable facility managers and real estate managers to benefit from an open IoT ecosystem that enables the integration of different technologies (building management systems, sensors, connectivity), making building management more intelligent. In particular, the IoT adoption allows to identify and implement new advanced strategies for FM operators. These strategies mainly refer to the following FM macro-areas of interest: a. Strategies for monitoring and controlling service performance; b. Operation and Maintenance strategies; c. Strategies for stakeholders management.

4.1 Strategies for Monitoring and Controlling Service Performance

The issue of monitoring the performance of FM services, thanks to the adoption of new technologies, is now based on two main topics: predictivity and control. Performance must be obtained and maintained through analysis, failure predictability and optimized and standardized processes as much as possible. Digital tools make it possible to make decisions through data that are beyond our perceptions, making us see real trends and allowing us to "anticipate" problems, even before we have to find them to solve them. The new frontier of real-time monitoring is represented by the centralized remote control and monitoring trough dynamic dashboards and visualization tools that show the collected sensor data on a user-friendly digital interface which allow facilities managers to observe at data to see in real-time exactly how employees and customers are using the building and its components and equipment. Indeed, the IoT enables to access in real-time data about several aspects of the building by easily controlling and monitoring the facility remotely. The data collected by the IoT allow facility managers to be more effective in preventing issues and reducing the time spent for "on site" inspections. This process has to be meant as circular and iterative, allowing a continuous improvement. In this way, the technological tools and intelligence can inform and improve facility management policies overtime, leading the FM industry forward.

4.2 Operation and Maintenance Strategies

The IoT application to the maintenance management allows to delineate new strategies, such as Condition-based and Predictive Maintenance, useful to contain costs, reduce the waste of resources and limit downtimes. These benefits can be achieved thanks to the continuous and dynamic monitoring of the state of operation and use of building components and plants. In particular, the maintenance strategies that recur in the traditional management practices of Operation and Maintenance are mainly the corrective maintenance and the planned preventive maintenance. Particularly, corrective maintenance is performed after the occurrence of a fault. This strategy often involves high costs, production downtimes and long intervention times and intervals. To overcome these issues, traditionally the scheduled preventive maintenance is used. It is performed at predetermined regular intervals, according to an established time schedule (calendar), in order to detect faults before they occur. However, usually in this case a high number of maintenance interventions is programmed independently of the condition of thc building components themselves. This often involves execution of interventions on components that are not actually affected by degradation/failure and still far from the end of their useful life, with consequent increased maintenance costs. Instead, Operation and Maintenance IoT-based strategies allow:

- to promptly intervene when abnormal conditions exist, that could lead to a drop in performance, or situations of deterioration or failure of the component (Condition-based Maintenance). In this way periodic on site inspections of building components conditions will be no longer needed (remote real-time monitoring) and interventions will be carried out according to the actual operating profile of the components, hence work orders will consequently be generated real-time. In this way, the maintenance activity will be aligned to the real and actual building needs avoiding unnecessary costs and limiting the use of resources;
- to predict the operating conditions of building components (Predictive Maintenance). Interventions will no longer be performed at regular and periodic intervals, but the frequencies will be defined using proper mathematical models useful to identify the time remaining before the failure. Therefore, the maintenance program is no more determined by a prescribed timeline but by analysis algorithms that use the data collected by IoT sensors in order to recognize in advance the occurrence of degradations, malfunctions or breakdowns, thus allowing to plan interventions in time, avoiding interference with on-going core activities.

By allowing facility managers to detect present trends and to forecast the values of the parameters of interest, making it possible to foresee the future behaviors of the building components, these new Operation and Maintenance strategies inform and support decision-making processes both at strategic and operational level.

4.3 Strategies for Stakeholders Management

The traditional FM practices are affected by the information silos approach, which is characterized by a lack of collaboration and information sharing between the different departments, as well as between the different stakeholders. Each department implements its own methods, procedures and tools to satisfy very specific operational requirements. Furthermore, these systems are often poorly implemented due to budget constraints and often suffer from a competitiveness aimed at demonstrating departmental efficiencies rather than global organizational ones. For example, different systems are often used to perform audits, plan maintenance, manage the help desk, track equipment, manage suppliers, check safety notices, organize cleaning, etc. This highly individual approach of the departments not only increases the complexity of the work but also the operating costs associated with these activities as well as the personnel training costs and administrative costs. IoT technology offers solutions aimed at overcoming this "data islands" sectorial approach that often leads to a supply of disconnected services and poor customer satisfaction. In particular, the IoT allows the creation of a connected ecosystem of people, devices and systems, in which it is possible to centrally manage the data, making the information bases of each department a common capital for all the departments and stakeholders. Thus, IoT has the potential to connect all the functional silos to make them horizontally integrated, favoring the stakeholders' engagement and collaboration and the integration of all the available knowledge bases. Moreover, the implementation of a widespread sensors network can detect data useful to different departments. If the IoT infrastructure, tools, methods and procedures are shared between departments and stakeholders—with a view to horizontal integration of tools, know-how and skills—it is possible to achieve considerable benefits also in terms of economic efficiency.

References

Ahmed V, Tezel A, Aziz Z, Sibley M (2017) The future of big data in facilities management: opportunities and challenges. Facilities 35(13/14):725–745

Barkham R, Bokhari S, Saiz A (2018) Urban big data: city management and real estate markets. New York, NY, USA, GovLab Digest

Fukada T, Huang W, Janssen P, Crolla K, Alhadidi S (2018) Field survey system for facility management using BIM model

Konanahalli A, Oyedele L, Marinelli M, Selim G (2018) Big data: a new revolution in the UK facilities management sector

Lee HJ, Kim M (2018) The Internet of Things in a smart connected world. In: Internet of Things-technology, applications and standardization. IntechOpen

Paganin G, Talamo C, Atta N (2018) Knowledge management and resilience of urban and territorial systems. TECHNE-J Technol Archit Environ 15:124–133

Talamo C, Atta N (2018) Invitations to tender for facility management services: process mapping, service specifications and innovative scenarios. Springer

Talamo C, Atta N, Martani C, Paganin G (2016) The integration of physical and digital urban infrastructures: the role of "Big data". TECHNE-J Technol Archit Environ 11:217–225

Wong JKW, Ge J, He SX (2018) Digitisation in facilities management: a literature review and future research directions. Autom Constr 92:312–326

Standard and Laws

ISO/IEC 20546:2019 Information technology—Big data—Overview and vocabulary

ISO/IEC 20924:2018 Internet of Things (IoT)—Vocabulary

ISO/IEC 30141:2018 Internet of Things (loT)—Reference Architecture

Websites

IoT Analytics 2018. https://iot-analytics.com/state-of-the-iot-update-q1-q2-2018-number-of-iot-devices-now-7b/. Accessed April 2019

Statista (2019). https://www.statista.com/statistics/471264/iot-number-of-connected-devices-worldwide/. Accessed April 2019

BIM Digital Platform for First Aid: Firefighters, Police, Red Cross

Alberto Pavan, Cecilia Bolognesi, Franco Guzzetti, Elisa Sattanino, Elisa Pozzoli, Lara D'Abrosio, Claudio Mirarchi and Mauro Mancini

Abstract This document explains the use of BIM in emergency response scenarios, considering both the current rescue procedures and the innovative development of a digital platform able to support first aid, firefighters, civil protection, police, traffic wardens, etc. This ongoing study has a practical aim: to allow all rescuers to intervene promptly and safely in places where there are alarm conditions. Thanks to the BIM methodology, linked to indoor navigation of the building and Google Maps tools, in case of an intervention users will be able to query a digital model for information. Starting from these premises, this paper assumes and analyses the methodology and tools used to develop the project of a BIM digital platform for first aid.

Keywords BIM digital platform · First aid · Indoor navigation of buildings · IFC parameters

1 Introduction

Despite the wide range of occupations, professional rescuers share a unique responsibility: saving lives. This type of work implies a continuous training of the staff, and a continuous update of the techniques and tools to be adopted in order to face different scenarios. The efficiency of a building rescue operation is often slowed down by the timing of receiving data in real time, concerning traffic congestion, the state of an ongoing construction site or the presence of some unexpected event. For instance, in the case of a fire, a main role is played by the firefighting team because they

A. Pavan · C. Bolognesi · F. Guzzetti · E. Sattanino (✉) · L. D'Abrosio · C. Mirarchi
Architecture, Built Environment and Construction Engineering—ABC Department, Politecnico di Milano, Milan, Italy
e-mail: elisa.sattanino@polimi.it

M. Mancini
Department of Management, Economics and Industrial Engineering—DIG, Politecnico di Milano, Milan, Italy

E. Pozzoli
Milan, Italy

B. Daniotti et al. (eds.), *Digital Transformation of the Design, Construction and Management Processes of the Built Environment*, Research for Development,
https://doi.org/10.1007/978-3-030-33570-0_25

have to collect all the main information about the building while scouting the site. In the majority of cases, the knowledge of the how a building is planned out or, for instance, the placement of lifts, safety exits or specific rooms are difficult to source in real time. Currently there are no platforms suitable on the market to help rescue services, including ambulances technicians, firefighters, police officers, to identify the most suitable path for their destination, allowing in the same application for the chance to visualise the useful information of a building for emergency operators, navigating a 3d model in real time. In this scenario the Italian emergency sector is becoming aware of the numerous advantages of the *BIM (Building Information Modelling)* method linked to *Indoor Navigation* and to *GIS (Geographic Information System).*[1]

The indoor navigation approach to public buildings is a developing reality that offers a considerable potential, not only in the retail sector, but also in emergency situations, increasing people's safety. On the other hand, the GIS tools plays a fundamental role in the analysis of an area, providing information such as the size of a street, the height of a viaduct, the presence of crowded public spaces and so on.

In this paradigm, the Politecnico di Milano recognises the importance of realizing a Digital Platform, where BIM and GIS coexist, in order to provide a mobile app (for tablet, smartphone or 3d visualizers) that helps rescuers.

2 3D BIM Project

Financed by *Regione Lombardia*, the 3D BIM project is part of the "Smart Living" context[2] and sees the partnership of the Politecnico di Milano with the two Italian companies Noovle spa and Fasternet spa.

The 3D BIM Project aims to create an innovative system for storing and displaying geographic information of an area and its buildings (public and private), such as to allow rescuers (First Aid, Fire Brigade, Police, City Patrol, etc.) the chance to intervene promptly and safely. The objective of the project is the realization of a cloud platform that will allow the professionals (architects) to insert a simplified and georeferenced survey of the buildings. The platform will give users the chance to store BIM models, initially modelled with an Authoring software, enriching them with all the information deemed fundamental by the safety and rescue organs to act promptly in case of an emergency.

---

[1]The GIS (Geographical Information System) is a geographic information system thanks to which it is possible to create maps, extract data, visualize scenarios three-dimensionally, look into specific elements and extrapolate geometries with data. The contribution of the GIS in this area is relevant because there is a large amount of information related to the geometries of objects.

[2]*(…) The "Smart Living" call is aimed at supporting development and innovation projects carried out by partnerships of companies in the construction, wood, home furnishing, household and High-tech sectors in collaboration with the university system, finalizing the introduction of new or improved products, processes/services from a technological, productive and organisational point of view, in order to enhance the theme of "intelligent Living.(…)"* (Regione Lombardia).

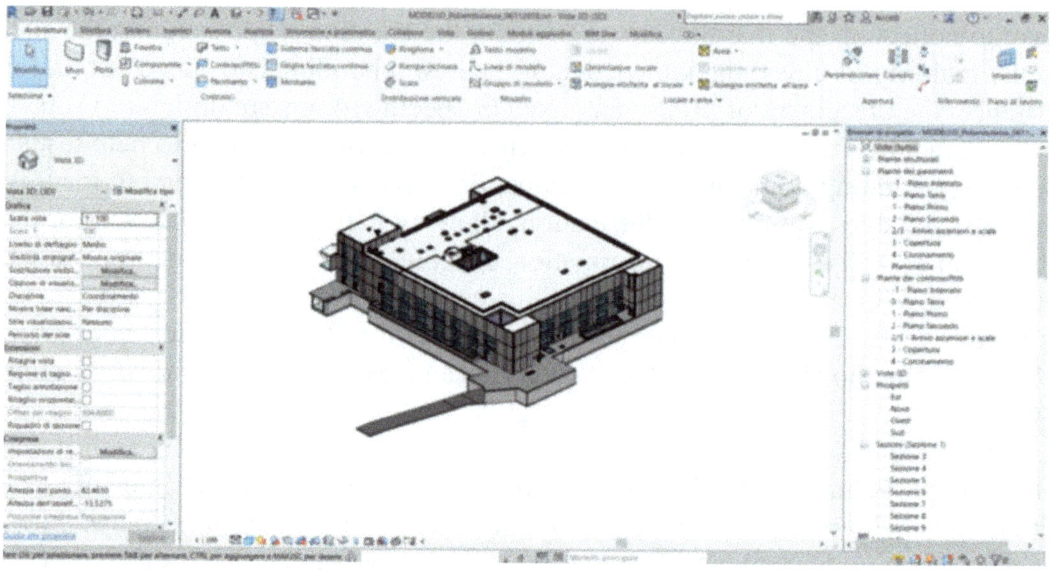

Fig. 1 3D model of the Poliambulanza Pavilion. Made with Revit (2018)

The main features of the platform are summarised as follows: a web portal for loading and managing BIM 3D models and its informative set; mobile application/tablet; search map of the building where users can view useful information; calculation of the shortest route to reach a particular building; display of the BIM model of the building and its information; indoor location of operator position and that of users.

In order to be able to test the functionalities of the research platform, in the analysis and development phase, the new hospital Pavilion of the Poliambulanza hospital of Brescia was modelled through BIM Authoring software (Fig. 1). The hospital did not have, even if recent, a BIM-oriented modelling design, so it has been analysed in all its components, from the georeferencing of the building in a GIS environment, to the geometric information up to the definition of the information attributes in view of 3d modelling.

3 Method

The development of the project is articulated in three main phases:

- The first phase concerns the analysis and aims to interview the end users of the platform, identifying the technologies to be used for the development of the platform;
- The second phase concerns the implementation of all the requirements collected during the previous phase, realizing the 3d model of the case study and the platform itself;
- In the final phase the tests are carried out.

3.1 Interviews and Questionnaires

In the initial phase, it was necessary to perform preliminary interviews with the subjects mentioned above: the heads of the Red Cross of Milan (Lombardy's Regional committee), Milan's section of Firefighters and the State Police of Milan and Brescia were interviewed. Based on the information gathered in these meetings, we structured questionnaires to obtain useful information for the realization of the technical data sheets (Fig. 2).

In order to facilitate the subjects involved in the comprehension of the indicated requests, we prepared a number of examples of graphic representations and visualizations of the symbols, colours and annotations of architectural objects. Representation of doors, elevators and rooms for firefighters and the Red Cross were chosen in order to make the visualization of the elements in the platform as intuitive as possible. We decided to give the following indications:

- *element accessible and recommended along the path*: the architectural element, for example a door, is highlighted green and graphic symbols are displayed with the most important information (opening, REI, fitted with alarm, emergency exit,...);
- *inaccessible element*: model elements that are inaccessible, such as ports, are displayed in red with the symbol of the access ban;

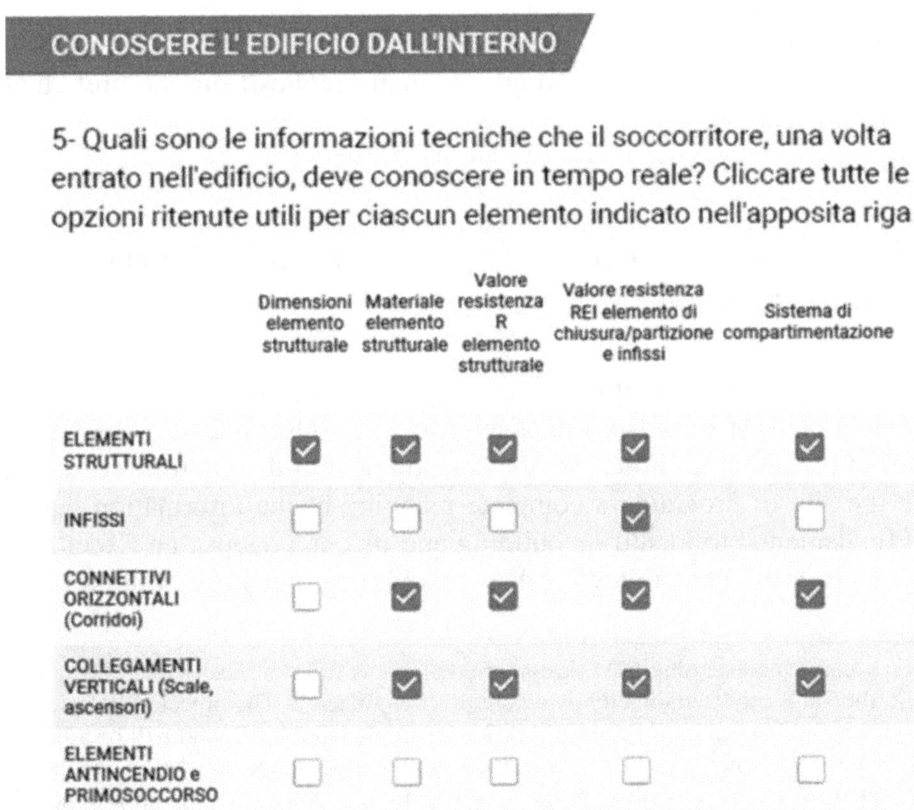
CONOSCERE L' EDIFICIO DALL'INTERNO

5- Quali sono le informazioni tecniche che il soccorritore, una volta entrato nell'edificio, deve conoscere in tempo reale? Cliccare tutte le opzioni ritenute utili per ciascun elemento indicato nell'apposita riga

	Dimensioni elemento strutturale	Materiale elemento strutturale	Valore resistenza R elemento strutturale	Valore resistenza REI elemento di chiusura/partizione e infissi	Sistema di compartimentazione
ELEMENTI STRUTTURALI	☑	☑	☑	☑	☑
INFISSI	☐	☐	☐	☑	☐
CONNETTIVI ORIZZONTALI (Corridoi)	☐	☑	☑	☑	☑
COLLEGAMENTI VERTICALI (Scale, ascensori)	☐	☑	☑	☑	☑
ELEMENTI ANTINCENDIO e PRIMOSOCCORSO	☐	☐	☐	☐	☐

Fig. 2 Example from the survey for the Firefighters

Fig. 3 Example of doors representation with the information for the rescuer

- *elements in the model*: all the elements (of a category) present in the model that the user wants to see are coloured yellow.

The route to follow is indicated in green with an arrow, allowing the operator/rescuer to arrive at their destination in the shortest possible time, along the most suitable and safe path (Fig. 3).

Following the representation of the objects, a visualization of the paths for the end users was defined. We received questionnaire replies from the firefighters and the Red Cross surveys.

3.2 *Structuring Technical Data Sheets and BIM Object Management*

Starting from the definition of the recurring project scenarios in the modelling and management of BIM models by the design studios (Osello 2015), the informative attributes of the buildings useful for the development of the platform were defined.[3]

With the aim of providing a complete mapping of the information that is considered fundamental to identify a building and its components, an Excel table was structured with attributes,[4] following these macro-areas: territory, building, real estate

[3]Common scenarios concerning BIM design are defined as follows. Scenario 1: The object to be inserted in the 3D is modeled directly by an architect/engineer. 2: The object to be inserted in the model is downloaded from online object libraries (example Bimobject) and will be adapted to the needs of the designer. Choosing the most recurrent type of modelling, namely the one that includes objects created ad hoc by specialists in the sector and those downloaded from Internet sites, we have proceeded to define the information attributes tab of a generic building.

[4]An *informational attribute* is standardly defined as "any property of an area, a work or its physical entity (subject, object) or procedure (action), of a geometric or non-geometric nature. Informative

	FAMIGLIE DI ATTRIBUTI INFORMATIVI NON GRAFICI				
OBIETTIVO/ USO	CATEGORIA	NORMA	PRESTAZIONE	GRAND. FISICA	OPERATORE
ARCHITETTURA	Edificio				Progettista architettonico
STRUTTURE	Edificio	Inquadramento normativo (DM 16/01/1996)			Progettista struttutre
ANTINCENDIO	Edificio	Inquadramento normativo (D.M. 246 del 1987)	Categoria	Altezza antincendio	
BARRIERE ARCHITETTONICHE	Edificio				
SICUREZZA	Edificio				Coordinatore della sicurezza in fase di esecuzione

SITO | OPERA-edificio | AFO-collegamenti | ASO-u.i. | SPAZIO-locali | SISTEMA-strutture | SISTEMA-impianti | SISTEMA-muri e solai | SUBSISTEMA-serran

Fig. 4 Example of a technical table of a building, in line with UNI11337-1:2017

units, premises, connections, structures, plants, walls and ceilings, windows and doors. In the final analysis, the classes corresponding to the technical data sheets of the previous Excel table were defined according to the regulation UNI 11337-1:2017, as follows: Site, OPERA-building, AFO-links, ASO-real estate unit, SPACE-premises, system-structures, system-systems, subsystem-windows and doors (Fig. 4).

Following the structuring of the technical data sheets, they were processed with the IFC[5] study (ISO 16739; buildingSMART; Autodesk IFC standard manual) to define the most suitable export method of the case study modelled in Revit.

At an early stage, taking as a reference point the IFC scheme (ISO 16739:2013), the composition of the elements and their cataloging according to the *ABS Building Element* was analysed. On the basis of the classification given, a table of building elements was formulated, including beams, columns, continuous façades, doors, components, foundations, parapets, access ramps, rooves, floors, staircases, walls, windows, to which the respective IFC, attributes and reference *Psets* were associated.[6] The model export tests were then performed both in the standard IFC 2 × 3 Coordination View format and with different information display settings (Table 1).

property of a territorial class (or superclass) or of the construction sector, of a model, of a constituent digital object or of a composition". (UNI 11337-1:2017).

[5]The Industry Foundation Classes (IFC) is an open and standardized data format useful for sharing model-related information, including different modeling software. Specifically, it consists of a complex structure of data, grouped according to a system of classification of the physical components of a building (walls, doors, floors etc.), their attributes (volume, quantity, transmittance etc.) and the relations between them. Without this framework there can be no sharing of building information.

[6]The classes identified are valid for all models and are defined as: rooms, walls, continuous façade, doors, stairs, floors, rooves, countertops, columns, beams, railings, walls, ramps, fire alarm devices, fire extinguishers, safety devices, special equipment (lifts, elevators, hoists, etc.), mechanical equipment, electrical appliances, electrical equipment, windows.

Table 1 Extract of the table concerning the different IFC exportations with Revit

Export Revit setup	Class of visible attributes IFC (wall)	Class of visible attributes IFC (door)
Export IFC common property sets	Identification, Location, Quantities, Material, Profile, Relations, Classification, Hyperlinks, Pset_WallCommon	Identification, Location, Quantities, Material, Relations, Classification, Hyperlinks, IfcDimensions, IfcDoorPanelProperties, Pset_Door Common
Export IFC common property sets, export Revit property sets	Identification, Location, Quantities, Material, Profile, Relations, Classification, Hyperlinks, Other, Building, ID Data, Size, Phases, Graphics, Analytical Properties, Pset_Wall Common, Structural, Restrictions	Identification, Location, Quantities, Material, Profile, Relations, Classification, Hyperlinks, Other, Other(Type), Building (Type), ID Data, ID Data(Type), Size, Size (Type), Phases, IfcDimensions, IfcDoorPanelProperties, Materials and finishing (Type), Analytical Properties (Type), Pset_Door Common, Restrictions
Export IFC common property sets, export Revit property set, export schedules as Pset	Identification, Location, Quantities, Material, Profile, Relations, Classification, Hyperlinks, Other. Base Quantities, Building, ID Data, Size, Phases, Graphics, Analytical Properties, Pset_Wall Common, Structural, Restrictions	Identification, Location, Quantities, Material, Profile, Relations, Classification, Hyperlinks, Other, Other (Type), Base Quantities, Building (Type), ID Data, ID Data (Type), Size, Size (Type), Phases, Ifc Dimensions, Ifc DoorPanelProperties, Materials and finishing (Type), Analytical properties (Type), Pset_Door Common, Restrictions

The IFC files obtained from the Solibri viewers, BIM Model Viewer and usBIM. Viewer + V. 7.00 and (Acca Software) were imported to verify the management and loss of data. Based on the needs of respondents and the information visible in the IFC viewers, an Excel table was defined with the attributes to be shown in the platform and those to be hidden. For each attribute, users (VVF, CRI, Police) have been defined to display the information on the platform (Fig. 5).

UTENTE	ATTRIBUTI da aggiungere ex-novo	VALORE
VVF, Polizia,CRI	Porta allarmata	YESNO
VVF, Polizia,CRI	Chiusura porta con badg	YESNO
VVF, Polizia,CRI	Chiusura porta con codice	YESNO
VVF	Altezza netta	NUMBER
VVF	Tempo di chiusura porta	NUMBER
VVF, CRI	Larghezza netta	NUMBER

Fig. 5 Extract from the excel table. Attributes and users

3.3 Geographic Information System (GIS) Use in the Project

The topographical database, basis of the Webgis of the province of Brescia, has been used for two different phases of the work: on the one hand to provide the coordinates for the correct positioning of the model with respect to the system of Reference, on the other to derive all useful information relating to points of interest within the city or otherwise intersecting the path identified and viewable by the rescue organs. The topographical DB standard is strictly recognized on a regional and national level. This study then focused on considering, deriving from the DBT, a likely recovery (and thus replicable) model for a regional and national level. Starting from the DBT of Brescia, in a first phase, the shape of the buildings' points of interest and shape of the road were selected.

Within each category data was selected, where present, which provides useful information for knowing the influx of people in given moments of the day. This part of the analysis was also realized with the help of Google Maps which allows users to visualize inside information about the opening and closing times of the activities.

The shapes thus recovered will be positioned and highlighted within the interactive map of the application, while the shapes relating to road traffic will be used for the possibility of identifying closed roads and dead ends by setting a selection considering at the geometric level the segments that are open.

4 Results

The goal of this research project is the creation of a graphic database and an information database, poured into a cloud platform where the different users can view, through a web and mobile interface (on tablets/smartphones or 3D viewers), the information coming from the BIM models (Mirarchi et al. 2018) overlapped by specific information for assistance. The approach of modelling for the rescuers' path will then take place with traces on the road graph already enriched with information from Google the database, which will allow users to reach the building. The travel tracking is done by GPS connected to the smart tool in the hands of the rescuer. Inside the building, it will be carried out using Beacon[7] technology, capable of locating the user with precision in the exact spatial location within the building itself and thus in the model. The research will soon reach its conclusion with the testing of the platform that Noovle s.p.a. is developing and with the final approval of the users.

Below is a high-level description of the components used for the developing platform:

- *Frontend*: this component will be responsible for processing all requests from the desktop and mobile clients and for communicating with the upload component when creating the information.
- *Upload*: it has the task of processing the IFC file format in order to extrapolate the most relevant information and to prepare it to be processed by the BIM Service.
- *Database*: This is a classic relational database that can save geographic and non-geographic information that will be analysed later.
- *Static Content*: This platform component will contain all static information such as images, configuration files and static BIM models loaded into the platform.
- *BIM Server*: This has the task of processing the IFC BIM files in order to make them usable and viewable on all devices.

5 Conclusion

The research described traces a path starting from generic models of buildings (i.e. modelled by any BIM-oriented software) to outline them as navigational tools for the specific use of rescue services.

Thanks to geo-localization and three-dimensional modelling, although simplified, of the masses of buildings and their common parts (gardens, atriums, staircases, elevators, technical rooms, electric control units, roofs, etc.) the rescue services will have all the information needed to optimize their time of intervention.

With this system of mapping and geolocation of environments, the information will be able to reach them through any mobile device (smartphone) or even directly, in augmented reality, on portable viewers (digital glasses).

For public buildings a BIM modelling of greater detail and the use of Beacon sensors for georeferencing will guarantee the localization also in internal up to the single environment besides the possibility of using the same device regarding an individual civilian as a direct source of the information necessary to the rescuers.

[7]Bluetooth beacons are small radio transmitters that send signals in a radius of 10–30 m (interior spaces). The benefits of beacons are: that they are affordable (three to thirty euros), can be installed with minimal effort, you can determine a position with accuracy up to 1 m and are supported by many operating systems and devices. Beacons can be used for both client-based and server-based applications. The new standard BLE (Bluetooth Low Energy) is also very energy efficient, in fact the choice fell on BLE technology, which allowed us to achieve excellent tracking results by keeping extremely small, cheap (almost disposable) tags.

In the first processes of Augmented Reality (realized by overlapping clear information for paths on the real environment) the emergence of information relating to the surrounding environment can be seen both through wearable devices and by smart devices in our case held by the rescue operator identified within the Beacon sensor network. After its trial the project of course will benefit from the possibility of spreading into several environments and on a large scale: starting from socio-health care to building large private properties.

References

AAVV (2013) Realtà Aumentate. Esperienze, strategie e contenuti per l'Augmented Reality. Apogeo Education, Milano

Autodesk (Nov 2018) Detailed instructions for handling IFC files. https://abcdblog.typepad.com/abcd/2018/Success_Stories/IFC-Manual-2018-ENU.PDF

Battini C (2017) Realtà virtuale, aumentata e immersiva per la rappresentazione del costruito. Altralinea, Firenze

Costin A, Pradhananga N, Teizer J (2014) Passive RFID and BIM for real-time visualization and location tracking. In: Proceedings of the construction research congress: construction in a global network, ASCE, Atlanta, GA, USA, p 169–178

Eastman C, Teicholz P, Sacks R, Liston K (2016) Il BIM: Guida completa al Building Information Modeling per committenti, architetti, ingegneri, gestori immobiliari e imprese. Hoepli, Milano

http://www.regione.lombardia.it/wps/portal/istituzionale/HP/DettaglioBando/servizi-e-informazioni/imprese/filiere-eccellenti/smart-living/smart-living

https://www.buildingsmart.org/

https://sit.provincia.brescia.it/Metadati

Lee Y-C, Eastman C, Solihin W (2016) An ontology-based approach for developing data exchange requirements and model views of building information modelling. Adv Eng Inf 30(3):354–367

Mirarchi C, Pavan A, De Marco F (2018) Supporting facility management processes through end-users' integration and coordinated BIM-GIS technologies. Int J Geo Inf 7(5):191, 1–10

Osello A (2015) Building information modelling—geographic information system—augmented reality per il facility management. Dario Flaccovio Editore, Palermo

Pavan A, Mirarchi C, Giani M (2017) BIM: metodi e strumenti, Tecniche nuove

Standards and Laws

ISO 16739 Industry foundation classes (IFC) for data sharing in the construction and facility management industries—Part 1: Data schema

Level of development (LOD) specification (2019)

NBS BIM object standard

UNI 11337-1:2017 Edilizia e opere di ingegneria civile—Gestione digitale dei processi informativi delle costruzioni—Parte 1: Modelli, elaborati e oggetti informativi per prodotti e processi

UNI 11337-4:2017 Edilizia e opere di ingegneria civile—Gestione digitale dei processi informativi delle costruzioni—Parte 4: Evoluzione e sviluppo informativo dei modelli, elaborati e oggetti

The Effect of Real-Time Sensing of a Window on Energy Efficiency, Comfort, Health and User Behavior

Tiziana Poli, Andrea G. Mainini, Alberto Speroni, Juan Diego Blanco Cadena and Nicola Moretti

Abstract Sensing technologies integration in buildings has grown rapidly. Most of them are connected to platforms for monitoring data and notifying anomalies, but few are integrated within building elements (either for data collection or response). These technologies help regulate HVAC, or detect building systems' failures, but few enable passive sustainable strategies or space maintenance. It is possible to enhance building responsive operation by gathering granular data on passive systems' operation and space occupation by setting a building component, that houses a sensor network. That is, if plugged into Building Information Models, it permits adaptation to unusual climate conditions or abrupt space use [Linked with the following correlated research projects:

Research title: 2017, ELISIR. R&D+S&I Smart Living, R.L 26/2015 Widespread creative and technological manufacture 4.0 (ongoing)
Research type: Funded by Regione Lombardia
Responsible: Tiziana Poli (research unit, Politecnico di Milano).].

Keywords Sensors · Interactive buildings · Energy efficiency · Indoor thermal comfort · User centered approach · Building information modelling · Digital twin

---

T. Poli (✉) · A. G. Mainini · A. Speroni · J. D. Blanco Cadena · N. Moretti
Architecture, Built Environment and Construction Engineering—ABC Department, Politecnico di Milano, Milan, Italy
e-mail: tiziana.poli@polimi.it

B. Daniotti et al. (eds.), *Digital Transformation of the Design, Construction and Management Processes of the Built Environment*, Research for Development,
https://doi.org/10.1007/978-3-030-33570-0_26

1 Smart Buildings Operation

Industry has shown that data availability can boost the product performance (Brynjolfsson et al. 2011). In the building sector, this can be noted in the application of BIM methodology for managing design information at first instance, however its application in asset management has shown remarkable improvements in terms of operational savings and efficiency (Re Cecconi et al. 2017). To do so within the building operation, in terms of energy and well-being, granular data acquisition is required for monitoring the trend of all or few parameters defining the indoor conditions and provoking human-building interaction.

Shen et al. (2014) have shown energy performance improvements on building energy performance when monitoring in parallel occupancy, illuminance and air temperature at different locations of the room. Additionally, Konis and Annavaram (2017) have studied the influence of studying the occupant interaction when a comfort disruption occurs, which can reach cooling energy savings ranging from ~27 to 90%. Agha-Hossein et.al. (2013) carried out numerous surveys on buildings as post-occupancy evaluation, trying to identify the main reasons for comfort disruption among building occupants, and their magnitude; highlighting visual, thermal, and acoustic comfort plus air quality as highly relevant.

There has been considerable research in identifying what parameters shall be monitored for rating the building performance, the required sensors for acquiring them (together with their sensibility), and how adequate they are as control criteria; however, few researches have been done for understanding their integration within the indoor environment and/or building elements. Hereby, a study is presented on the possibility of predicting the occupant indoor satisfaction and maintaining a healthy and safe environment, by using granular data collected from sensors installed in a window unit. This innovative window unit is intended to work not only as a sensor, but also as a building system actuator able to adjust the local indoor environment, providing to a traditional window unit new functionalities and services.

1.1 Building Smartness Degree

A building can be as smart as any other device, it is perhaps a matter of sensitivity, connectivity and interactivity. The building shall be able to sense any alteration, predict any favorable response and interact with the user for enhancing the response efficacy. Therefore, it shall be equipped with sensor nodes (i.e. smart building elements), and a control algorithm able to dictate its behavior.

For doing so, it's important to know: (1) which type of data is need for providing a comfortable, healthy and safe environment for the building user; (2) how the needed sensor nodes, or network, shall be installed to gather useful data for the devised control algorithm; (3) how these data should be integrated to work in a holistic and unique control logic; and (4) which type of actions it will be able to carry out, or how would it interact with the building user.

The degree of smartness of the new window unit would be established based on the amount of data gathered, the knowledge produced with data collected, and the extent of the action produced by the installed smart system (i.e. extent of the interactivity and/or connectivity with the rest of the building). This smartness has been planned, designed and embodied within the building element by SEEDlab.ABC, in collaboration with Italserramenti, Schneider Electric and University of Brescia.

1.2 Data Collection and Processing

For the data collection it was necessary to understand the main factors for each aspect (i.e. comfort and well-being), that is: (1) for thermal comfort—air temperature and solar radiation (2) for visual comfort—illuminance (3) for respiratory health—concentration of pollutants (e.g. CO_2). Nevertheless, the control logic of the building will function differently if the occupants are present or not, requiring from the building sensitivity to acknowledge this fact.

An initial proposal of the sensor integration is presented in Fig. 1b for obtaining the required data on a south-oriented office space in Milan, Italy. The sensors have a latency of approximately 1 min. Direct measurements are useful, however for some parameters, the sensor becomes expensive and/or impossible to integrate within a building system given its dimensions, requiring some re-engineering for extracting these values from correlation of simple measurements (i.e. the case of radiation). From extracted granular data, correlations can be established from the use of temperature difference between irradiated and non-irradiated surfaces, in Fig. 1a, a matrix correlation diagram is presented with the Pearson correlation method for understanding their similarities.

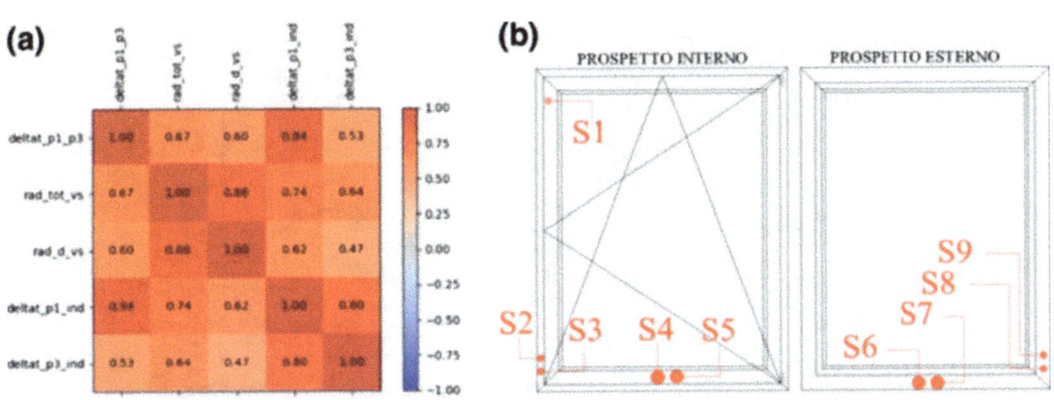

Fig. 1 Results for **a** Pearson correlation test (+1 and −1 corr. strength) between data obtained from temperature probes installed within the window frame; **b** sensor location in window unit

After gathering the required parameters to be measured and sufficient granular data, it was necessary to evaluate the convenience of the sensors and their position, by attempting to give an added value to the collected datum by possible correlated information or avoiding numerous overlapping effects. Thus, different tests were carried out measuring sensor precision and accuracy at different positions and conditions. For instance, analysis on the amount of radiation falling on the window frame, to consider possible alterations due to excess of solar radiation falling on the sensors, the variation of the air temperature values according to the sensor location and the sensibility of these values given the frame material (see Fig. 2).

Not only indirect measurements allow the implementation, and/or complexity reduction, of a sensor node within a building system, the interpretation of the datum is crucial to avoid noise hampering the building performance (Wu and Clements-Croome 2007). For instance, using the results from the Pearson test from Fig. 1a, the data and evidence found from analysis such as the one presented in Fig. 2c, it was possible to neglect the use of a radiometer to determine the amount of radiation falling on the window from external temperature probe readings with reasonable accuracy (see Fig. 3).

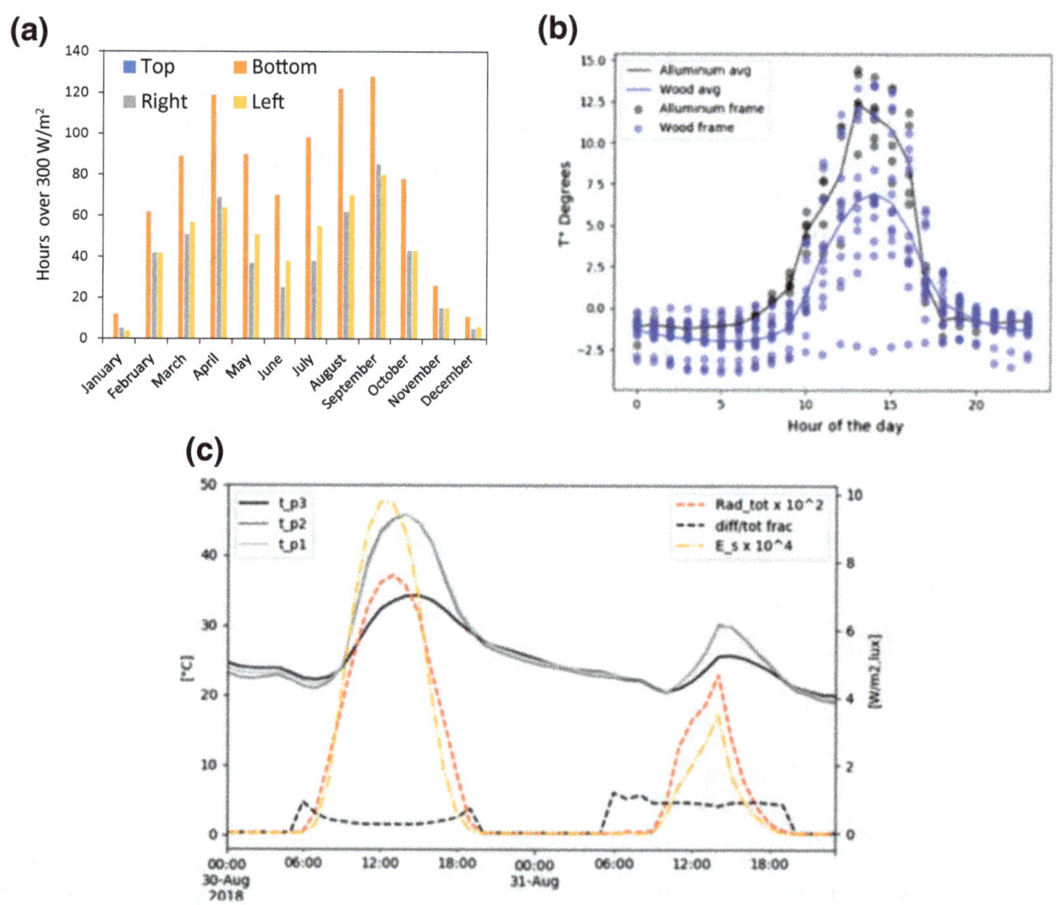

Fig. 2 Data acquisition comparison for **a** frame irradiance intensity; **b** temperature with different window frame; **c** temperature probe at different location, external illuminance and radiation

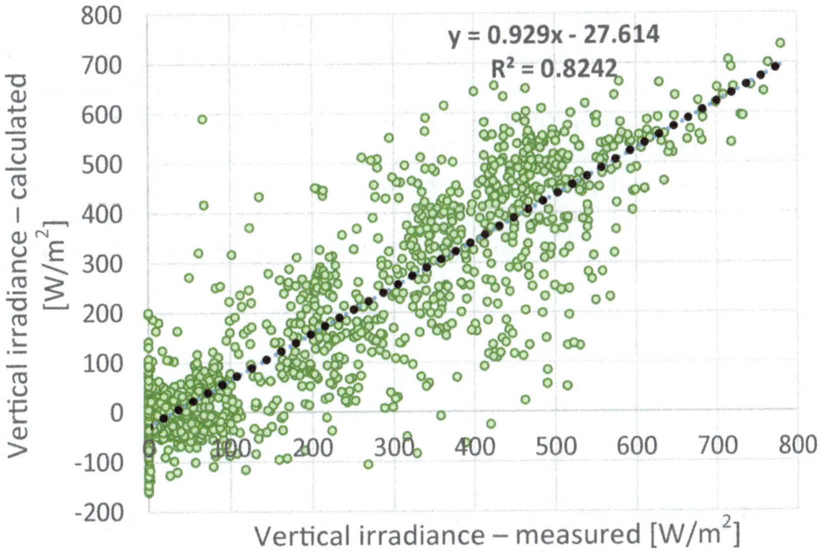

Fig. 3 Comparison between vertical irradiance values on the window obtained from weather files and the ones computed from the correlation found with external temperature measured at the window frame

The use of the minimum quantity of data types for obtaining all required parameters, increases the smartness of the device, and maximises the productivity of the system. That means, faster or immediate connectivity through IoT devices for generating building element adaptions.

2 Conclusions

Sensor integration allows real-time building performance monitoring, enabling immediate intervention when any disturbance occurs. Nevertheless, different data interpretation can be performed by the facility manager or the designer who configures the control algorithm. This will affect the building adaption to climate and the occupant interaction, but the control algorithm could try to benefit from this feedback and adjust the initial control settings.

The integration of sensors within building elements eases the proper data acquisition, data interpretation and, if wanted, reduces the need of an external online platform for managing and storing data, because all the system controller could be integrated within the building element.

Further studies are foreseen to reduce the need for other sensor, such as the passive infrared (PIR) motion sensor, by extracting the room occupancy from air pollution readings as has been attempted by Meyn et al. (2009).

Acknowledgements This work is supported by Regione Lombardia, with the research grant funding the project ELISIR Energy, Life Styled and Seismic Innovation For Regenerated Buildings. Business Models e innovazioni puntuali per la rigenerazione urbana nell'ambito della digitalizzazione dell'ambiente costruito, R&D +S&I Smart Living, L.R. 26/2015 Manifattura diffusa creativa e tecnologica 4.0 (P.I. Prof. A.L.C. Ciribini and Prof. G. Plizzari).

Is acknowledged also the support coming from Italserramenti by providing the window units for experimentation and Schneider Electric for supplying the sensors. The authors thank the research group of DICATAM from University of Brescia for valuable suggestions and for managing the data collected.

References

Agha-Hossein MM, El-Jouzi S, Elmualim AA., Ellis J, Williams M (2013) Post-occupancy studies of an office environment: energy performance and occupants' satisfaction. Build Environ 69:121–130. Elsevier Ltd

Brynjolfsson E, Hitt LM, Kim HH (2011) Strength in numbers: how does data-driven decisionmaking affect firm performance? SSRN

Konis K, Annavaram M (2017) The occupant mobile gateway: a participatory sensing and machine-learning approach for occupant-aware energy management. Build Environ 118:1–13. Elsevier Ltd

Meyn S, Surana A, Lin Y, Oggianu SM, Narayanan S, Frewen TA (2009) A sensor-utility-network method for estimation of occupancy in buildings. In: Proceedings of the 48h IEEE conference on decision and control (CDC) held jointly with 2009 28th Chinese control conference, pp 1494–1500

Re Cecconi F, Maltese S, Dejaco MC (2017) Leveraging BIM for digital built environment asset management. Innovative Infrastructure Solutions

Shen E, Hu J, Patel M (2014) Energy and visual comfort analysis of lighting and daylight control strategies. Build Environ 78:155–170. Elsevier Ltd

Wu S, Clements-Croome D (2007) Understanding the indoor environment through mining sensory data-A case study. Energy Build 39(11):1183–1191

Digital 3D Control Room for Healthcare

Liala Baiardi, Andrea Ciaramella and Ingrid Paoletti

Abstract The building process is in an evolutionary phase dictated by the constructive innovations and the digital revolution that has involved the tools and the technical and design contents of the entire life cycle of buildings. In this context, the operators of the sector need to develop organizational models capable of protecting and managing the conceptual and scale transition, between the conceptual framework of the architecture and the subsequent ones of construction, use and management. In the development of complex projects the elaboration of the organizational model is in fact fundamental to reach adequate figurative, performance and qualitative levels by incorporating the necessary contents of environmental, economic and management sustainability of buildings. This text illustrates the development of a horizontal organizational model for the smart and dynamic control of complex buildings through the creation of an innovative digital Web-Based platform capable of integrating Building Information Modeling (BIM) technology with a "Facility Management platform". The project involves experimentation applied to a real case involving the restructuring of a complex building.

Keywords Process approach · Building automation · BIM · Facility management · Interoperability

1 The Role of Digital Technologies in the Construction Process

Digital technologies have fundamentally transformed our interactions with the built environment. Mobile devices provide a means to rapidly access and share information. Within the facilities context, these technologies impact both capital project delivery and day-to-day operations (McArthur and Bortoluzzi 2018).

L. Baiardi (✉) · A. Ciaramella · I. Paoletti
Architecture, Built Environment and Construction Engineering—ABC Department, Politecnico di Milano, Milan, Italy
e-mail: liala.baiardi@polimi.it

B. Daniotti et al. (eds.), *Digital Transformation of the Design, Construction and Management Processes of the Built Environment*, Research for Development,
https://doi.org/10.1007/978-3-030-33570-0_27

In the management of advanced construction processes, the BIM (Building Information Modeling) can play an active role throughout the entire life cycle, giving the possibility of having a single integrated system able to collect and manage information concerning the building or infrastructural object.

One of the topics that the research and innovations focus on in the field of BIM (Building Information Modelling) systems is the creation of tools capable of moving the management and control of design projects onto interoperable platforms. Based on this principle is the ability to manage, by means of the exchange and sharing of complex data and multidisciplinary knowledge, the activities and roles of the various operations involved in the programming, design and management process of these interventions. In particular, within the context of new construction or retrofitting interventions in complex buildings (as is the case for the large healthcare districts), the possibility of developing design models capable of simulating alternative conditions and gauging the effects, guaranteeing a shared feedback, is an essential condition for optimizing management activities during the entire lifecycle of buildings (Di Giulio et al. 2017).

Now the application of methodologies, protocols and BIM tools in the design stage (and even more so in the subsequent phases of the building process) is to be considered no longer an innovation but rather a key element of the whole process (Simeone 2018).

BIM concerns issues of information sharing, interoperability, and efficient collaboration throughout the life cycle of a building, from feasibility stages to the demolition and recycling stages (Isikdag 2015) and now it is seen as a solution for sharing data among multiple systems (Utica 2010). At the same time, the 3D digital approach provides support to the synergic management activity through the high accessibility of a 3D environment.

Recent research efforts have focused on investigating building information modeling (BIM) implementation and its potential expected benefits in project management operations and maintenance (O&M) (Eastman et al. 2011), including facility management, maintenance, and energy management (Becerik-Gerber et al. 2012; Eastman et al. 2011; Motawa and Almarshad 2013; Teicholz 2013).

Based on these premises, the text, starting from the principles underlying the development, experimentation and innovation project, selected and financed by the Lombardy Region within the Smart Living program, describes the project "digital 3D control Room for Healthcare".

The principles of the research project are in line with the H2020 community strategy, aimed at intelligent, sustainable and inclusive synergies with the research activities carried out in this area by Universities.

1.1 The Digital 3D Project

BIM plays a fundamental role in improving current design and building management practices, contributing to the sustainable implementation of performance levels in terms of creation and management of the built environment.

BIM systems make it possible to draw great benefits in the wide-ranging actions of programmable interventions on public and private building heritage.

The potential of the system is the ability to elaborate complex actions even during the planning of the interventions, contributing to the selection of the most suitable operational paths. Moreover it allows for the model's implementation during the management phases thus the library of the elements can also grow exponentially.

Thanks to BIM coordination it is possible to implement systematic actions aimed at identifying any project problems (with their respective coordinates and associated data) following their corrections until they are resolved and the main deliverables are reached:

- design models related to individual disciplines (e.g. structural, systems, etc.) coordinated with a predetermined level of detail;
- space and circulation programs;
- cost plan and financial information;
- collection and verification of concessions and authorizations;
- assessment of possible procurement strategies and supply chain management issues;
- acceptance and approval of the project by the client.

During the life cycle of a project, the BIM model is progressively enriched with data and information, it is possible to update information relating to design, engineering, construction and economic aspects along with costs.

Fundamental importance is also given to the integration with the sensor network and the building management system (RICS 2014).

The Digital 3D project was developed with the aim of outlining an operational model for intelligent and dynamic building control by relating the executive project to process innovation relating to the Facility Management (FM) operating mode. The model is based on the interoperability of data, in a coordinated form, starting from the design stage and involving the entire process, on several levels.

The innovative aspects are augmented by the building management and personal services model (property and facility and Building automation) which obliges us to rethink the entire building process from design, to construction, to management with a systemic approach.

The interoperability of the management and control tools of the design process is one of the themes on which research and innovations in the field of BIM modeling systems (Building Information Modeling) are focused.

The experimentation and validation of the new model was carried out by applying it to a real case of restructuring of a complex building, a hospital building, and involved the creation of an innovative digital Web-Based platform that integrates BIM technology with an FM platform.

2 The Elaboration of the Digital 3D Model

The development of the Digital 3D model takes its cue from the international principles inherent to the methods of maturation and use of BIM in the design, construction and management phases of built environments. Among the main ones there is the BIM Policy (BIM report 2012) of the British government which introduced the concept of GSL (Government Soft Landing) in order to favor a closer alignment of the design and construction phases with that of management and asset management. GSL guarantees the centrality of the property from the planning and construction phases up to its delivery and management.

The ramp representation of the BIM Maturity Diagram (Richards 2010) shows a systematic transition of BIM maturity levels (Fig. 1).

At "level 0", the project is fundamentally based on paper which of course provides two-dimensional (2D) information.

"level 1" marks a transition from a paper environment to a 2D and 3D environment.

At "level 2" sometimes called "pBIM", we move to a common method of producing, exchanging, publishing and storing information. At the same time, the inclusion in the smart model and additional metadata begins. Being proprietary models focused on individual disciplines. Model integration is based on a Common Data Environment (CDE).

At "level 3" a fully integrated "iBIM" is reached, marked by the use of a single model accessible by all team members. This level of BIM uses 4D, 5D and 6D (management of the life cycle of the building, Facility Management). The 4D environment includes the planning and management of construction phase times,

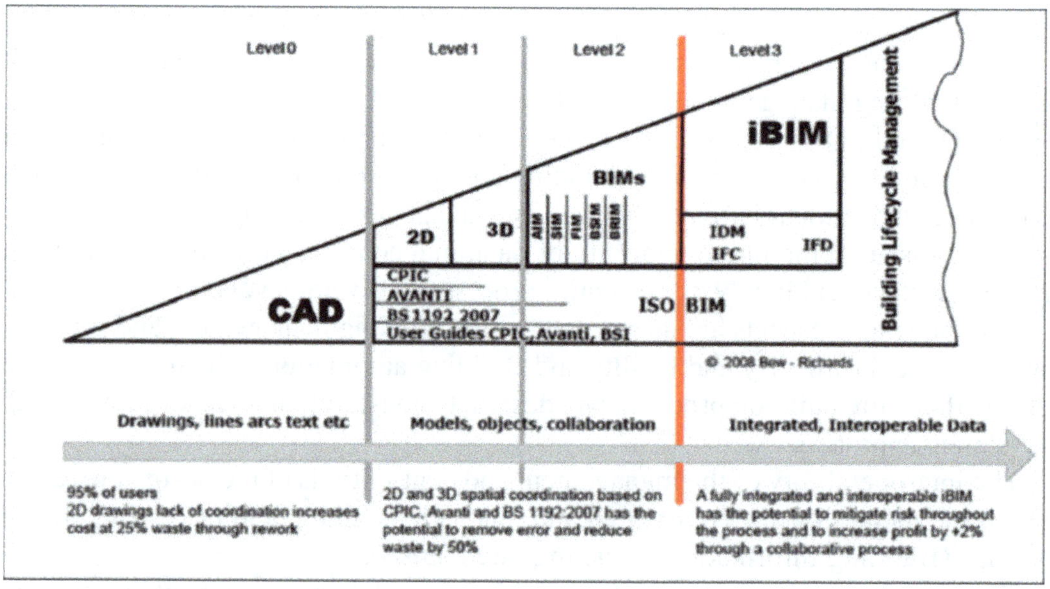

Fig. 1 The BIM maturity model by Bew and Richards (Richards 2010)

including logistics and site operations. The 5D adds cost management and the possibility of performing economic forecasting analysis and contains technical/economic information such as quantity, unit costs and total materials. The 6D environment extends the model to Facility Management.

If updates are correctly carried out based on developments in the work phase, the "as built" model is obtained upon completion of the work. This contains all the management and maintenance specifications, useful for future maintenance. It will also be possible to monitor the costs of a building's life cycle and optimize its efficiency and evaluate the costs/benefits of any proposed updates.

An example of application of BIM systems to the management of public real estate assets is the model developed in Finland by Senate Properties, a government body run by private individuals under the patronage of the Finnish Ministry of Finance and responsible for the management of real estate of the government and of the rent of the constituent premises of a patrimony of approximately 10 thousand buildings, for a total of approximately 7 million m^2.

Since 2001, Senate Properties has developed a fair number of pilot projects for the adoption of BIM for property management. For construction work on their buildings and for new constructions (for operations of recovery or restoration), the company has decided to request models in IFC[1] format (Industry Foundation Classes). The implementation of the projects through parametric systems has improved the level of compliance of the overall costs, and the feasibility of the projects to their set objectives.

3 Interoperability System of the Model

Of fundamental importance, for the success of the project, is the interoperability of BIM with the FM management system and with the network of sensors and the building given by the software and hardware capacity, on multiple IT platforms (of different origin) in order to exchange information in a useful and reliable way. The efficiency of BIM therefore lies in its interoperability, that is, the ability to allow for the management and communication of electronic data of different types (graphic data, text data and related data) between the various participants in the design, engineering and construction activities, maintenance and related business processes.

To promote interoperability we chose to use an open and publicly managed scheme (dictionary) with a standard language. A common example of a scheme is that conceived by BSI (Building SMART International) and by COBie (Constructions Operations Building Information Exchange) in order to allow for the representation and the public and open exchange of data, in the built environment sector.

[1] Industry Foundation Classes (IFC) for data sharing in the construction and facility management industries is the international openBIM standard, registered by the International Standardization Organization (ISO).

COBie is a standardized open-ended approach aimed at facilitating the interoperability of essential information in the BIM process (Fig. 2).

The approach focuses on entering data when created during the design, construction and commissioning phases of the structure. The acquired data is recorded in neutral format and can be exchanged between the various players in IFC format (Industry Foundation Classes).

The basic structure of Digital 3D solves the limitation of the COBie standard given by the object-oriented work environment (Object Oriented Modeling). In this perspective the definition of the minimum and maximum units of components may not coincide with the structuring of the model in the maintenance phase, which, in the case of complex structures, may require groups of objects.

Through the "Open Control Room" interface, the following is available:

- mapping of maintenance events over time, scheduled or one-time ex-post;
- possibility to attach documents;
- reading of the direct properties of the object;
- mirroring of the main data set on local parameters, to allow the generation of display filters;
- construction and writing of values in parameters shared between the two BIM-FM platforms;
- construction of "maintenance routes";

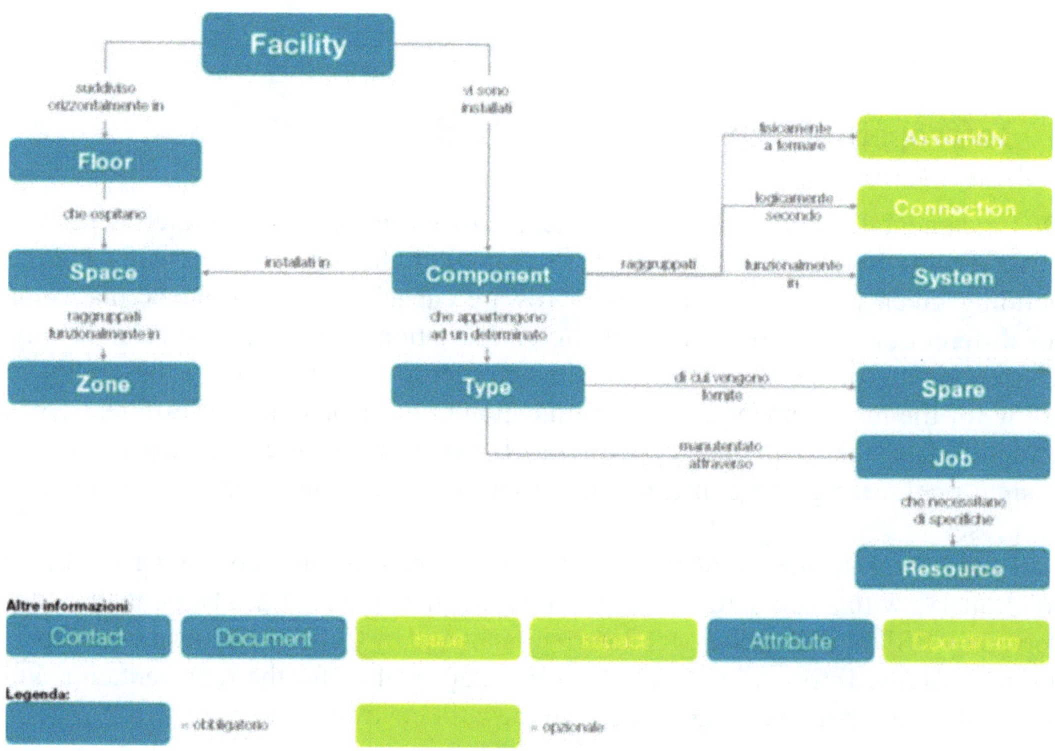

Fig. 2 Example of data structure of the COBie standard. Authors: BIS-LAB, Contec Verona [Further information available on http://www.gruppocontec.it/ricerca−per−la−progettazione−bis−lab/]

- mapping of inspections as memorization of sequences of rooms over time, directly in the timeline.

The results of the BIM + FM integration are represented by the "Open Control Room", an evolution of what in today's evolved construction sites is the "BIM Room". The "Open Control Room" represents a "model reference" where access to the Big Data model is mediated by "Agents".

Agents are professionals capable of activating the relationships between data and providing continuous feedback to internal and external operators.

The platform used by the Agents will record, in a rapid, structured and interactive manner the technical/energetic, cadastral, accounting, tax, maintenance and guarantee data, with the following specifics:

- cloud integration of a "networked" 3D shared environment and the concept of an "Internet of Computer" (IoC), which aims to make each building "a sender of itself", networked with others;
- web consultation (via browser without specific plug-ins/software);
- navigable and three-dimensional view of the building: this will allow us to isolate information on consumption, costs, maintenance histories, scheduled maintenance, potential for efficiency, etc. with 3D Data Visualization displays;
- monitoring of people's accesses and behaviors through geolocation;
- semantic query system;
- data return by type of user.

The integration of the BIM and CAFM software has involved the elaboration of a single software architecture able to return the combined and navigable data on the building, with a different level of detail and interaction.

An analysis was then carried out on the two sides (BIM and FM) to perform a "match" between the classification of executive project objects and the classification of "maintenance elements", referring to the long history of standardization and study of the sector, and to the COBie model. The control room can be connected to a large amount of data, improving the decision-making power of the actors that make up the project and management team. The project can be facilitated by real time data sources, marketing data, data coming from the sensors, data on the cloud, stratifications on safety and user preferences.

The key object of the developed interface is to form a "timeline", with a customizable shape and extension, which allows it to "map" over time and spatially a series of events that can be external or internal to the system and that synthetically produce the planned or dynamically mapped life. Moving in the timeline means moving in the building that becomes the spatial reference of events. The chosen graphic wants to highlight the concept of dynamic "database building", which obviously also allows for the use of information of various kinds, for users more directly oriented towards a textual display.

To the functions of integration between databases, the system therefore adds those of direct online interaction, today largely excluded by viewing interfaces, to which the Digital 3D Control Room project adds the possibility of dynamic generation

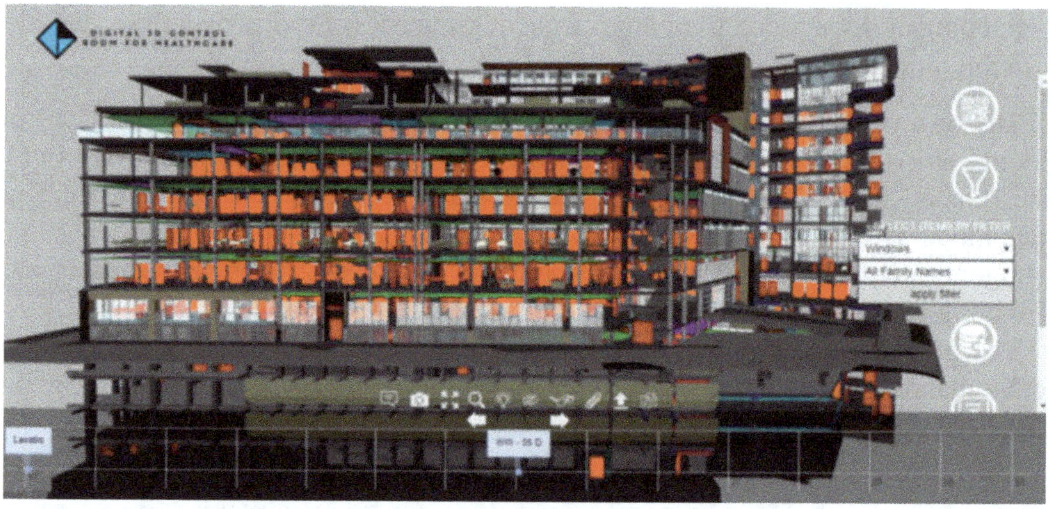

Fig. 3 Example of processing the case study on the Open Control Room online model. Authorship

of maintenance elements, allowing for multiple assignments and the creation of maintenance elements starting from the model (Fig. 3).

Functions of management and a simulation of inspections are then added, made viewable in the timeline both as single events and in sequence. In this sense, a work of mapping the rooms in the floors allows them to be used as a key for physical positioning in 3D vision by framing the identified maintenance object. By clicking on the interested part, the model "moves" to go there showing, in addition to the local data, the detailed photos, memorized according to its timeline, of the reference room/environment.

The system aims to consider two more aspects of the maintenance activity: the synthetic and synergic visualization of the data, and the physical inspection, in sequence, of "visual mapping" (Fig. 4). In this sense, the referencing of photos over time is associated with the maintenance tour for rooms, which allows reading by time key (the tour) and by spatial key (all the photos of this room over time, etc.).

From the Control Room the Operators have access to a customized Web Application that governs the structures, monitors them via webcam, commands or controls the automations, launches the applications and supports the customers.

The web application stores data in a database. Via PC or tablet users can interact with the control room with a dashboard for the control of electronic functions.

The Internet of Things (IoT) makes it possible to measure objects (information provision) but also detect and (or) remotely control them through the existing network infrastructure, creating opportunities to regulate or activate or deactivate systems remotely.

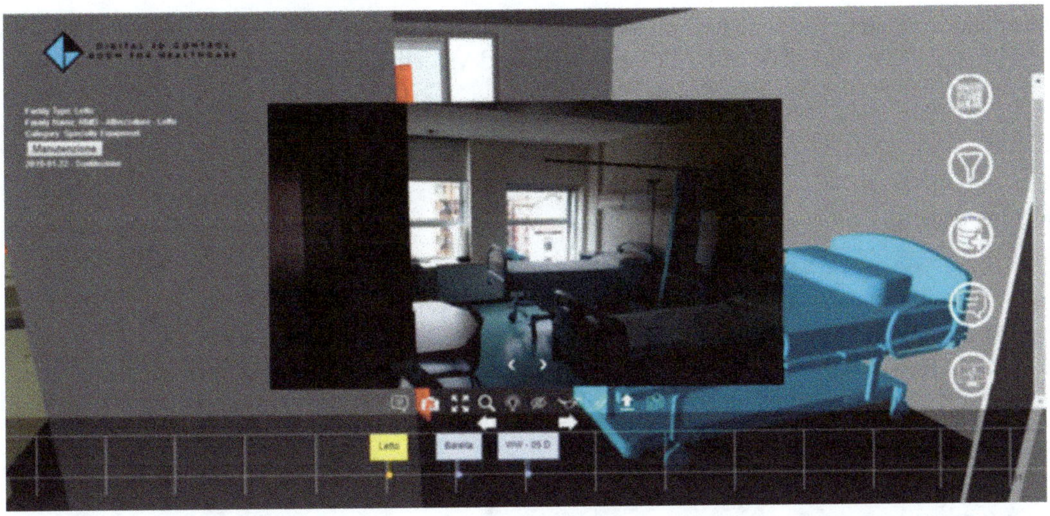

Fig. 4 Example of direct online integration on the Open Control Room model of the "maintenance tour". Authorship

4 Final Considerations

The innovation of the management model (facility and property) involves the rethinking of the entire building process, from design, to construction, to maintenance. The approach of the initiative is systemic and sees all the elements participating in a systemic and interdependent way.

The combined synergy between intervention on an existing building with attention to the principles of circular economy (considering for example carbon dioxide emissions, waste, etc.) promotes and guarantees the implementation of the principles of inclusion, safety, well-being, health, eco-sustainability, smart supply chain, recovery and re-use of buildings.

With the emergence of technologies associated with the Smart Built Environment (SBE) concept, in which intelligent objects are connected and interact with integrated installations, it is necessary to study the role that BIM can play in this context, helping to increase the efficiency, safety, and user comfort of future intelligent buildings.

A further integration with mobile applications can orient the building process towards cloud based management where the information of a project is made available at any time, addressing specific applications for Field Management.

This way it is possible to improve the efficiency of the process through suitable IT tools that allow operators to generate a circular flow that supports the control and management reducing repetitions, redundancies and manual transcriptions by the professionals involved.

Acknowledgements The development, experimentation and innovation project entitled "Digital 3D" were selected and financed by the Lombardy Region under the Smart Living program. Project partners: Politecnico di Milano, ABC Department, Rimond S.r.l., Tecnosys Italia S.r.l. Politecnico di Milano work group: Andrea Ciaramella, Ingrid Paoletti, Liala Baiardi, Giulio Barazzetta, Stefano Bellintani, Domenico Chizzoniti, Marzia Morena and Valentina Puglisi.

References

Becerik-Gerber B, Jazizadeh F, Li N, Calis G (2012) Application areas and data requirements for BIM-enabled facilities management. J Constr Eng Manag 138(3):431–442

Di Giulio R, Turillazzi B, Marzi L, Pitzianti S (2017) Integrated BIM-GIS based design for high energy efficiency hospital buildings. Techne (13):243–255. https://doi.org/10.13128/techne-19728

Eastman C, Teicholz P, Sacks R, Liston K (2011) BIM handbook: a guide to building information modeling for owners, managers, designers, engineers and contractors. Wiley, Hoboken, NJ

Isikdag U (2015) BIM and IoT: a synopsis from GIS perspective. In: Joint international geoinformation conference, vol. 10, Kuala Lumpur. International archives of the photogrammetry, Kuala Lumpur, pp 33–38

McArthur JJ, Bortoluzzi B (2018) Lean-Agile FM-BIM: a demonstrated approach. Facilities 36(13/14):676–695

Motawa I, Almarshad A (2013) A knowledge-based BIM system for building maintenance. Autom Constr 29:173–182

NBS (2012) National BIM report 2012, NBS, Newcastle. www.bimtaskgroup.org/wp-content/uploads/2012/03/NBS-NationalBIMReport12.pdf. Accessed 21 February 2019

Richards M (2010) Building information management: a standard framework and guide to BS 1192. BSI Standards, London, UK

RICS, (2014) International BIM implementation guide, Produced by the International BIM Working Group of the Royal Institution of Chartered Surveyors. London, UK

Simeone D (2018) BIM and behavioural simulation for existing buildings re-use design. Tema 4(2). https://doi.org/10.17410/tema.v4i2.188

Teicholz P (2013) BIM for facility managers. Wiley, Hoboken, NJ

Utica G (2010) Tecniche avanzate di gestione dei progetti. McGraw-Hill, Città di Castello, IT

Guidelines to Integrate BIM for Asset and Facility Management of a Public University

Giuseppe Martino Di Giuda, Paolo Ettore Giana, Marco Schievano and Francesco Paleari

Abstract This research work aims at providing a methodological approach tested on a real case study, where a public university applied guidelines to structure BIM and IMM procedures. The research allows the public client to procure works and services to manage the portfolio. The approach is initially tested on the ABC department to be gradually extended to the entire Politecnico di Milano.

Keywords Portfolio information management · Public administration client · Data-driven process · Asset information model (AIM) · Organizational information requirements (OIR)

1 Introduction

The effective implementation of a new methodology such as BIM has repercussions on the entire value chain of an industrial sector: for this reason, it requires a planned and structured intervention by the legislator to encourage its adoption. European Union BIM Task Group aims at aligning the national programmes in order to adapt national programmes and spread information modelling among European community, boosting public administration's productivity.

The EUBIMTG (2018) supports the need for collaboration in the development of national programmes by sharing the results achieved by individual nations. Thus, the most effective and efficient measures can be repeated in countries that share a uniform market, speeding up the implementation process. According to the literature and to main consulting company (McKinsey Global Institute 2017; The Ellen MacArthur Foundation 2012; Barbosa et al. 2017), the public administrations have an important role in the implementation of BIM methodology in playing in the digital transition process, defining guidelines to regulate the BIM methodology and address the sector. In a process where the path is not well defined, international standards have to set

G. M. Di Giuda (✉) · P. E. Giana · M. Schievano · F. Paleari
Architecture, Built Environment and Construction Engineering—ABC Department, Politecnico di Milano, Milan, Italy
e-mail: giuseppe.digiuda@polimi.it

B. Daniotti et al. (eds.), *Digital Transformation of the Design, Construction and Management Processes of the Built Environment*, Research for Development,
https://doi.org/10.1007/978-3-030-33570-0_28

principles to follow rather than impose strategies. It is up to the single Appointing Parties to define the strategies that drive procurement processes. The EUBIMTG (2018), through the strategic framework for BIM implementation process, recognizes the task of the public client to decrease the frequency and mitigate the impact of factors that act negatively against the success of the BIM implementation process. In the strategic development and guidance framework for the AECO sector, it identifies the following four areas of intervention: (i) definition of the public strategy to increase support for the BIM methodology; (ii) communication of the vision and promotion of communities to increase interest and participation; (iii) development of a collaborative framework to improve the sharing and transparency of information; and (iv) growth of the expertise of clients and the sector to improve the process in technical aspects.

The EUBIMTG identifies the need to reform the regulatory framework of building processes to support the collaboration principle of the BIM methodology by encouraging the information exchange among the parties involved. The public client has the function of identifying critical issues in the data sharing and the process transparency to support the legislative bodies in the drafting and publication of appropriate laws or transpose EU directives to reduce problems. The public client is entrusted with the task of increasing the BIM skills and competencies of the sector through pilot projects and direct adoption of the BIM methodology in public contracts.

2 Information Management According to International Standard

Information management (i.e. the integration, distribution and coordination of information) plays a key role in modern production processes. The literature (McKinsey Global Institute 2017; The Ellen MacArthur Foundation 2012; Barbosa et al. 2017) shows that it is possible to improve the final project result through the development of the information management skills inside the organization and of the information outside the company. In fact, the information management improvement results in greater management effectiveness, efficiency, quality and flexibility (Liu et al. 2015).

The British Standards Institution (BSI 2018) recognizes the added value of the BIM methodology, summarized in the ability to reduce the project time and cost and the work and the management costs recreation during the entire asset life cycle, through a standardized approach to information management.

Effective information management within the client's processes favours their optimisation and operational performance during the asset use phase. Structuring the information production of a client allows producing coherent information between the different projects and operators involved. The information production outside the organization assures better management of internal processes, activating greater collaboration and transparency that is reflected in the overall projects improvement, both internal and external. Transparency and collaboration during the activities performance allow organizations involved to assess the risks arising from their activities and the activities of employees. Namely, this process reduces the risk level in the production of information. It is, therefore, necessary that the contracting authority structures a BIM process in order to improve the information produced by the different economic operators. On these grounds, the new EN ISO 19650 part 1 and 2 are the new international guidelines to define the process of organizing and digitizing information through the BIM methodology for the management of civil and construction engineering works (Fig. 1). The objective of the concepts and principles defined is to reduce risks costs associated with the asset management of buildings, complexes and infrastructure. The functions of these organizations, in the context of a public procedure, are covered by the contracting authorities acting on behalf of the public administrations they represent. The BIM methodology requires even before starting a project process to make explicit the information requests related to the final objective, i.e. the information to be obtained in order to achieve this objective.

In this context, the client plays a fundamental role as the person in charge of defining the objectives of the project and the requirements to be met. For this reason, the drafting of strategic objectives whose requirements are expressed in the OIR. It influences both maintenance decisions whose requirements are expressed in the AIR and all project decisions whose requirements are expressed in the EIR. The expression of needs at different levels allows, to standardize the approach that an organization has to Asset Management strategies.

Once the contracting station has selected the different Lead Appointed Parties, the project process begins. As mentioned above, this process must take place within an

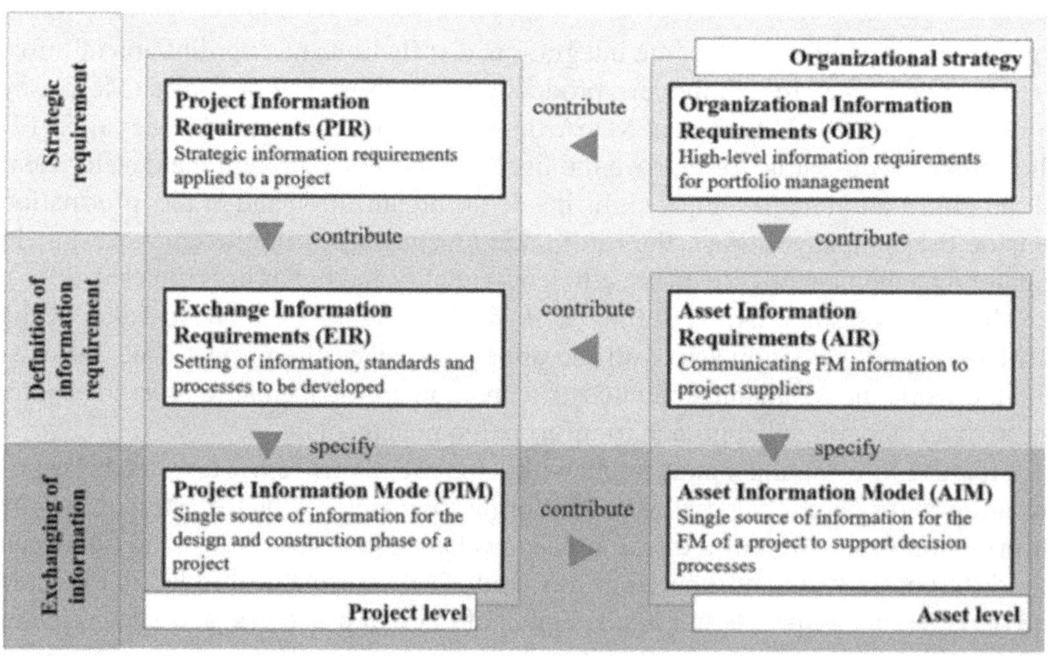

Fig. 1 Information workflow according to ISO19650-1

information-sharing platform Common Data Environment. The BIM methodology, reported in the ISO 19650 series, requires the contracting authority to be the first party to produce information. The contracting authority is, in fact, responsible for objectives defining and information requirements at the starting point of a project. The information requirements determine the types, times and methods of information that must be produced, necessary for the contracting authority to support the decisions set in each milestone. There is, therefore, a significant change whereby the design phases are no longer only conceived as levels of definition to be achieved, but as key moments when certain decisions must be taken. These moments are defined as project milestones.

As it is necessary to clarify from the very beginning the objective of applying the BIM in a building intervention the definition of an application strategy to achieve this aim is crucial. The definition of the BIM use required for the project under analysis represents the starting point for the definition of the phases and documents to be produced later. The definition of the project intended use is based on the concept of 'starting with the end in mind': a profound paradigm shift linked to the transition to BIM.

3 State of the Art of International Guidelines for Information Modelling

Considering the advantages provided by a BIM approach, different Countries have adopted different strategies in implementing the methodology. The most advanced clients have a structured programme for the BIM implementation, and are equipped with standards for data management and information modelling. The creation of proprietary guidelines is to be considered positive as it allows clients to develop project information requirements in a structured way, according to processes driven by a specific organizational requirement (Di Giuda et al. 2017).

The most advanced Countries have engaged in the definition of proprietary guidelines for specific clients (e.g. Finland (European Construction Sector Observatory 2016), Norway (Statsbygg 2011), Singapore (BCA 2017) and USA (US Dept of Veterans Affairs 2017; Bloomberg et al. 2012; Los Angeles World Airports 2017; USC Capital Construction Development and Facilities Management Services 2012; Massachusetts Institute of Technology 2016)).

Namely since 2007, Finland has imposed a progressive obligation on BIM for public works, and in 2014 one of the first proprietary guideline for the construction and management of bridges was published. This guideline, published by the Finnish Transport Agency, incorporates what is presented in the CoBIM standard (European Construction Sector Observatory 2016) and specifies the process and the BIM requirements for the government body activity in charge of the Finnish infrastructure management.

On the other side, the General Service Administration (GSA) published in the early century a guideline for the management activities of public offices in the USA. The guidelines have been partially updated and conformed to the standard published by the National Institute of Building Science (NIBS) (BuildingSMART alliance 2015).

Later on, also the government body Building and Construction Authority (BCA) (2017) of Singapore publishes the guideline for the design, construction and management phases with the related objectives of use. This document gives an overall view of the issues to be developed in the BIM implementation, defining in general terms the results to be achieved.

The Statsbygg in Norway draws up the first guideline for the implementation of the BIM methodology in projects managed by the institution. This guide aims at defining the information requirements, generic and specific disciplines, to facilitate the control and management of the process by the customer.

Unlike the Countries mentioned above, others have undertaken to define national regulations which, however, do not present the operating instructions for achieving the benefits deriving from the methodology (Hooper 2015). In this sense, the internal management of an organization, defined by standardized approaches, does not allow to obtain the most from the BIM approach, in terms of efficiency and attitudes of professionals. The definition of proprietary guidelines allows to standardize and guide processes based on the needs of a specific client.

4 Research Methodology

This research describes the process of digital transition of a public university in managing its own asset through a BIM approach (Fig. 2). The methodology started with the definition of the requirements of the Appointing Party, ABC Department of Politecnico di Milano, and develops with the definition of client proprietary BIM guidelines for asset management. This approach guides the specific client in the transition from a traditional approach to a BIM methodology. It activates a computational transition process to improve the design, construction, maintenance, and management quality: using BIM models and connected databases, accessing and obtaining data from any building and creating a dynamic archive of documents for each settlement are transition key steps. These models have been connected to external databases to better manage historical data about buildings and their components.

The first initial effort is devoted to the analysis of procedures and data workflows to identify main requirements, objectives, and useful information of the organization, aiming at improving data exchange and storage.

The next step has been the analysis of existing assets, based on drawings and documents provided. A Working Breakdown Structure (WBS) has been set up to identify, for each building system, all objects and components connected, with a hierarchical structure. From the database and the drawings supplied, the spaces and buildings codification has been taken. The soft-landing approach provides a smooth transition linking new procedures with the old ones, keeping the same codification

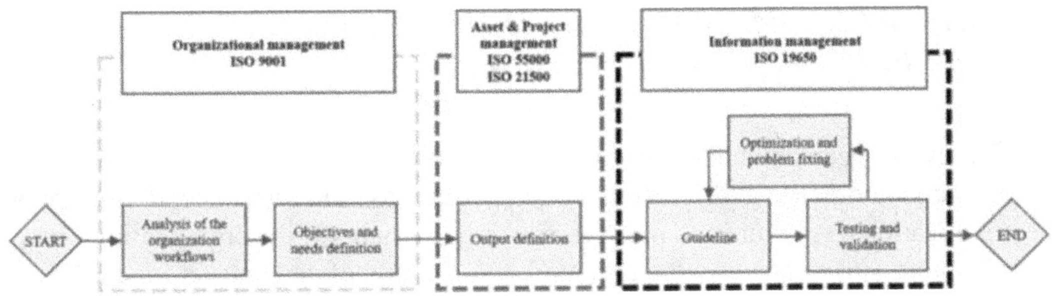

Fig. 2 Methodology approach

rule has also been required in the past, even though setting a complete dataset to every object.

A BIM Library has been created for architectural, structural and MEP system. The dataset were mainly divided in categories: (i) codification; (ii) generic information (description, model, pictures, etc.); (iii) specifications (current, voltage, frequency, etc.); and (iv) maintenance. The attributes come from COBie (British Standards Institution 2014) dataset, organization's and maintenance requirements and consultant experiences gained in past projects.

The guideline has been written for the following major existing guidelines and the ISO 12911:2012, in order to have a guideline tailored to client's needs, but is also valid at international level. The guideline was tested on a real case study, as described in the following paragraph.

5 Case Study

The BIM guidelines have been applied to the ABC Department (Architecture, Built environment and Construction engineering) building, representative of the Politecnico di Milano portfolio. The three-storey building is divided into two main wings with a connection, for a total of 4300 m^2 of gross floor area. The case study is a pilot project for the guideline validation. After the validation phase, the main goal is to gradually extend the guideline to the entire Politecnico di Milano building asset.

The guideline flexibility allows the building to be modelled without comprehensive data from all three disciplines: architecture, structure and systems (MEP). Since the structural designs were not available, only visually detectable structures were modelled in the architectural BIM model, without changes in the guideline structure and in the modelling workflow. This information can be added during further investigations in the structural component.

5.1 Geometric and Non-geometric Data Survey for the Creation of a BIM Library

The survey has been conducted following rules set in the guidelines. The objectives, methods and geometric and informative data collected structure depend directly on the organizational structure and management requirements of the organization's asset management. Survey sheets were prepared with the predefined dataset for each asset elements to be recorded.

The building has been surveyed in order to update and correct the as-built and design drawings provided: a detailed survey has been performed to gather all the finishing of each space. Information related to floor, wall, and ceiling finishes is a key data for the asset management. A detailed survey has been performed to validate number and type of security, fire alarm systems, light bulb and every maintainable asset. The equipment maintenance and use of information were collected at the same time as the geometric survey.

After the survey and identification of architectural, structural and MEP disciplines elements, the BIM library was built and organized. The information requirements for management and maintenance were linked to the modelled objects. Attributes are inserted in objects thanks to shared parameters file, to allow the implementation and application in other new objects. Attributes are standardized and organized into following categories: (i) project information; (ii) codification; (iii) general data; (iv) fire prevention equipment; (v) electrical equipment; (vi) mechanical equipment; (vii) spaces; and (viii) maintenance.

5.2 Master Model and DataBase

The master model's main purpose is the geometric information coordination and visualization, while the data management is carried out in the individual discipline BIM models. Discipline models are linked to the master model through a system of shared coordinates; this guarantees the correct spatial alignment of the models.

The links between the discipline models affect the database structure which is identical to the model's structure, so there is a Database (DB) of the architectural, structural and MEP data connected in a master DB.

The BIM model-DB connection is bi-univocal to link each object instance to historical data. For instance, there is the possibility to easily: (i) attach photos taken during the survey to the instances of the objects; (ii) attach data sheets or other documents to object types; (iii) attach documents to the entire building or settlement, so as to have a digital archive; (iv) manage data on scheduled and performed maintenance; and (vi) manage data about objects condition, installation and end life date. This connection also allows us to dynamically update data, filling them in the database without a BIM authoring tool, and graphically checking them in the model, thanks to thematic plans and dedicated filters in the BIM model (Fig. 3).

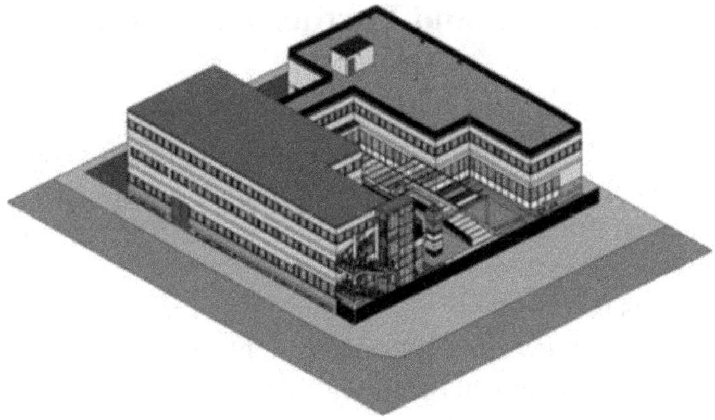

Fig. 3 Master BIM model

5.3 *Model Output*

The BIM models output, such as thematic plans, schedules, views (sections, plans, prospects, etc.), has been set up to meet client objectives. The databases also have outputs and thanks to the software, they used can be customized according to users' needs: queries, tables and data elaboration can be extracted from the database.

An interesting output is the finishing materials bill, that can be used as the starting point for conducting a tender for cleaning services. All the output have been defined in the guidelines, where also the operational procedures to produce new ones are also described.

Existing buildings frequently need to be modified, so the guideline has been set up to regulate the information exchange in case of internal layout modification.

The office 012 layout changes have been modelled to test the procedures, workflows and tools developed. Geometrical and non-geometrical data were consistent and updated at all stages (Fig. 4).

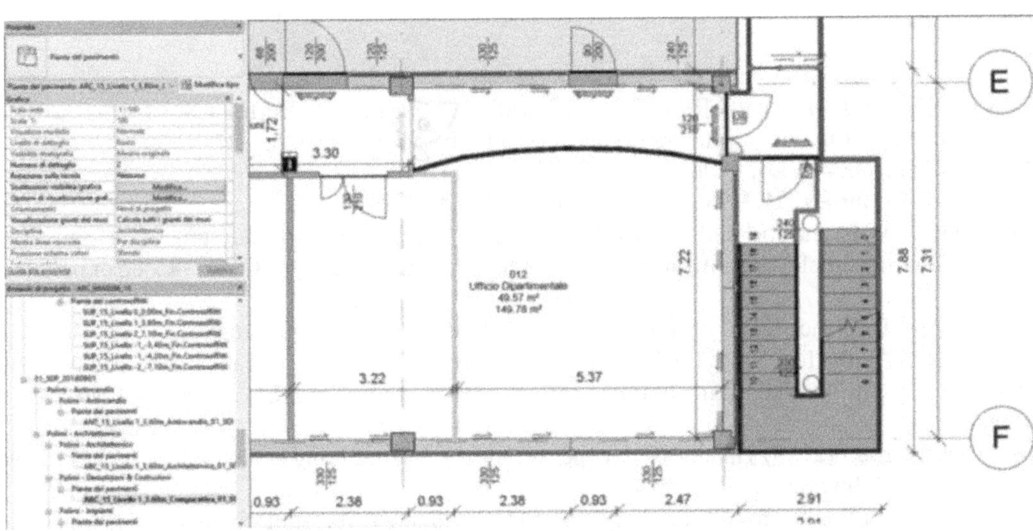

Fig. 4 Example of a layout modification

6 Discussion and Further Development

This paper presents a research aiming at creating a guideline for managing existing assets of Politecnico di Milano portfolio, and for easing design of new construction. To reach this goal, OIRs have been divided into packages, easier to handle, and then translated into AIRs and eventually into attributes of the BIM models and of the connected databases.

This methodology allows to control a series of fundamental processes, such as (i) managing relevant information about existing buildings and new constructions; (ii) managing documents of the entire assets; (iii) monitoring maintenance and operation phases; (iv) standardizing projects information thanks to a detailed LoIN; (v) controlling data exchange among parties thanks to robust procedures; and (vi) extracting output useful portfolio management. The authors planned to revise and improve the guideline with Appointing party's feedbacks.

The customization of procedures, according to client's requirements, allows a soft-landing approach to the BIM methodology that otherwise is rarely and reluctantly accepted by the structured organization. The guidelines are used as a tender specification to manage different procurement according to common strategy of the organization achieving the consistency among data produced in different situations.

The future development of the research project is the definitions of an optimization method for the management of real estate assets, based on the usage model of the building, through a Post-Occupancy Evaluation, that could make it possible to collect data and define the usage model. The presence of cameras connected to a neural platform transforms images into data useful to define the usage model. Once the model usage of the spaces has been defined, the data collected into the BIM model can be used to define and optimize cleaning operations and contracts.

References

Barbosa F, Woetzel J, Mischke J, Ribeirinho MJ, Sridhar M, Parsons M, Bertram N, Brown S (2017) Reinventing construction: a route to higher productivity. McKinsey Co, p 168

BCA (2017) Singapore VDC Guide

Bloomberg MR, Burney DJ, Resnick D (2012) BIM Guidelines

British Standards Institution (2014) BS 1192-4:2014 Collaborative production of information Part 4 : Fulfilling employer's information exchange requirements using COBie—Code of practice

BSI (2018) Gov.uk 1

BuildingSMART alliance (2015) National BIM Standard for The United States V3

Di Giuda GM, Giana PE, Villa V (2017) Comparison between different collaborative procurement methods: a system for integrating building information modelling in AEC processes. Constr Eng Manag

EUBIM Task Group (2018) Handbook for the introduction of building information modelling by the European Public Sector

European Construction Sector Observatory (2016) Finland: coBIM Requirements

Hooper M (2015) BIM standardisation efforts—the case of Sweden. J Inf Technol Constr 20:332–346

Liu H, Al-Hussein M, Lu M (2015) BIM-based integrated approach for detailed construction scheduling under resource constraints. Autom Constr 53:29–43. https://doi.org/10.1016/j.autcon.2015.03.008

Los Angeles World Airports (2017) Volume 2: Requirements for building information modeling

Massachusetts Institute of Technology (2016) MIT Design Standards—BIM and CAD Drawing Standards v6.0

McKinsey Global Institute (2017) Reinventing construction: a route to higher productivity

Statsbygg (2011) Statsbygg BIM manual 1.2. Bim 1.2.1:98

The Ellen MacArthur Foundation (2012) Towards a circular economy—economic and business rationale for an accelerated transition. Greener Manag Int 97. doi:2012-04-03

US Dept of Veterans Affairs (2017) VA BIM Standard

USC Capital Construction Development and Facilities Management Services (2012) University of Southern California BIM Guidelines for Design-Bid Build Contracts

BIM and Post-occupancy Evaluations for Building Management System: Weaknesses and Opportunities

Giuseppe Martino Di Giuda, Laura Pellegrini, Marco Schievano, Mirko Locatelli and Francesco Paleari

Abstract The goal of this work is to provide a state of the art about POE fields of use, opportunities, weaknesses and tools, which is currently used to perform POE combined with BIM methodology. The application of POE on existing buildings can provide a large amount of data on actual uses, supply needs and users' behaviour: the main aspect is to explore the potential application of IoT sensors and Machine Learning techniques to POE.

Keywords Post-occupancy evaluation · Digitalization · Performances optimization · Operational phase · IoT sensors · Machine learning

1 Introduction

Several analyses carried out since the 1990s established a ratio of 1:5:200 over the life of a 30-year-old office building, in relation to construction, maintenance and operating costs respectively (Wu and Clements-Croome 2007; Evans et al. 1998): costs associated with the operational phase have a significant impact on the total cost of the building life cycle. It is necessary to optimize the process of building management in this phase to ensure functionality and efficiency. Actual uses of spaces, supply needs and users' behaviour have strong impacts on functionality and consumptions (Bento Pereira et al. 2016; Zimmerman and Martin 2001) and can cause higher consumptions of energy, space or resources. This, in turn, can result in additional costs or lower quality of available services and, in any case, in lower satisfaction of users. Despite its importance to optimize building performances and consumptions, users' feedback is not even investigated (Royal Institute of British Architects (RIBA) 1965; Cooper 2001). Unless these aspects are managed during the operational phase, it is hard being aware of possible issues in terms of quality or higher costs.

G. M. Di Giuda (✉) · L. Pellegrini · M. Schievano · M. Locatelli · F. Paleari
Architecture, Built Environment and Construction Engineering—ABC Department, Politecnico di Milano, Milan, Italy
e-mail: giuseppe.digiuda@polimi.it

B. Daniotti et al. (eds.), *Digital Transformation of the Design, Construction and Management Processes of the Built Environment*, Research for Development,
https://doi.org/10.1007/978-3-030-33570-0_29

In this context, post-occupancy evaluations (POEs) can lead to an improvement in the above issues. POEs, that are also called 'building-in-use-studies' (Preiser 2010), were defined by RIBA as systematic studies 'of buildings in use to provide architects with information about the performance of their designs, and building owners and users with guidelines to achieve the best out of what they already have' (Royal Institute of British Architects (RIBA) R.S.G. 1991). POEs can provide predictive data to improve buildings' use and management (Leaman et al. 2010).

First applications of POEs started in the 1960s, but the main development of POEs' theory and strategy has been carried out since the 1980s, as a tool for facility management and design phases.

This work provides a state of the art about POEs, including POE's opportunities and weaknesses, an overview of subjects involved and a resume of available tools to perform POEs. A systematic approach was adopted to review related publications and research gaps and further developments are finally presented.

2 Opportunities

POEs can improve the management of existing buildings and the design of new ones, providing more efficiency and user satisfaction. POEs benefits are shown in the following flow (Fig. 1).

Moving from short-term benefits to medium- and long-term ones, the effort required increases, but the effects are spread from the operational phase to the whole life cycle. Benefits and drivers provided by POEs are reviewed in Table 1.

Drivers show the usefulness of POEs to improve existing and future buildings. They could provide savings in terms of resources, costs, and time, as well as increased user satisfaction. Referring to feedback for the design process, however, it may take a long time to obtain valid and consistent data to define databases. Increasing user satisfaction can even be a hard result to achieve with very fast user turnover. In addition, there are some weaknesses relating to POEs highlighting several limits to the spread of this kind of analyses that are presented in the following paragraph.

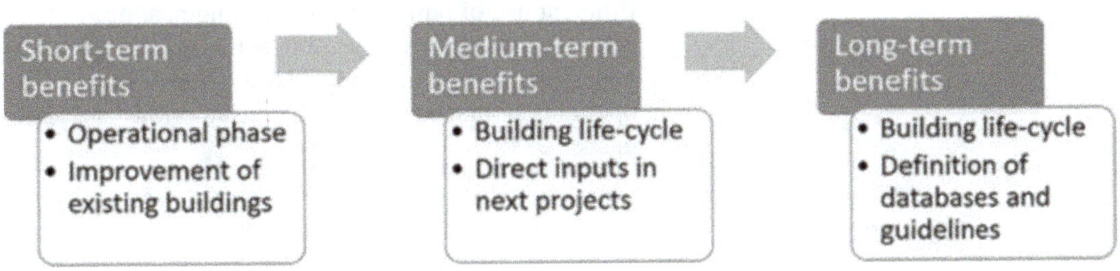

Fig. 1 Definition of short-, medium- and long-term benefits of POEs (Preiser 2010 adjusted)

Table 1 Opportunities of POEs

Continuous improvement	The large amount of data of actual uses can provide valuable information for continuous improvement of building operational stage, the short-term benefit of POEs (Zimmerman and Martin 2001; Preiser 2010)
	Analyses can provide predictive information (Leaman et al. 2010) for building operational phase in order to define accurate and specific management plans
Increased users' satisfaction	POEs help to determine whether the initial conditions ensure comfort and user satisfaction in the operational phase, and results can be used to both rectify weaknesses and inform next projects (medium-term benefits of POE application (Preiser 2010))
Feedback for design process	POEs allow to improve future designs, removing or modifying spaces that are not producing the expected function in existing facilities, according to actual needs (Zimmerman and Martin 2001). As a result, user satisfaction and building efficiency would increase
	POEs collect robust data on usage patterns (Leaman et al. 2010) producing valid, reproducible and generalizable results, leading to the definition of design criteria and guidelines relating to the building function and type of users (long-term benefits of POE application (Preiser 2010; National Research Council 1987))
Reduction of energy consumption	POEs help defining whether the building's performance is negative or positive
	The influence of users' effective behaviour on usage patterns and consumptions of the building can be defined (Straka and Aleksic 2009)
	Benchmarks and standards are available for energy performance and indoor air quality monitoring, resulting in easier data analyses and evaluation
Reduction of operational phase's costs	Lower need for adjustments once buildings are occupied results in lower costs
	Efficient use of buildings, i.e. limiting heating, cooling, resource use, cleaning activities and spaces management only to occupied areas and during operating hours, can optimize consumptions and lead to cost savings

3 Weaknesses

Traditionally, POEs are not part of the standard in common practice, despite the advantages introduced in the previous section. Some obstacles to the dissemination of POEs are investigated as follows (Table 2).

The barriers stated above involve most of the actors of the building design and management process, limiting the application of this type of analysis.

The lack of awareness of the main goal and benchmarks, on one hand, causes the collection of useless data or to over/low-detailed analyses. The result is the increasing of costs and complexity of POEs. The lack of indicators referring user's needs, on the other hand, precludes the definition of whether POEs show positive or negative results.

Table 2 Barriers to POEs

Standard practice	There is institutionalized pressure in AEC industry to carry on with standard practice and not to innovate, to avoid any delay in financing, approvals or design processes (Lovins 1992)
	AEC industry specialists are neither interested in actual usage patterns nor in the analyses of building performances and weaknesses, once occupied (Leaman et al. 2010)
Benchmarks	Users' needs depend on their age, features, occupation, etc. (Bento Pereira et al. 2016) resulting in their difficult definition
	Each building specialist taking part in the building life cycle has its own targets, performance requirements, outlook, technical language and incentives to achieve their goals. Clients, in turn, are placing ever-increasing demands on buildings (Preiser 2010). Besides this, in many cases clients' and specialists' requirements are different from users' needs (Zimmerman and Martin 2001). This fragmentation of the AEC industry and lack of shared goals results in huge efforts to define benchmarks
	Thresholds should even be defined in relation to each type of building, whether residential, service or commercial buildings
Liability	POEs can highlight current issues of existing buildings: low user satisfaction, usage patterns producing waste of energy and resources, poor indoor environmental conditions and inefficient energy performances. Liabilities resulting from an awareness of the real conditions of buildings create resistance of AEC industry in the use of POEs (Zimmerman and Martin 2001; Leaman et al. 2010)
Users' reluctance	POE performing appears as a discomfort and a restriction on user privacy
	Users are reluctant to take part in this kind of analyses as a result of the effort required
Implementation costs	Monitoring by means of sensors and devices can be expensive especially to large buildings
	High levels of details in the analyses cause even increased costs

Referring to liability, several rented out building owners will be reluctant to carry out analyses such as a POEs that could reveal weaknesses of their building compared to similar ones. As a result, Zimmerman and Martin pointed out that tenants would move out, producing a reduction in revenue. They define this mentality as 'ignorance is bliss' (Zimmerman and Martin 2001): building managers and owners reject innovative methods that generate better or more complete information, since they can result in lower profits.

It is clear that costs of application are a further obstacle to the spread of POEs and, as mentioned above, researchers should consider a level of detail according to the criticism detected with a basic survey. This avoids collecting unnecessary and oversized data on lower critical areas or topics. Indeed, the lower the costs and the shorter the application time, the better. This, in turn, can increase the cooperation of users and occupiers and may limit users' opposition to carrying out the analyses.

4 Actors Involved

Once drivers and obstacles to POEs were discussed, positive and negative aspects related to the actors involved in the design and management process were investigated. Each AEC industry specialist and stakeholder has its own goals to achieve, as well as targets in terms of building performance and negative impacts to avoid, as shown below (Table 3).

The observed aspects point out the need for attention in carrying out this kind of analysis considering the possible negative consequences on the different actors involved.

5 Tools and Levels of POEs

Users' reluctance towards POEs, one of the barriers outlined above, has a strong impact on the selection of how a POE is performed. In this sense, the three levels of detail of POEs are defined (Preiser 2010) as

- Indicative POEs;
- Investigative POEs;
- Diagnostic POEs.

The more in-depth and invasive the analyses are, the more reluctant users will be (Fig. 2).

Indicative POEs can be indeed used for an overall analysis of the building to identify the main issues. The aim is the identification of the most critical areas and aspects, which are therefore object of more in-depth analyses. Observations, on-site photographic surveys and interviews are used in indicative POEs; they are

Table 3 Actors involved and pros–cons of POEs

Actor	Pros	Cons
Facility owner	Better end product creates value for money invested (Meir et al. 2009)	Overemphasizing on malfunctioning and hazardous buildings (Meir et al. 2009)
Building manager/owner	Optimization of existing and future buildings (Leaman et al. 2010): more efficient buildings, lower consumptions and maintenance costs	Overemphasizing on malfunctioning and hazardous buildings (Meir et al. 2009)
Users	Increasing healthy conditions, comfort and productivity	Analyses can be invasive such as interviews
		Hard to provide objective feedback
Design teams	Feedback from previous occupied facilities can improve future projects and gain a competitive edge over other specialists, resulting in increased fees/additional work and better future designs	Reluctance to use POEs since early planning due to the complexity of managing large amounts of data and the need for coordination between disciplinary teams
Institutional stakeholders	Promotion of better design and building practices (Meir et al. 2009)	Costs of performing POEs especially in large buildings
	Increasing longevity of buildings minimizing the need for changes (Meir et al. 2009)	

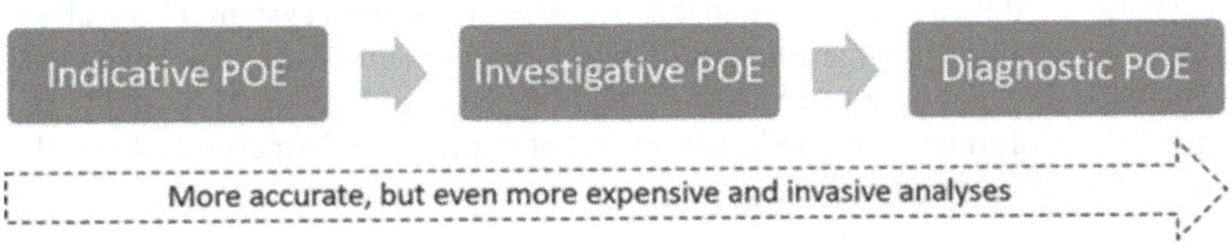

Fig. 2 Indicative, investigative and diagnostic POEs

non-invasive and less expensive analysis tools (Preiser 2010), despite they can be inaccurate. Another problem concerns interviews, i.e. subjective feelings expressed by users.

Investigative POEs are more in-depth analyses adding questionnaires, video recordings and local measurements (Preiser 2010). These tools can be invasive referring to user's privacy. Questionnaires show the same problem as interviews, subjectivity, therefore objective indices to evaluate the building performances should be set.

Diagnostic POEs are detailed analyses involving widespread and continuous monitoring, defining consistent databases (Preiser 2010). Monitoring by means of sensors systems, mostly used to verify building energy consumptions (Marzouk and Abdelaty 2014; Costa et al. 2015; Demian et al. 2018), can provide objective data concerning the building usage pattern. Sensors technologies allow to compare actual use and consumptions with the building performance as obtained through simulations during the design phase (Straka and Aleksic 2009). Design documents' analysis allows to identify discrepancy between the way a building should be used and its actual use defined with the above tools, especially referring to sustainable buildings, whose improper use can be the cause of potential waste. Despite this benefits, diagnostic POEs are invasive, expensive analyses and provide large amounts of data to be analysed.

The application of IoT sensor systems can be useful to collect huge amounts of data, a key aspect to ensure the effectiveness of POE. This is a rapidly expanding technology in many contexts, and is now consolidated with several research and applications in the optimization of energy performances.

An in-depth analysis of several publications (Demian et al. 2018) dealing with the integration between sensors and BIM methodology showed these results: it was identified that a limited number of publications dealt with the use of sensors in the operations and maintenance phase (23%) and in the tracing of people and facilities (12%). It is clear that structured data can be then integrated into the building information model; sensors data, saved in the building model, become valuable to keep information accurate and up to date (Underwood and Isikdag 2011).

Nonetheless, a deep analysis with sensors, besides being limited to critical areas, could involve a huge quantity of data, as a result of the integration of POEs analyses with IoT sensors systems. To date, there is an imbalance between data acquisition and analysis (Ahmed et al. 2017), so it is necessary to define a system for the classification and analysis of large amounts of data. These structured data can be fed to an artificial intelligence based on Machine Learning methods (ANN). This could bring to predictive information for an improvement of building performance and use, that is one of the main goals of POEs.

6 Conclusions

The state of the art stated above aims to identify the main obstacles to the dissemination of POEs and to the integration of these analyses into the traditional management and design process. Furthermore, a review of available tools and their negative and positive sides can allow an informed choice of the best approaches to adopt, according to the level of detail needed. The aim should also be to create fewer negative effects on actors.

The research gaps identified can be resumed in the following aspects, underlining potential developments:

1. Define accurate thresholds and benchmarks to compare collected data.
2. Make owners and managers aware of the considerable savings that result from the application of POEs. At the same time, there is a need to reduce the perceived fear of POEs producing a decrease in profits. Analyses, indeed, may initially highlight shortcomings and problems, but once solutions are identified, increased user satisfaction and predictive information relating to building management during the operational phase can lead to benefits such as
   - Improving users' perception of carrying out POEs.
   - Increasing users' morale and even productivity at work.
   - Cost savings in maintaining and operating facilities, resulting from an efficient use of spaces and resources.
3. Find the best method of carrying out the analyses depending on lower discomfort and limitation of users' privacy. The optimization of the costs of POEs implementation can also be achieved according to desired outcomes and weaknesses detected.

POEs could represent an answer to existing issues, both through questionnaires and interviews to evaluate users' satisfaction, and by means of sensor for environmental condition analysis. This approach could bring to a complete definition of an existing building's conditions. The integration of POEs, IoT sensors technology and Building Information Modelling could result in a better structuring, archiving and continuously updating of data coming from POEs.

Further developments may regard the use of Machine Learning techniques to accurately analyse data and to define predictive information for the operational phase of a building. The application to case studies can lead to the definition of advantages, disadvantages and issues of the outlined method. At the same time, databases from POEs on existing buildings will be used to define guidelines for future buildings implementation.

References

Ahmed V, Tezel A, Aziz Z, Sibley M (2017) The future of Big Data in facilities management: opportunities and challenges. Facilities 35:725–745

Bento Pereira N, Calejo Rodrigues R, Fernandes Rocha P, Bento Pereira N, Calejo Rodrigues R, Fernandes Rocha P (2016) Post-occupancy evaluation data support for planning and management of building maintenance plans. Buildings 6:45

Cooper I (2001) Post-occupancy evaluation—where are you? Build Res Inf 29:158–163

Costa AA, Lopes PM, Antunes A, Cabral I, Grilo A, Rodrigues FM (2015) 3I Buildings: intelligent, interactive and immersive buildings. Procedia Eng 123:7–14

Demian P, Liu Z, Deng Z (2018) Integration of Building Information Modelling (BIM) and sensor technology : a review of current developments and future outlooks. In: CSAE

Evans R, Haryott H, Naste N, Jones A (1998) The longterm costs of owning and using buildings. London

Leaman A, Stevenson F, Bordass B (2010) Building evaluation: practice and principles. Build Res Inf 38:564–577

Lovins A (1992) Energy-efficient buildings: institutional barriers and opportunities. E Source Inc., Boulder, CO (United States)
Marzouk M, Abdelaty A (2014) BIM-based framework for managing performance of subway stations. Autom Constr 41:70–77
Meir IA, Garb Y, Jiao D, Cicelsky A (2009) Post-occupancy evaluation: an inevitable step toward sustainability. Adv Build Energy Res 3:189–220
National Research Council (1987) Post-occupancy evaluation practices in the building process: opportunities for improvement. National Academy Press, Washington, DC
Preiser WFE (2010) Post-occupancy evaluation: how to make buildings work better. Facilities 13:19–28
Royal Institute of British Architects (RIBA) (1965) Handbook of architectural practice and management. RIBA Publications, London
Royal Institute of British Architects (RIBA) R.S.G. (1991) A research report for the architectural profession. In: Duffy F, Hutton L (eds) Architectural knowledge: the idea of a profession. Taylor & Francis, London
Straka V, Aleksic M (2009) Post-occupancy evaluation. Three schools from Greater Toronto. In: PLEA2009—26th conference passive and low energy architecture 5
Underwood J, Isikdag U (2011) Emerging technologies for BIM 2.0. Constr Innov 11:252–258
Wu S, Clements-Croome D (2007) Ratio of operating and maintenance costs to initial costs of building services systems. Cost Eng 49:30–33
Zimmerman A, Martin M (2001) Post-occupancy evaluation: benefits and barriers. Build Res Inf 29:168–174

Digital Technologies for Multi-Scale Survey and Analysis

Introduction

Bruno Daniotti, Marco Gianinetto

Surveying a historical building, monument or place means measuring and analyzing its geometries, structural elements and connections to evaluate its state of conservation, to make structural analysis or to plan a proper project of conservation, consolidation or reuse. Thus, the survey is the very first step for their knowledge investigation.

This section describes the outcomes of some relevant reseach activities related to digital survey technologies of the built environment at different territorial scales, from that of large historical monuments to that of their surroundings and landscape. Specifically, the case histories of the Cathedral of Milan (Duomo di Milano) and Saint Mark's Basilica (Basilica di San Marco) in Venice describe the technical development of the digital survey techniques for cultural and historical heritage during the last decade. The sinergy of traditional and cutting-edge survey approaches made possible to build a digital replica of two of the most famous churches in the world, which are today used for both documentation and conservation of these exceptional monuments.

Moving to the urban scale, this section shows how the cross-referencing of geo-databases and national census data can be used to map the energy demand of city districts and how this information then becomes the input for modeling energy upgrading scenarios towards smart energy district. Finally, aircraft and satellite survey technologies are typically used when analysing very large areas. In this case, geo-information made of big data and massive measures should be managed, thus automatic algorithms for their analysis becomes a must. In this regard, this section describes from a theoretical point of view how to implement automatic geocoding workflows of digital data and some case studies for mapping at the territorial scale complement the dissertation.

The common threads of this section are the technologies of Geomatics, such as web-based Geographic Information Systems and Building Information Modeling for cultural and historical heritage. Both use a set of tools to allow the handling, updating and analysis of different digital data to monitor their conservation state, planning the intervention of restoration and conservation, storing the data and sharing all these information among the experts involved. This section outlines the results of the several research activities carried out on these topics, including the significant case histories of the Sacri Monti of Piedmont and Lombardy (North Italy), devotional paths part of the UNESCO World Heritage list, and Basilica of Collemaggio in L'Aquila and Basilica of Saint Ambrose in Milan.

From a Traditional to a Digital Site: 2008–2019. The History of Milan Cathedral Surveys

Cristiana Achille, Francesco Fassi, Alessandro Mandelli, Luca Perfetti, Fabrizio Rechichi and Simone Teruggi

Abstract Since 2008, an intense survey has been underway in the Milan Cathedral. The operations, over the years, have been conducted with laser scanner and photogrammetry choosing or integrating the different methodologies according to the environment, the necessary "drawing" to support the different sites' maintenance operation but always concurrently with the evolution of the method and the software. The "principal actor" of conservation activities is the "marble block." The Veneranda Fabbrica organizes the activities identifying the areas that are in need of intervention and identifying which blocks will be affected by replacement operations, tessellation, or consolidation. Thus, the objective of the survey activities and the subsequent modeling phase was to build a detailed 3D model in which the marble blocks are easily recognizable in terms of their shape, size, position, and texture (only for the outer part). All the elaborations produced, the 3D models, the two-dimensional representations, and the orthophotos, allow for the identification of the blocks, as to provide proper technical support for site operations. In parallel with the survey and modeling activities, an ad hoc online information system was created to support the construction site activities in a smart manner. The system allows for the consultation and the sharing of the 3D models and all the data necessary for maintenance operations of the entire Cathedral.

Keywords Survey · Photogrammetry · Reality-based model · Online information system · VR/AR applications

C. Achille (✉) · F. Fassi · A. Mandelli · L. Perfetti · F. Rechichi · S. Teruggi
Architecture, Built Environment and Construction Engineering—ABC Department, Politecnico di Milano, Milan, Italy
e-mail: cristiana.achille@polimi.it

B. Daniotti et al. (eds.), *Digital Transformation of the Design, Construction and Management Processes of the Built Environment*, Research for Development,
https://doi.org/10.1007/978-3-030-33570-0_30

1 The Milan Cathedral and the Veneranda Fabbrica: A Brief Historical Introduction

The construction of the Duomo began in 1386. The promoter of the construction was the archbishop Antonio da Saluzzo and the first patron Gian Galeazzo Visconti (Moschini 2012). Simultaneously with the start of the construction of the Cathedral, Gian Galeazzo Visconti established the Veneranda Fabbrica del Duomo in Milan, a historical institution in charge of preserving and restoring the Cathedral. This institution has been a key player for over 630 years in all the activities of the Cathedral: from maintenance to custody and liturgical service and the valorization and crowdfunding activities to provide for the necessary resources for its continuous maintenance through the century (https://www.duomomilano.it).

The dimensions of the Gothic Cathedral are impressive: it is the third highest, after the Cathedral of Beauvais in France and St. Peter's in Rome, and one of the largest in the world, second only to St. Peter's in the Vatican.

The entire Gothic Cathedral is covered in marble while all decorations, medium/little spires, statues, and its tallest structures are made by marble entirely. The marble itself comes from the quarries of Candoglia and Ornavasso in Val D'Ossola, a white–pink marble deposit, located in the high Alpine foothills between Monte Rosa and Spluga. The marble deteriorates in a short time due to the run-off of rains, pollution and temperature changes and over the years the marble becomes very friable. It could lead to potentially dangerous situations for the structure and the people living in the Cathedral. So one of the essential activities of the Veneranda Fabbrica is a periodical inspection to identify the damaged parts. The necessary maintenance tasks consist of repairing cracked or chipped marble blocks up to their complete replacement in more damaged situations. The activities of the site are organized and carried out block by block, identifying the type of necessary intervention (consolidation, cleaning or replacement) on the single marble block (Monti et al. 2013). This means that all technical representations, 2D or 3D, must accurately and precisely describe the shape and position of the individual blocks of marble (Fassi et al. 2018).

2 2008–2019: The Survey Activities of the Cathedral

Among the essential "working tools" of the Veneranda Fabbrica site, are the architectural 2D representations as plans, sections, profiles, and 3D models. They are necessary for the correct documentation and planning of the activities. The technical office of the Fabbrica periodically deals with updating the technical drawings, from the representations that describe the whole structure, in part still on paper, to the accurate representation of smaller parts presenting complex maintenance operations. In order to respond to this need, a survey campaign was launched in 2008 focused on the "Guglia Maggiore", in preparation for the later extraordinary restoration of the following years. Initially, the planned activities included the representation of the Main

Spire traditionally with predetermined plans and sections of the structures. However, the constant relationship with the Fabbrica site had immediately highlighted the limitations linked to a classic approach. In other words, to follow the activities of a long and complicated construction site, a fixed number of general representations were very limiting. The need of continuous different surveys and representations before and during the restorations has increased the need to overturn the classical processes (2D survey and restitution) in favor of a survey aimed at the production of complete 3D models, from which it is possible to extract any two-dimensional representation when needed. A classic survey and 2D restitution approach always finds a solution, also starting from a complex shape; on the contrary, to create a three-dimensional reality-based model means to solve theoretical and practical problems that are both theoretically and practically challenging. Since the beginning of the activities in 2008, the main issues faced with a technical research approach are as follows:

- the extremely complex object shape, considerably rich in decorations, in narrow passages, and height of out-of-reach structures;
- the management of an impressive amount of data;
- the reality-based modeling phase in terms of time necessary for the realization of accurate and precise 3D models but also in terms of data organization and modeled representation accuracy and details;
- the management of big 3D models both during the modeling phase and also later for the data fruition.

The first part of the activity was aimed at solving the problems that arise in situ: the presence of an active site, scaffolding, narrow spaces, people, inaccessible areas, covered and non-covered spaces, different lighting conditions, times for execution of the surveys dictated by the site (Achille et al. 2012).

It is good to remember that the start of these activities dates back to 2008, so the hardware and software limitations were not trivial. Today, the geometric description makes use of tools that allow users to measure objects of any shape (linear or free form), producing in real time or near real time a three-dimensional portion of data (point clouds). At the beginning of the project the survey technique that allowed us to solve the multiple issues (shape–position–material) was photogrammetry: starting from the traditional digital photogrammetric approach (manual bundle adjustment and stereo plotting) in 2008, up to the experimentation of the emerging semi-automatic image-based techniques in the following years. The photogrammetric survey, although possible, was initially limited by the size of the object itself, which forced users to use an impressive number of images elaborated through manual procedures both in the orientation phase and in the modeling phase for the visual recognition of homologous points. This was the most significant limitation of the process since it was extremely time consuming.[1] Today, the tested automatic image-based techniques allow for the orientation of a considerable number of images automatically, for the auto calibration of the sensors and for the extraction of dense Digital Surface Models (DSMs) in the form of point clouds comparable (if not superior) to those obtainable with scanner instrumentation (both in terms of maximum resolution achievable and expected accuracy).[2]

The photogrammetric methods are in this way very flexible, allowing users to survey both large architecture and minute decorations, using the same instruments. Another reason why the methods are necessary on the cathedral is because it can adapt itself quickly to the many different situations and "architectural environments". Nevertheless, not always can the methods be used effectively due to architectural or environmental constraints. It is the case of vast interior spaces that are poorly lit by large windows and that are always occupied by liturgical activity and the presence of tourists or, on the contrary, very narrow secondary passages where any survey method couldn't give satisfying results. In order to solve these issues, many tests were conducted both with unconventional range-based (Zeb1, DOT DPI 8, Heron backpack) and image-based methods (fisheye photogrammetry, camera rigs, and panoramic cameras) to identify the optimal strategy for (i) detecting Candoglia marble, (ii) speeding up the survey phase, and (iii) detecting narrow and poorly illuminated spaces. The tests were conducted both inside and outside the Cathedral (Perfetti et al. 2017). The final 3D point cloud format model of the Cathedral derives from the contribution of different survey systems.

Candoglia marble is a material that is difficult to detect with instruments that use laser light. The crystalline structure of the marble allows the laser light to penetrate the surface, giving an incorrect shape (Fassi et al. 2011). After different tests, we identified the Leica C10 scanner as the only instrument capable of scanning marble without penetration. It was used inside the Cathedral to survey all the naves, substituting the photogrammetric technique because time consuming. The designed scans were acquired both from the ground and two upper levels (totally 850 scans), using a lifting platform, in order to avoid shadow areas and guarantee a uniform resolution

[1]At the beginning of 2008, the laborious photogrammetric pipeline (image acquisition, manual orientation, and restitution) produced a wireframe model. For each stereoscopic pair, the operator had to identify common points (orientation) and then—on the stereoscopic model—proceed with returning the shape profile manually. Several days of work were required to get all 3D lines and to construct the three-dimensional wireframe representation. To create the final surface model, yet another processing step via an external piece of software was needed.

[2]These features and automations are the evolution of the classic and consolidated photogrammetric rules combined with the developments proposed by the world of Computer Vision. An example is the Amadeo spire, dimensions: base 3.10 m × 3.10 m, height 20.5 m. Time required for the survey: 1 week; for processing data: 5 weeks. Elaborated products: dense point cloud, orthophotos (eight exterior sides and eight inner sides). Total images acquired: 8160 (6180 from the scaffolding, 163 from the level of the roofs, 23 from the elevator, and 1793 from the interior of the staircase). Digital cameras used: Nikon D810, 36MP, f: 12 mm, GSD = 0.8 mm; f: 50 mm, GSD = 1.5 mm; f: 8 mm fisheye. Canon 5DmkIII, 24MP, f: 85 mm, GSD = 2.2 mm. 1:10 for external facade, 1: 20 for interior facade.

of 5 mm and consequently a fixed accuracy suitable for a 1:50 and 1:20 representation scale. Currently—in 2019—the survey of the whole interior Cathedral is almost complete, so it is possible to explore the 3D model of the Duomo in a point cloud format. Due to the size of the Cathedral, traditional topographic measurements have always been necessary to accurately align the point clouds within a single reference system and to check all the photogrammetric elaborations. The cleaning phase of the scans was completely manual. The final size of the point cloud does not allow for management through a single file; the data was segmented following the logic of the site areas of the Veneranda Fabbrica (inferior nave, middle nave, central nave, apse, and choir). This allows to efficiently use the data during the restitution phase. The survey of the external facades, of the roofs and the falconature,[3] has been realized only through photogrammetry. It is easy to understand as it was not possible to use the scanner to survey the highest parts due to the impossibility of positioning the equipment in a stable position. Instead, it was possible to use the elevating platform for a photogrammetric survey. In this way, roofs and the highest spires and decorations of the facades were also surveyed.[4] The main goal was to produce high-resolution orthophotos of the facades with a restitution scale equal to 1:50; while the sectors of the roof, delimited by arches and falconature, were elaborated at a 1:20 scale.

All point clouds produced bit by bit are generating the complete 3D dense point cloud model of the Duomo (Fig. 1). This raw model can be used for visualization, measurement operations, features extraction or as a base for orthophoto generation.

In order to obtain a 3D surface model, a time consuming and arduous reality-based modeling and simplification phase is required. It is necessary to produce BIM-like models that can be easily modified or used for classical CAD applications (Achille et al. 2014).

For the first research activity of the Main Spire, a complete and detailed 3D model was created with the double aim of

(i) automatically extracting all the metric information useful for the restoration of the Cathedral with a useful accuracy for a 1:20 and/or 1:50 scale,
(ii) being the basis of an information system supporting site activities. For this reason, the 3D reality-based models describe all the individual objects, each block of marble, every decorative element, or statue.

The modeling phase was not easy, so different methods and software were tested to find a satisfactory method in terms of achievable accuracy and time needed for its realization.[5] In addition to the nurbs models elaborated with SW Rhinoceros (the

[3]Crowning elements of the Duomo (Benati et al. 2001).

[4]An example. South façade, digital orthophoto 1:50 scale, obtained by mosaicking 33 ortho images. 7.586 images (4.511 from elevator; 205 from ground level; 1.936 from roof level; 934 for apse). Five different full-frame camera configurations (Canon 5DSR, 51MP, f: 35 mm, Ground Sample Distance = 1,8 mm; f: 85 mm, GSD = 2,4 mm; Canon 5DmkIII, 24MP, f: 85 mm, GSD = 1,6 mm; Nikon D810, 36MP, f: 50 mm, GSD = 1,5 mm; f: 24 mm, GSD = 2,0 mm.

[5]In addition to direct modeling, a number of tests of parametric modeling were made. This approach is possible only when the element is repeated with the same shapes, but with different sizes and can be used only where the simplification respects the precision requirements. Some tests were done

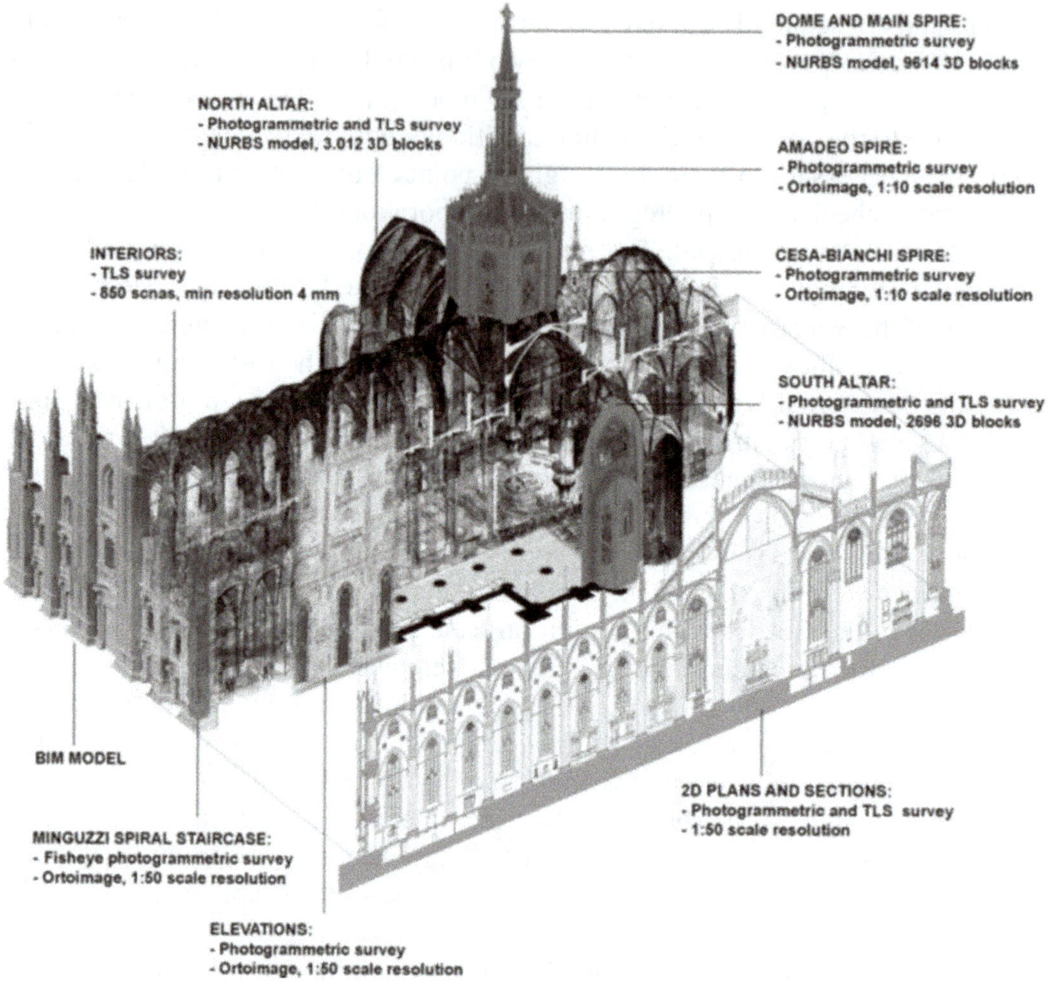

Fig. 1 The image aims to represent the types of survey and elaborations carried out in these years for the Duomo research projects. The scanner and photogrammetric survey produced a huge point cloud of the Duomo; data processing is progressively generating classical two-dimensional representations (plans, profile sections) and 3D models (mesh, nurbs, BIM)

Main Spire, the Dome Cladding and the vault, the internal facades of the transept, with the altars and portions of the roof), other modeling tests are underway using BIM software (Revit and Archicad). Some areas were chosen, consisting both of simple and free-form objects (the façade, a portion of roofs, the pillars and the vaults of the lower nave) in order to assess accuracy versus time, the level of interoperability from and to other software (Tommasi et al 2016; Achille and Tommasi 2017), the possibility of linking information (vector, raster and textual) and the possibility of managing site activities using the specific tools that this software provides.

The use of all the different elaborated 3D models, even only in terms of visualization, requires specific skills and enormous computational resources. In order to

for the series of gothic windows and for the pyramid on top of the main spire. In the case of the windows of the transept altars, the parametric approach (Rhino + Grasshopper) led to a significant reduction in the time, from an expected 330 h for direct modeling to 126 with parametric modeling.

facilitate the use of 3D models for everyday users as well, to allow the parallel modification of the 3D inside the modeling software and to provide a further potential to the model itself, a dedicated BIM-like system was implemented.

3 The Web-Based Informative System

The research had to immediately deal with the problems relating to the management and use of the three-dimensional models produced. The main issues identified were:

(i) a smart *management system*: The complete 3D model is vast in terms of "quantity of modelled objects" and resources needed for loading and visualization. Even today (2019) they cannot be managed simultaneously in any software;
(ii) an easy fruition: Because such a complex model requires simple and intuitive consultation methods to be also used by the general public;
(iii) continuous updating: It is necessary to guarantee a constant updating of the model itself, in support of future maintenance activities.

The information system (Fassi and Parri 2012; Fassi et al. 2015) specifically developed for the Veneranda Fabbrica[6] is composed of three parts. The first part refers to the modeling environment[7] where the model is created, modified, updated, and it is also possible to link it to the information database. The 3D models, once completed, are exported and are prepared to be used through the web interface, the second part of the system. It is sufficient to have access to a web browser (via PC, notebook, tablet, or smartphone) to view models at different levels of detail.[8] Models built with any modeling software can be displayed and can have a simple shape (structure) or complex forms (free-form models) or both possibilities integrated with one another. The system has no restrictions on the number of objects that can be viewed at the same time, as well as on the level of detail of the same.[9] The third part of the online information system consists of a database that contains both 3D models and associated information. The database[10] (PostgreSQL) allows operators

---

[6]The system created for the Milan Cathedral is to be considered as a system made specifically to support the Veneranda Fabbrica activities. A general system that adapts to the world of Cultural Heritage is under development with the name BIM3DSG (patent pending MI2014A002016).

[7]The choice of modeling software had a relapse (2009) in the Rhinoceros software to guarantee reality-based modeling (from a point cloud). The software allows for the creation of scripts in C language and therefore to build ad hoc tools according to specific needs. The plug-ins developed were written in C# and .NET while a windows form was used for the user interface (in the latest version of the system the plug-ins use WPF Windows Presentation Foundation, Xceed WPF Toolkit, a version of Avalon-Dock and ITinnovationsLibrary).

[8]It is possible to use devices even with limited hardware resources because the system allows objects to be displayed at different levels of detail, automatically calculated by the system.

[9]Powerful caching mechanisms have been implemented to ensure ease and speed of use of the system; the download is almost instantaneous, even if the available connection is slow.

[10]The DB is installed on a remote server and is accessible from a variety of devices via Internet connection. Only if necessary, the DB can be installed locally and managed within a local network.

to share and synchronize changes in real time; it contains all the information relating to the internal structure, textual information, and model files entered by users. The system can be used both by a specialized user (who also accesses the first part of the system) and a non-specialized one. It allows users (i) to view all or part of the models; (ii) to automatically calculate metric information of each object such as area, volume, and position; (iii) to display/add/modify information and files (raster, text, and vector) associated with one or more objects. The system also allows users to create (both inside the modeling software and the web interface) records of maintenance activities (cleaning, tessellation, or replacement), whether they be planned, ongoing or already completed. A series of functions have been designed to add/modify/see ongoing maintenance activities with all the related information (photos, texts, videos, documents, dwg, etc.) using visualization with different colors directly on the web viewer (Achille and Fassi 2016).

The designed system not only allows for records of information on interventions but also helps the real-time updating of 3D parts. In this way, the visualized 3D model always represents the latest state of the art of the object after the last intervention (Fig. 2). Using the system with portable devices (tablets, smartphones, etc.) the operator can update the information directly on site. The information, once entered, is saved in the system database and visible to any other user (Fig. 3). The management of 3D models through the WEB system guarantees: (i) the low-cost essence of the system (free browser); (ii) simultaneous access by multiple users (usability); iii) interoperability between different actors (participation); (iv) the immediate updating and synchronization of data entered.

4 Future Research

The three-dimensional survey started in 2008 and which is now almost complete, aims at achieving the geometrical knowledge of the Duomo, a starting point for conservation, diagnosis, monitoring, and therefore for safeguarding the Cathedral. For this reason, all survey activities are conducted with a metrological point of view guaranteeing the uniformity of the resolution, the completeness of data, and a mean accuracy under the centimeter. The metric representations (2D or 3D) required for the Duomo (high-resolution orthophotos and reality-based models) need a design phase, long elaboration processes and check phases. The new data acquisition and processing tools lead to a significant reduction in the time required, but, particularly, they make it possible to achieve the once unthinkable both in terms of general feasibility and available resolution and precision .

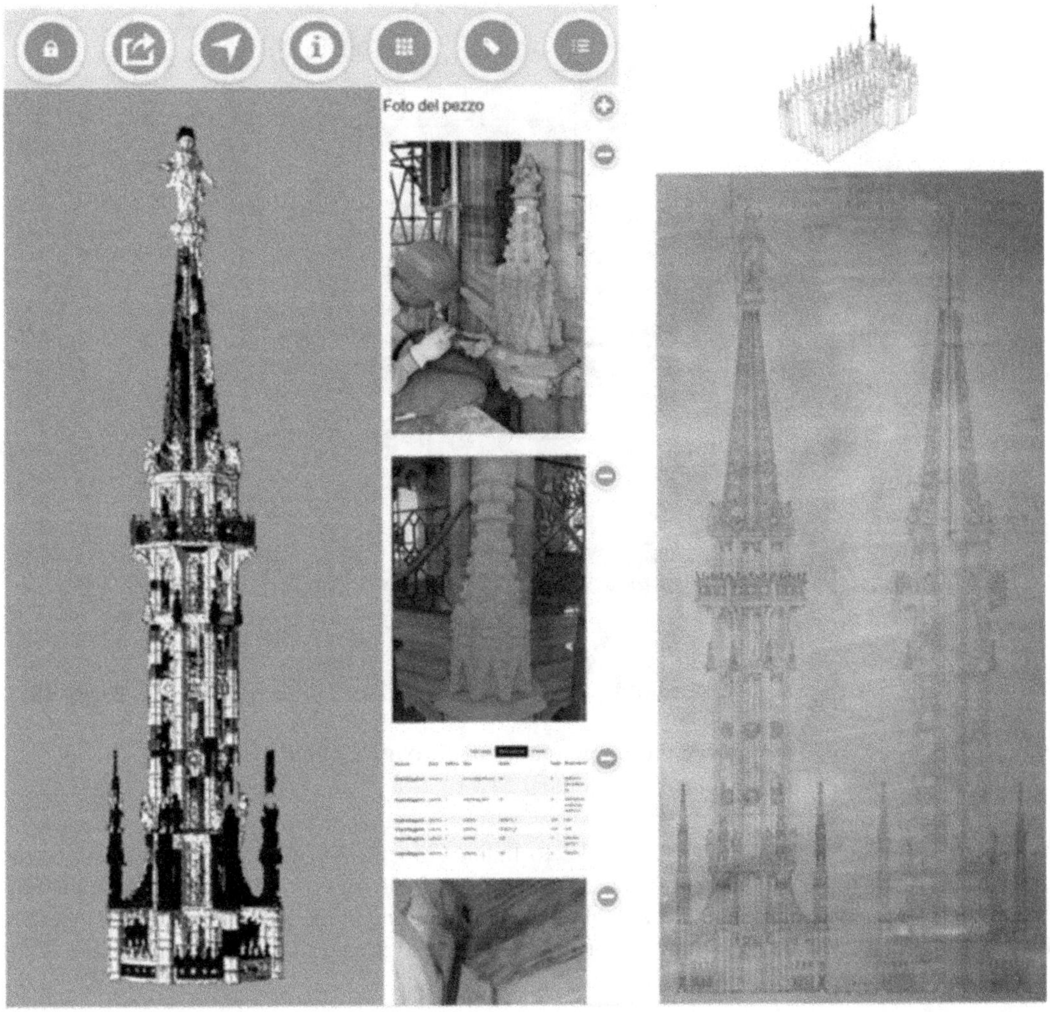

Fig. 2 On the left, a detail of the final part of the Main Spire visualized on the online system. The blocks subject to intervention (different colors according to the type of intervention) with some example images are provided; on the right the same detail in the original drawing made by Veneranda Fabbrica, on 1:20 scale, January 7, 1941

The bottleneck of the whole process is the modeling phase. For this reason, the Nurbs modeling activity was interrupted after the "Main Spire research project". The future research topic is to skip this stage and create a 3D system able to use the point cloud directly as a 3D reference model. Nowadays, the survey techniques produce, in one way or another, a point cloud, which already constitutes an archive of the geometry of the object that can be used conceptually in professional processes. For these prospects, the contribution of tools designed and developed ad hoc is decisive.

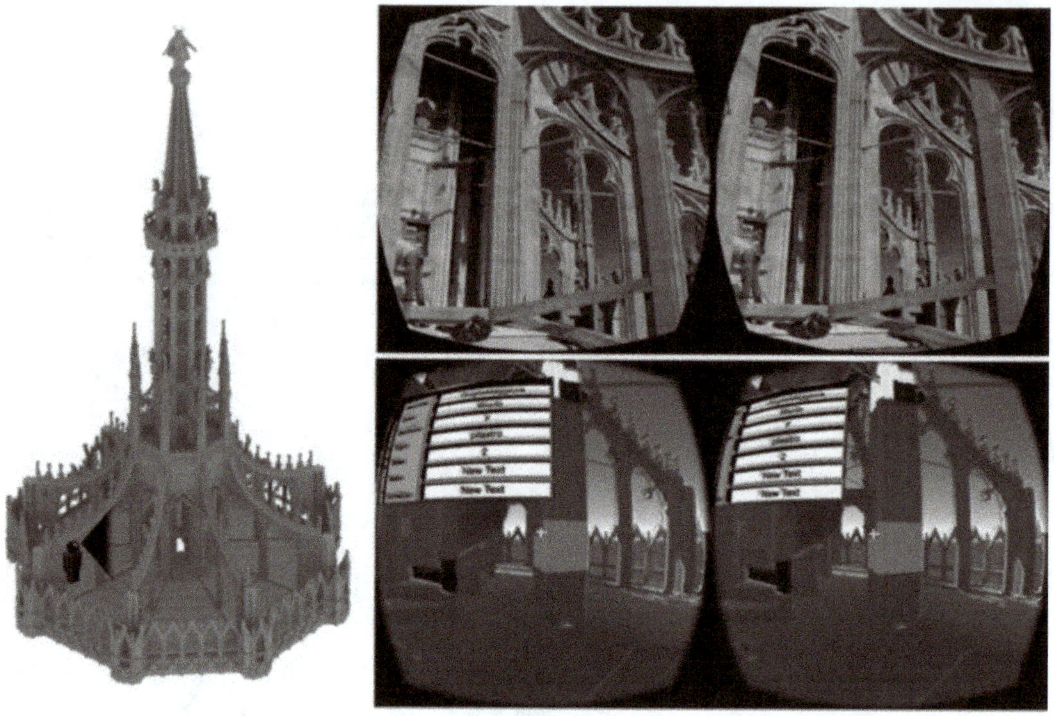

Fig. 3 Main Spire BIM system, stereoscopic rendering through Oculus Rift DK2

Acknowledgements The authors would like to thank Francesco Canali, the "Architect" of the Duomo, and the whole staff, in particular, Francesco Aquilano and the Massimiliano Regis. Furthermore, the authors would like to thank all the colleagues and students who have contributed to this project to date. Thanks to Benigno Mörlin Visconti Castiglione, the former "Architect" of the Duomo at the time when the research started.

References

Achille C, Fassi F, Fregonese L (2012) 4 Years History: from 2D to BIM for CH. The main spire on Milan Cathedral. In: Guidi G, Addison AC (eds) Proceeding of VSMM 2012. Virtual Systems in the Information Society, IEEE, pp 377–382. ISBN 9781467325622

Achille C, Fassi F, Mandelli A, Moerlin B (2014) The yards of the Milan Cathedral: tradition and BIM. In: Proceedings of the international conference preventive and planned conservation Monza, Mantua, 5–9 May 2014. ISBN 978-88-404-0318-2

Achille C, Fassi F (2016) Rilievo, modellazione e gestione BIM dei cantieri del Duomo di Milano. In: Eastman C, Teicholz P, Sacks R, Liston K, Di Giuda G, Villa V, Hoepli (eds) 'Il BIM Guida completa al building information'. ISBN 9788820367053

Achille C, Tommasi C (2017) Interoperability matter: levels of data sharing, starting from a 3D information modeling, vol XLII-2/W3. https://doi.org/10.5194/isprs-archives-xlii-2-w3-623-2017

Benati G, Roda AM (a cura di) (2001) Il Duomo di Milano Dizionario storico artistico e religioso Nuova edizione rivista e aggiornata Boniardi. ISBN 88 7023 270 0

Fassi F, Parri S (2012) Complex architecture in 3D: from survey to web. Int J Herit Digit Era 1:379–398. ISSN 2047-4970

Fassi F, Achille C, Mandelli A, Rechichi F, Parri S (2015) A new idea of BIM System for visualization, web and using huge complex 3D model for facility management. Int ISPRS XL-5/W4:359–366. https://doi.org/10.5194/isprsarchives-xl-5-w4-359-2015

Fassi F, Achille C, Fregonese L (2011) Surveying and modeling the Main Spire of Milan Cathedral using multiple data sources. Photogram Rec 26:462–487. https://doi.org/10.1111/j.1477-9730.2011.00658.x

Fassi F, Achille C, Mandelli A, Perfetti L, Polari C (2018) Recording the marble blocks of the Milan Cathedral: synergy between advanced survey techniques. In: Proceedings of the international masonry society conferences 2-s2.0-85062311862

Monti C, Moerlin B, Fassi F, Achille C, Mandelli A (2013) From the digital to the physical model. Case study in the yard of the Milan Cathedral. Mo.Di.Phy Modeling from digital to physical innovation in design languages and project procedures Maggioli Editore. ISBN 9788838762741

Moschini C (2012) a cura di, Il cantiere del Duomo di Milano Dai maestri del lago di Lugano a Leonardo, Silvana Editoriale

Perfetti L, Polari C, Fassi F (2017) Fisheye photogrammetry: tests and methodologies for the survey of narrow spaces. ISPRS XLII-2/W3. 3D Virtual reconstruction and visualization of complex architectures. https://doi.org/10.5194/isprs-archives-xlii-2-w3-573-2017

Tommasi C, Achille C, Fassi F (2016) From point cloud to BIM: a modeling challenge in the Cultural Heritage field. ISPRS XLI-B5. https://doi.org/10.5194/isprsarchives-xli-b5-429-2016

The 3D Model of St. Mark's Basilica in Venice

Luigi Fregonese and Andrea Adami

Abstract The San Marco 3D project had the ambitious goal of building a digital replica of the famous venetian basilica. Architectural surveying and modelling are very widespread procedures, but the complexity of the basilica, its decorative apparatus, in mosaic and marble, and its liveliness made this project a real challenge. Thanks to geomatics, from the most traditional topographic approach to cutting-edge methods of digital photogrammetry, it has been possible to build an information system of the basilica, a geometric database from which to continuously extract new and correct information. Even the mosaics, a main characteristic element of the basilica, have been documented through very high-resolution orthophotos, therefore providing useful and effective tools for the conservation of the basilica itself. Thus, the research project allowed for a better and deeper knowledge of the basilica, expressed through a very accurate 3D model where the geometry and the very rich decorative apparatus are merged into a single product.

Keywords Survey · Complex architectures · Photogrammetry · Orthophoto

1 Introduction

Many elements combine to constitute a research project: the innovativeness of methods, the use of the most advanced and recent technologies, the development of new software, the social impact, the applicability and many others. An element that is always present, but sometimes neglected, is the application case: the object (in this case the architecture) on which the objectives of the project are to be verified. For the San Marco 3D project, the exceptionality of the theme tackled strongly characterizes the research, to the point of becoming almost the theme itself; it is no longer just an application case, an occasion on which to make tests, but the element that determines the choices of the research itself.

L. Fregonese · A. Adami (✉)
Architecture, Built Environment and Construction Engineering—ABC Department, Politecnico di Milano, Milan, Italy
e-mail: andrea.adami@polimi.it

B. Daniotti et al. (eds.), *Digital Transformation of the Design, Construction and Management Processes of the Built Environment*, Research for Development,
https://doi.org/10.1007/978-3-030-33570-0_31

The construction of a 3D replica of an architecture is now a widespread activity because it lends itself to various activities and insights. It is the fundamental element of many specialized applications such as those that provide for the design, simulation and management in architectural and engineering of a building. 3D replication is also, however, a fundamental element for all communication and edutainment applications developed, for example in the field of cultural heritage. If therefore, 3D replication, from the survey of information to its geometric modelling, is a very widespread process, the same cannot be said of the methodology applied. As it is easy to understand, the different applications require 3D replicas with different characteristics, more or less detailed, more or less corresponding to reality in accordance with the purpose of the application itself.

In the context of the San Marco 3D project, the main need highlighted by the Procuratoria di San Marco, the managing body of the basilica and partner of the project, was to build a three-dimensional replica of the basilica, with very high resolution and precision, to be used for technical applications related to the knowledge and conservation of the building itself. This need, however, was not only linked to the shape of the basilica, but also to its appearance: golden mosaics and polychrome marbles cannot be overlooked in the phase of creating a 3D copy.

This article does not describe the technical aspects of the work, already described in Adami et al. (2018), Fregonese et al. (2017), but it highlights the choices made during the research and the reasons that led to those choices.

2 St. Mark's Basilica and Its Exceptionality

The uniqueness of St. Mark's Basilica lies without doubt in its historical vicissitudes and in its very high historical and cultural value. But for this research project, the elements that most determined the choices are linked to the use of the basilica and the construction techniques used.

The fruition of the basilica, in fact, means all the activities that take place in it: religious celebrations, concerts, cultural visits, tourist visits. It is estimated that tourists alone reach over 5.5 million each year and the basilica is open every day in fixed time periods. In this context, we should also consider all the activities necessary to ensure this enjoyment: in fact, the basilica always has a team of workers (electricians, cabinetmakers, mosaicists, stonemasons) engaged in ordinary and extraordinary conservation of the building. This intense programme of activities that takes place inside the basilica has caused many problems at the stage of architectural survey because, for obvious reasons, it was not possible to close the basilica to visitors and not even to work in the evening hours when the basilica was closed, so as not to weigh on the costs of management. All the measurement operations, therefore, were carried out during the normal period of opening of the church, with the tourists present and often with the need to move the work area to meet liturgical needs.

On the other hand, the construction techniques and materials that have guided the various choices are those of architectural surfaces generally decorated with mosaics or marble slabs. The stone slabs, of different marbles and positioned in the lower part of the basilica, confirmed the difficulties, already known in the literature (Godin et al. 2001; Boehler and Marbs 2001) and also tackled in other cases by the authors (Fregonese et al. 2018), in the acquisition through range-based techniques where the emitted signal seems to 'penetrate' the object because of the characteristics of the material. At the same time, mosaics are often made of golden tiles and even in this case, all this has led to serious problems of reflection in the use of range-based techniques such as laser scanners at flight time or phase difference.

The mosaics, in particular, highlighted another problem, only partially solved. In literature, there are many studies on colour and acquisition methods to be as faithful as possible to reality. All these studies, however, envisage working on small objects and being able to operate in controlled environments such as a laboratory. The case of San Marco is completely different because the architecture with mosaic cladding is very large (the upper part of the whole basilica measuring about 64 m in width and 78 m in height) and above all it is not possible to guarantee a constant lighting condition due to the fact that there are fixed times for the lights (to facilitate the visit of tourists) and the large openings on the south and west side allow a different light to enter (in colour, intensity and angle) depending on the time of day and season.

The decorative richness of the basilica, which can also be seen in the sculptural and moulding parts, as well as the deterioration due to time, also constituted a challenge in the field of modelling.

The last exceptional element of the Basilica is linked to the amount of data. The data collected (images and, after processing, point clouds), the geometric model, with and without texture, and the orthophotos represent a huge amount of data that must not only be stored but also used, thus opening new themes related to organization and accessibility of the data.

3 The Choices of the Project

The exceptional nature of the basilica has brought to light, as described, a series of issues to be addressed and which were encountered, in various respects, in the different phases of the work that, developed according to a shared approach, includes a phase of project survey, data acquisition, processing up to the moment of modelling and structuring of the data.

3.1 The Survey Project

The objectives of the project were known from the outset, so it was possible to define the quality of the final deliverables from the outset. The planned scale of return, 1:50, requires accuracy of 1 cm, but given the complexity and richness of the basilica, it is increased to 5 mm to be able to describe all the architectural details of value. The less obvious choice, however, concerned the survey techniques. The possibilities included the use of range-based (laser scanner) and image-based (photogrammetry) methods. The characteristics of the materials (marbles and golden tiles) led to a preference for photogrammetry as it was less affected by systematic errors. And it also responded to the need to acquire not only the geometry of the factory but also its surface finish (Chiabrando et al. 2015; Remondino et al. 2014). In this way, it was possible to obtain metrically correct and valid data both for the modelling of the geometries and for the construction of the orthophotos and the texturization of the models.

Once the photogrammetric approach has been chosen, the question to be answered was linked to the richness of detail of the basilica, an element to be considered, however, also in relation to the real need for the data to avoid falling into the risk of an 'unconditional' acquisition with the subsequent effect of exponentially increasing the amount of unnecessary data. In this way, we have operated with a multiscale principle, choosing to build different models, all georeferenced, but with different resolutions. Thanks to the use of different lenses (and consequently different quantities of images), general models and detailed models have been created by photogrammetry of the same area.

3.2 Data Acquisition

In data acquisition, the greatest difficulties were encountered in relation to lighting conditions and the need to work flexibly throughout the factory to comply with opening hours and the different activities that take place within.

To meet this need, the most efficient solution was to divide the entire factory into small areas and to conduct the survey area by area. This method also partly reflects the operating conditions of multi-image photogrammetry, which benefits from operating on closed and well-defined areas, as is the case with a topographic polygonal: the closure of the ring allows for a greater control over errors. Referring back to the Basilica, you can imagine the survey as a series of closed rings, connected together. The function of the topography, in this case, is precisely that of connecting the individual rings (referring to the individual areas or single elements) in a single reference system. The presence of an existing topographic network, with points well distributed and already compensated for previous works of survey carried out in the basilica (Fregonese et al. 2006), has therefore made it possible to make a non-linear process operational and thus meet the different needs encountered.

To solve, on the other hand, the problem of lighting, a number of tricks were used that have not allowed to completely solve the problem, but only reduce it. The need would have been to have a homogeneous light, for colour, intensity and time, but this is virtually impossible in the hypothesis of working during normal business hours of the basilica. For this reason, three Airstar lighting balloons were used to provide a homogeneous light. In addition to the balloons, four adjustable LED lamps (more manageable than the large balloons described above) were also used.

3.3 Geometric Modelling

The phase of geometric modelling also involved numerous choices, linked to the characteristics of the basilica, in this case, its decorative richness.

The decorative apparatus is very varied: San Marco, in fact, can be seen as a collection of pieces of art from the entire domain of Venice in the Mediterranean basin. Very often capitals, decorations and statues come from different areas of the world and different historical periods; each element has specific characteristics in terms of shape, size, position and geometric complexity. Therefore, each individual object has required specific modelling devices. The most evident example is the erosion of the bases of the columns of the central nave, for which it is difficult to lighten the original moulding.

Finally, the basilica is characterized, unusually for European architecture, by connected curved surfaces (for example, the connection between the domes and the arches of the aisles): so it is not always possible to identify the breaklines that generally 'build' the architecture (Fig. 1).

Considering all these difficulties, it was decided to use a modelling approach based on NURBS. This system combines the possibility to have a reliable representation of reality and easy to manage processes, at least in terms of memory occupation. Other alternatives such as parametric or mesh modelling, widely used in the industry, have in fact proved to be particularly effective only in one of the two areas (model accuracy, ease of use), but very weak in the other. The negative factor of the NURBS approach is the very low level of automation: all operations require a strong manual intervention of the operator and, therefore, make the entire construction phase of the model very expensive in terms of time. The work pipeline has provided researchers to start from the point clouds, obtained by photogrammetry, and then build basic and generating profiles for sweeping, extrusion and loft operations (Fig. 4). In some specific cases, such as domes or more complex elements, profiles were extracted at close range and then modelled using commands such as 'loft' or 'network of curves'. Even the most

Fig. 1 Example of modelling by NURBS. From the left: pointcloud, geometric model and final result with a single element highlighted. *Source* He.Su.Tech. group

Fig. 2 Examples of sculptural elements modelled by mesh approach. *Source* He.Su.Tech. group

complex elements, such as capitals, where possible were represented as NURBS (Fig. 4).

Only a few elements, usually of a sculptural type such as statues, bas-reliefs, were represented as meshes (Fig. 2).

3.4 Texture and Colour Mapping

A fundamental element of the research was the management of colour, understood as the need to provide not only a geometric representation of the Basilica but also, and above all, an accurate description of its appearance. This aim stems from the project's conservative objective: the Procuratoria wanted to equip itself, as with the pavement (Fregonese et al. 2006), with an effective tool for the restoration of mosaics. With this objective in mind, but without neglecting the need to navigate through the model, we worked on the texturization of the NURBS model by reprojecting the photographs oriented on the surfaces. This operation was carried out in the Agisoft Photoscan software. In order to proceed with the projection of the photographs, the working pipeline planned never to modify the reference system or to translate the object. The result of this procedure is that the geometric model, modelled by a manual process with the photogrammetric model previously developed in Photoscan, coincides completely. The result of the texturization phase is a three-dimensional model mapped with the real texture, in which the position of every single tile is topologically correct. It is, therefore, a mapping not only obtained for the purpose of immersive representation but guarantees the metricity of the result (Figs. 3 and 4).

Still, with regard to the treatment of colour, but with the aim of obtaining the most useful outcome for conservative purposes, the orthophoto, the next step was the projection of images oriented on a plane of reference. In this case, the resolution of the result is very high: the covering of the single pixel on the real surface is 0.5 mm

Fig. 3 The same object represented only with geometric information (left) and with texture extracted from georeferenced images (right). *Source* He.Su.Tech. group

Fig. 4 The area of the Ascension, in the centre of the Basilica, with the overlapping of textured areas. *Source* He.Su.Tech. group

(Fig. 5). This very small value is linked to the need to accurately represent the single card. To recognize the single tessera, it was necessary to set a pixel size smaller than the gap between two different tiles, to be able to display it in the final result.

In the most complex areas, domes, vaults and connecting elements, it was decided to use simplified reference planes.

Fig. 5 The orthophoto of the dome of Baptistery, (top), with some details (bottom) where it is possible to see the smallest details of the golden tessera in the orthophoto (pixel size of 0.5 mm)

3.5 Data Management

The exceptionality of the study, even in data management, is linked to several aspects. First of all, the very aim of the study was to elaborate a three-dimensional model from which to extract a lot of other information. This means that it is not necessary to decide in advance on the necessary sections but to postpone the choice to a later stage according to the different needs. In addition, the research project immediately involved not only surveying and modelling the geometry of the basilica, but also its image—the finishing of the surfaces. All this translates into the need to manage various types of data (geometry and images), to be able to organize them in such a way that access to the data is effective and immediate and, above all, to provide for different types of use linked to the operators involved.

The answer to all these needs came through the BIM systems. Designed for new buildings, they allow the design to be shared by the various operators of the building process, from the conception phase to the realization and management. The attempt made within the research project was to identify a system that could apply the same method to the existing heritage. The HBIMs differ from the BIMs precisely because they concern objects that are already made, in which the knowledge phase is not determined by a design choice, but implemented by a series of analyses, surveys and subsequent observations. The HBIM logic highlights many problems such as the construction time of three-dimensional models, which serve as a three-dimensional index of all information, the cost of realization and the difficulty in building reliable models of reality.

A step in solving these problems has been taken with the use of the BIM3DSG system, developed by the 3DSurveyGroup of the ABC Department of the Politecnico di Milano, already tested on the Duomo di Milano and in other cases (Fassi et al. 2015).

To allow for the greatest adaptability to the case studies, the system has not provided its own modelling environment but relies on external software of a commercial and non-commercial nature. This guarantees, therefore, that it can be integrated into an existing work process and avoids a very steep initial learning curve.

In the case of San Marco, for example, the use of the NURBS model made with Rhinoceros allows for the extraction of infinite plants and sections, where necessary, without having to choose them a priori. At the same time, the three-dimensional model is also the three-dimensional element (or rather the sum of the 3D objects) to which all the information can be linked.

Among this information, in particular, the orthophotos that constitute the most reliable and effective elaborate in the representation of the mosaic and stone surfaces are also included.

Finally, the third type of data is stored in the archive and that is the textured model. The latter, generated to offer a more immersive visualization of the entire basilica, is built through the import of the NURBS model into Photoscan (then with the transformation from NURBS to MESH) and the texturization through the reprojection of photographs on the model.

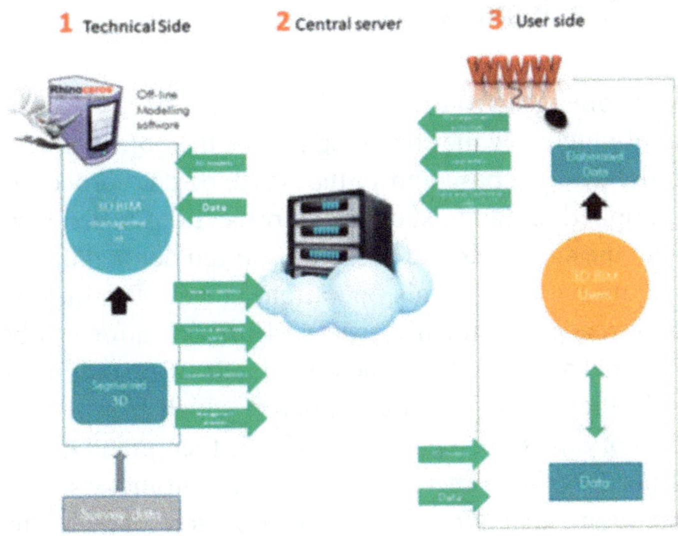

Fig. 6 Scheme of BIM3DSG system. *Source* Fassi, Rechichi

The system is structured in three parts that reflect three different actors of the operational process (Fassi et al. 2017): the modeller, the user and the data manager (Fig. 6).

The first part of the system provides the interface with a modelling software chosen by the user (in this case, the software used was Rhinoceros). The separation of this phase from the general BIM process responds to the need for specialized figures in data modelling starting from reality-based data and relieves users from the BIM and also from the heavy construction load and/or management of the three-dimensional model. The modeller can load the model, element by element, and eventually also load subsequent versions, which allow for keeping the initial ones.

The second part of the system, on the user's side, provides a simplified interface so that you can easily and directly access the georeferenced data on the 3D model. This part of the system is responsible for both data entry and querying the system through a database that can be adapted according to existing needs. This is done on a web-based basis and therefore does not require the use of any specific software. This aspect should not be underestimated as it ensures the multi-platform use of the system: you can access, enter data and perform queries from both desktop and mobile systems that can then be used directly on site (e.g. via tablet).

In this phase, the user can load all the objects of interest through a selection menu and, in addition, define the level of detail. The user can also decide whether to see the geometric or textured model. After loading the model, if you select an element of the model, a window will open containing all the files linked to the element itself: in the case of San Marco, the orthophotos of the individual elements will be linked, but other types of images could also be linked.

The third part of the system, transparent with respect to the other phases, is the true heart of the system: the central server, where all the information, both geometric and logical, reside, dedicated to the management of data through a database Postgres.

4 Conclusions

As it emerges from the critical issues discussed, the San Marco 3D project has highlighted how the characteristics of the object affect the choices of the survey, both in the design phase and in the phase of data processing and management. In cases of applied research, such as the project described here, the objective is precisely to bend the standards and guidelines, often defined in a theoretical way, to the concrete operational needs. This operation, however, is only possible if, behind the choices to be made, lies the awareness of the operations to be performed, the results to be obtained and especially the knowledge of methods and tools and their operation.

In the case of photogrammetry, which in recent years has reached a very high level of diffusion, the need for a basic theoretical knowledge is a particularly burning issue: the development of increasingly automatic software seems to go in the opposite direction—that of a total 'democratization' of the system, but it is actually an attitude that shows its limits especially when the object to be detected is complex and therefore not suitable for an automatic approach.

Finally, the research has shown the need for a multidisciplinary approach to the problem: the collaboration between the field of surveying, the computer world and that of conservation, final users of the system, has allowed those involved to focus on the results from the beginning with a certain degree of clarity.

Acknowledgements The HBIM part of the project was also developed with the support of the SIR project entitled 'Building Information Modelling for the planned conservation of Cultural Heritage: even a Geomatic question' (identification code RBSI144B5K) funded by the Ministry of Education, University and Research in the national research framework SIR (Scientific Independence of Young Researchers) 2014.

References

Adami A, Fassi F, Fregonese L, Piana M (2018) Image-based techniques for the survey of mosaics in the Basilica di S. Marco in Venice. Virtual Archaeol Rev (2018) (In press). ISSN 1989-9947. https://doi.org/10.4995/var.2018.9087

Boehler W, Marbs A (2001) The potential of non-contact close range laser scanners for cultural heritage recording. In: CIPA XIX international symposium, XVIII-2001, Potsdam and IAPRS

Chiabrando F, Donadio E, Rinaudo F (2015) SFM for orthophoto generation: a winning approach for cultural heritage knowledge. In: The international archives of the photogrammetry, remote sensing and spatial information sciences, vol XL-5/W7, pp 91–98. https://doi.org/10.5194/isprsarchives-xl-5-w7-91-2015

Fassi F, Achille C, Mandelli A, Rechichi F, Parri S (2015) A new idea of BIM system for visualization, sharing and using huge complex 3D models for facility management. In: The international archives of the photogrammetry, remote sensing and spatial information sciences, Avila, Spain, vol XL-5/W4, pp 359–366. https://doi.org/10.5194/isprsarchives-xl-5-w4-359-2015

Fassi F, Fregonese L, Adami A, Rechichi F (2017) BIM system for the conservation and preservation of the mosaics of San Marco in Venice. In: ICOMOS/ISPRS international scientific committee on [...] ES, vol XLII-2/W5, Bahnhofsalle 1e, Gottingen, 37081, Copernicus Gesellschaft MBH,

Germany, pp 229–236, Ottawa, 28 Aug 2017–01 Sept 2017. ISSN: 2194-9034. https://doi.org/10.5194/isprs-archives-xlii-2-w5-229-2017

Fregonese L, Monti C, Monti G, Taffurelli L (2006) The St. Mark's Basilica pavement: the digital orthophoto 3D realisation to the real scale 1:1 for the modelling and the conservative restoration. In: Abdul-Rahman A, Zlatanova S, Coors V (eds) Innovations in 3D geo information systems. Lecture notes in geoinformation and cartography. Springer, Berlin, Heidelberg, pp 683–693. https://doi.org/10.1007/978-3-540-36998-1_52

Fregonese L, Taffurelli L, Adami A, Chiarini S, Cremonesi S, Helder J, Spezzoni A (2017) Survey and modelling for the BIM of Basilica of San Marco in Venice, 3D Arch-2017, Nafplia, Greece. In: International archives of the photogrammetry, remote sensing and spatial information science, vol XLII-2-W3, pp 303–310. https://doi.org/10.5194/isprs-archives-xlii-2-w3-303-2017

Fregonese L, Taffurelli L, Adami A (2018) BIM application for the Basilica of San Marco in Venice: procedures and methodologies for the study of complex architectures. In: Remondino F, Georgopoulos A, González-Aguilera D, Agrafiotis P (eds) Latest developments in reality-based 3D surveying and modelling. Basel, Switzerland, MDPI, pp 348–373

Godin G, Rioux M, Levoy M, Cournoyer L, Blais F (2001) An assessment of laser range measurement of marble surfaces. In: Proceedings of the 5th conference on optical 3-D measurement techniques, Vienna, Austria, pp 49–56

Remondino F, Spera MG, Nocerino E, Menna F, Nex F (2014) State of the art in high density image matching. Photogram Rec 29(146):144–166. https://doi.org/10.1111/phor.12063

Automatic Processing of Many Images for 2D/3D Modelling

Luigi Barazzetti, Marco Gianinetto and Marco Scaioni

Abstract The era of big data requires increasing automation for the analysis of huge information in a short time and this need becomes critical when dealing with geoinformation. This chapter describes the automatic geocoding of digital images based on high-end Photogrammetric and Remote Sensing methods. In particular, the so-called Structure-from-Motion (SfM) technique is developed to handle image data sets in close-range applications, and here, it is generalized to deal with multi-scale applications. Some examples are proposed with panoramic images for the measurement of indoor narrow spaces, with smartphone cameras and UAV for the 3D reconstruction of complex monuments, as well as with airborne and satellite images for the survey at the territorial scale.

Keywords Aerial · Archive photos · Close range · Drones · Satellites · Geocoding · Image registration · Photogrammetry · Remote sensing · Structure from motion

1 Fundamentals of Image Geocoding

Digital imagery is nowadays an enormous source of qualitative informative content, which may also provide accurate 2D and 3D metric information when they are processed using Photogrammetric and Remote Sensing techniques (Luhmann et al. 2014). In the past decades, the research in this field has developed powerful and consolidated methods for the automatic alignment of multiple data sets and the integration with the Image Processing and Computer Vision communities has played a paramount role. Today, the boundaries between these disciplines is much more blurred than it was in the past (Hartmann et al. 2016).

Disregarding the specific processes to extract 2D or 3D geometries from different types of images, *geocoding* (or *georeferencing*) is a fundamental phase to associate a

L. Barazzetti · M. Gianinetto (✉) · M. Scaioni
Architecture, Built Environment and Construction Engineering—ABC Department, Politecnico di Milano, Milan, Italy
e-mail: marco.gianinetto@polimi.it

B. Daniotti et al. (eds.), *Digital Transformation of the Design, Construction and Management Processes of the Built Environment*, Research for Development,
https://doi.org/10.1007/978-3-030-33570-0_32

location in the geographical space to each geometric primitive. In general, geocoding may be obtained from one out of the following approaches:

1. *Direct georeferencing* of the sensor is adopted for data acquisition by integrated Global Navigation Satellite Systems (GNSS), Inertial Navigation Systems (INS) and other types of sensors (e.g., star trackers in spaceborne sensors);
2. *Indirect georeferencing* is based on Ground Control Points (GCPs), whose coordinates are known in either the images space and the object space;
3. *Mixed techniques*.

When a time series made of images collected on the same location from the ground (e.g., fixed camera stations) or above the ground (e.g., drones, aircraft or satellite) is analysed, it is not necessary that all the frames are georeferenced. As in Photogrammetric *bundle block adjustment*, GCPs are measured on a smaller subset of the whole block of images, while the others are *co-registered* among them and with respect to it.

If *indirect georeferencing* techniques are used, one image is manually *geocoded* using known GCPs (this image is usually called 'master' or 'reference' image) and all the other images of the series are then manually or automatically co-registered to it. On the other hand, when using *direct georeferencing* techniques, all the images are already georeferenced and only a few GCPs are needed to correct some residual biases. Unfortunately, this approach is not possible for any type of applications, such as for instance, in close-range Photogrammetry (Luhmann et al. 2014). In other cases, it may only provide approximate geocoding to be used for instantiating other georeferencing techniques. This is the case of most Photogrammetric blocks recorded using drones (Colomina and Molina 2014; Granshaw 2018a) or for the analysis of satellite imagery, where direct geocoding is not enough accurate.

When indirect georeferencing methods should be called for, provided that some external constraints are always needed, the option to measure GCPs only on a (small) subset of the images and then co-register the rest of the data is really strategic to reduce the processing time and to limit the operator workload. Consequently, in recent years, several automatic approaches have been developed to this purpose. The Department of Architecture, Built Environment and Construction engineering (DABC) of Politecnico Milano has contributed to this topic by channelizing the registration process of different types of images within a common framework. This can be referred to the Photogrammetric procedure addressed to as *Structure from Motion*, which is discussed in the next section.

2 Structure from Motion

Structure from Motion (SfM) is, nowadays, the most popular technique used in Photogrammetry for image co-registration (in such a case, the correct term is *image orientation*). The reader can find a review of its origins and development in Granshaw (2018b), including the most relevant literature. In the beginning, SfM was limited to

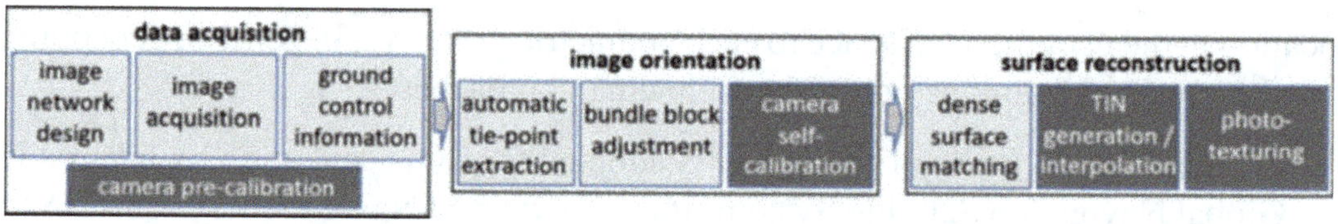

Fig. 1 Workflow of Structure-from-Motion basic processing pipeline

small–medium-size images (up to few megabytes) due to computational limitations of the implemented feature extraction algorithms such as SIFT or SURF (Barazzetti et al. 2009). Consequently, its use was limited to close-range blocks collected from ground-based stations or from drones.

Despite this, SfM was rapidly spreading out in the scientific community, where it reached a great popularity, especially in the field of Geosciences (Westoby et al. 2012; Eltner et al. 2016) and Cultural Heritage documentation (Barazzetti et al. 2011). The concurrent progress of the algorithms' performances and the use of efficient computing solutions (e.g., parallel computing and processing at GPU level) has recently opened the way to deal with large blocks of large-size digital photos, such as the ones adopted in aerial projects for topographic mapping or Remote Sensing projects for land mapping. Figure 1 summarizes the basic concept of SfM, though many variations may come up in specific software implementations. Besides, the term SfM should only refer to the orientation stage, however, in the common language, it frequently refers to the whole Photogrammetric pipeline, also including the dense surface matching stage to generate a colourized point cloud.

2.1 *Panoramic Images for the Survey of Indoor Narrow Spaces*

Consumer-grade cameras able to capture 360° photos and videos are becoming more popular for the opportunity to look in any direction as well as for exploiting immersive visualization with virtual reality headsets (Barazzetti et al. 2018). Today, many cameras are already available on the market and some of them are quite cheap (100–500 EUR), such as the Ricoh Theta S, 360fly 4K, LG 360 CAM, Kodak PIXPRO SP360 4K, Insta360, Kodak PIXPRO SP360, or the Samsung Gear 360. On the other hand, professional systems for 360° imaging such as the GoPro Odyssey, Sphericam V2, Nokia OZO or the GOPRO OMNI are still expensive today (dozens of thousands of euros).

As mentioned, 360° cameras capture the whole scene around the standpoint in a single shot and are becoming a new paradigm for Photogrammetry. Recalling that a 3D model of the scene can be created from images acquired from different points of view (at least two), multiple images can be processed following the typical workflow for image processing based on the spherical (equirectangular) camera model, whose equations in a camera centered reference system can be written as follows:

$$\begin{aligned} x &= f \arctan\left(\frac{X}{Y}\right) \\ y &= f \arctan\left(\frac{Y}{\sqrt{X^2+Z^2}}\right) \end{aligned} \tag{1}$$

where $f = \text{image\_width}/(2\pi)$.

Two examples of Photogrammetric surveys using panoramic cameras are shown in Figs. 2 and 3. The first example describes the modelling of a narrow corridor (Fig. 2). While this is a conventional application for Photogrammetry, nevertheless, traditional Photogrammetric or (static) laser scanning surveys would have required a lot of time for the data acquisition because of the tight space. In this case, the collection and processing of more than 200 panoramic images required only few minutes.

Fig. 2 Modelling of a narrow corridor with panoramic cameras

Fig. 3 Comparison of the shape of the staircase of the Pestagalli's spire in the Cathedral of Milan (on the left) with the model made from 360° panoramic images (on the right)

Figure 3 shows a non-conventional application. In this case, 360° panoramic images are used for surveying the narrow staircase inside the Pestagalli's spire of the Cathedral of Milan (Italy). Again, the tight space available makes spherical images a valid alternative to laser scanning or traditional Photogrammetric techniques.

These examples demonstrate that data processing is feasible not only for small images but also for large blocks. The geometry of a spherical image is more suitable than central perspective of traditional Photogrammetry when the field-of-view and image overlap are critical parameters. On the other hand, this solution may be problematic when considering very long image sequences, that could result in accuracy problems when images are progressively added without external constraints. In this case, the use of GCPs measured with a total station remains a primary tool to control the network geometry, as in the application examples abovementioned.

2.2 *Frame Images for the Survey of Complex Monuments*

If panoramic cameras are an emerging technology for 3D Photogrammetric surveys, frame cameras are a well-established technology for most applications. The SfM approach is an almost fully automated technique, nevertheless, in close-range projects, the image acquisition still remains under complete control of the operator. Yordanov et al. (2019) proposed some guidelines to drive this task, which is crucial to achieve final good results. Two examples with different types of photogrammetric blocks are presented to show how the same processing pipeline and the same software package (here Agisoft Metashape® ver. 1.5.0) may cope with the orientation of images based on SfM.

The first example (Fig. 4) describes the documentation of Cultural Heritage with a smartphone. In this case, 26 photos of a Medieval capital were collected during a visit to the 'Museum de los caminos' in the Episcopal Palace of Astorga (Spain). The internal camera of a Samsung Galaxy Grand Prime was used (focal length 3.3 mm, sensor size 3,264 × 1,836 pixels, pixel size 1.2 μm). Figure 4 (on the left)

Fig. 4 On the left: camera poses used to reconstruct a Medieval capitel. On the right: the final textured 3D model of the capital obtained from the projection of images

shows the camera poses after image orientation based on SfM. Here, we can see that most photos were captured from stations in front of the capital and were located at approximately the same distance. A few images were taken with the camera rolled approximately 90° with respect to others in order to help camera calibration during the bundle adjustment applied to estimate image orientation parameters (Luhmann et al. 2016). An average number of 1,700 tie points were extracted per image, with a Root Mean Square (RMS) of residuals on reprojected image coordinates of 0.5 pixels.

Figure 4 (on the right) shows the textured 3D model obtained after dense matching (Remondino et al. 2014). Despite the lack of any quality assessment, the 3D model shows that very interesting results for documentation of a small Cultural Heritage element (the capital may be contained in a sphere of radius 60 cm) may be obtained using a cheap camera (about 100 EUR).

The second example concerns the reconstruction of an old industrial heritage building in Sicily (Italy). The 'Fornace Penna' was built up at Scicli (Ragusa province) at the beginning of XX Century, and then severely damaged by a fire in 1924, when it was abandoned. Figure 5 shows the camera poses after image orientation based on SfM and the textured 3D model obtained from the projection of images.

The complex building geometry and texture were surveyed using images from a small drone (DIJ Phantom 2, equipped with a camera featuring: focal length 3.61 mm, sensor size 4,000 × 3,000 pixels, pixel size 1.6 μm) and some photos from ground-based stations. In total 252 images were recorded. This data set belongs to the SIFET benchmark described in Piras et al. (2017). Since the use of Unmanned Aerial Vehicles (UAVs) allows to predefine the acquisition points, thanks to the GNSS navigation system onboard, the geometry of camera poses can be carefully planned in advance (Pepe et al. 2018).

In such a case, the whole data acquisition was accomplished by a team of expert surveyors which also included the measurement of some GCPs for accurate georeferencing. These were realized by targets on the ground surface and natural features on the construction's facades. Consequently, in such a case study also some evaluations

Fig. 5 On the left: camera poses of the UAV block of 'Fornace Penna' (Scicli, Italy). On the right: the final textured 3D model of the building obtained from the projection of images

of the accuracy were possible. An average number of 800 tie points were extracted per image, with an RMS of residuals on reprojected image coordinates of 0.4 pixels. Residuals on 23 GCPs resulted in an RMS error of 2.7 cm.

This second case study proves how the SfM approach combined with the use of UAV data allows modelling complex buildings in accurate and complete way.

2.3 *Airborne and Satellite Images for the Survey at the Territorial Scale*

When moving from the local to the territorial scale, aerial Photogrammetry and satellite Remote Sensing are the technologies used for extensive mapping.

Referring to aerial surveys, Fig. 6 shows a topographic mapping project over a small village in Italy. In this case, 18 images recorded with a ZI DMC-II digital airborne camera were processed using SfM (focal length 120 mm, sensor size 7,680 × 13,824 pixels, pixel size 12 μm). Digital airborne cameras are usually integrated into a GNSS/INS unit which may directly provide the exterior orientation parameters for georeferencing, besides the known interior orientation parameters that are used as input in the bundle adjustment. On the other hand, automatic aerial triangulation based on corresponding tie points observed on the images is applied to have more control over the final results. An average number of 1,400 tie points were extracted per image, with an RMS of residuals on reprojected image coordinates of 0.5 pixels. Here, the SfM approach was tested to replace the standard methods for automatic aerial triangulation adopted in aerial Photogrammetry.

In a similar way to the previous case studies, Fig. 6 shows some intermediate products of the SfM process. Though the example considered here only concerns a small block, this procedure may be also applied to larger blocks like the ones used in mapping projects at regional scales.

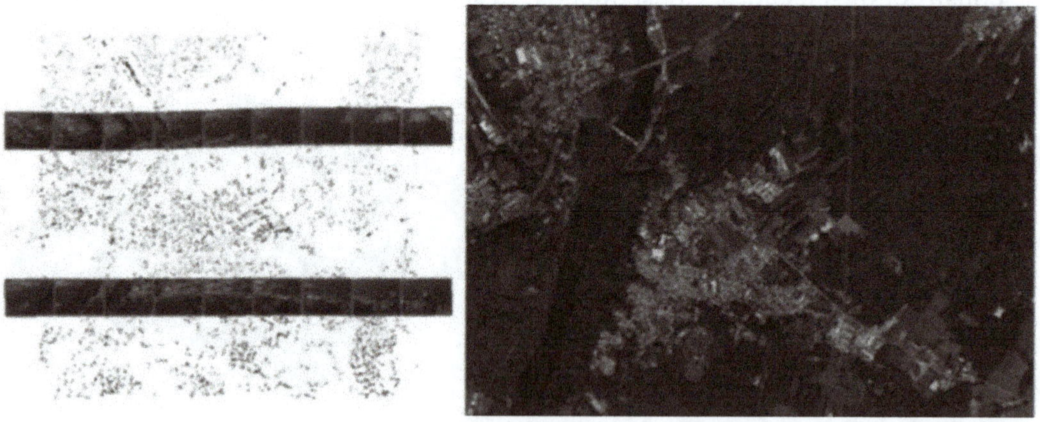

Fig. 6 On the left: camera poses used of the aerial block. On the right: the final orthophoto obtained from the projection of images on the obtained 3D model

When dealing with satellite Remote Sensing technologies, the exponential growth in the availability of traditional images and the revolution of microsatellites started with CubeSats pose several challenges. From the end-user point-of-view, commercial off-the-shelf software packages and cloud platforms able to process large number of images are becoming more accessible. As a consequence, new opportunities open up not only to experts but also to a wider public interested in digital survey and reconstruction at the territorial scale.

However, the improved availability of satellite data and products requires more efficient methods for data processing, in which the combined use of many images than the ones traditionally exploited is quite attractive. This task requires novel approaches and algorithms. Big data has, therefore, become a popular word in the Remote Sensing community and, at the same time, an opportunity and a challenge.

With reference to this matter, the Department ABC of Politecnico Milano has developed an alternative approach (MIRA—Multi-Image Robust Alignment) for the co-registration of large multitemporal data sets of images (Barazzetti et al. 2014) that extends the traditional pairwise approach (i.e., 'one-to-one') to the simultaneous processing of the whole time series (i.e., 'one-to-many'). The basic concept is to use not only corresponding features between the 'master' image and the other images to be processed, but all the corresponding features shared between all the images of the time series. The coordinates of these correspondences (tie points) are used to instantiate a system of redundant equations to be solved within a Least Squares framework for the determination of the unknown registration parameters in a given geodetic datum. In such a way, also the images without corresponding features shared with the 'master' image can be registered, thus exploiting all the available images. In addition, this method increases the inner reliability of the observations and thus gain robustness against gross errors and limits the error propagation (Scaioni et al. 2018). Finally, with the growth of the time series, the network geometry improves, with overall benefits for the geometric alignment.

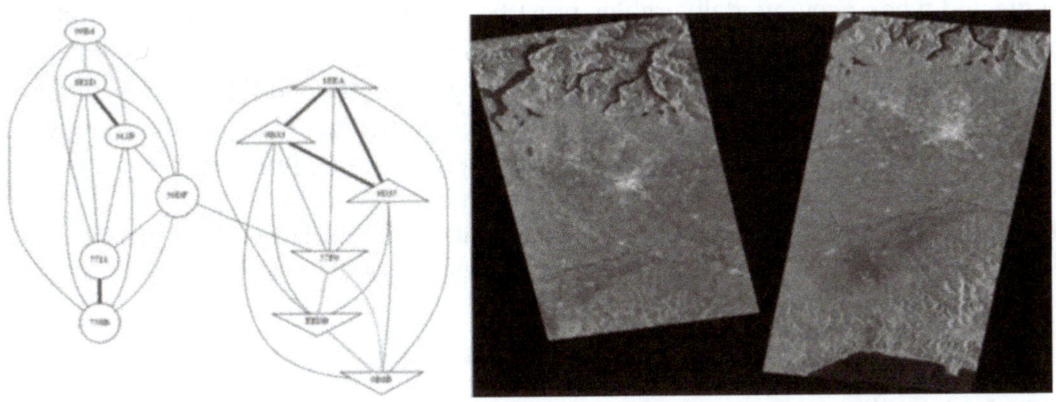

Fig. 7 On the left: connection graph of the Sentinel-1 SAR images. On the right: corresponding features between the ascending image (56DF) and the descending image (57F0) found by applying MIRA method

Figure 7 shows an example of data processing of Sentinel-1 (SAR) satellite time series with the MIRA method described above (Gianinetto et al. 2016). The surveyed region spans over an area of approximately 15,500 km^2 in the North-West of Italy, from the Lake of Como to the Gulf of Genoa. Twelve Sentinel-1 images (6 ascending and 6 descending) were collected in Stripmap mode (10 m geometric resolution), with dual polarization (HH/HV for ascending, VV/VH for descending), and with large differences in incidence viewing angles. This is a very challenging acquisition scheme for image registration.

Looking at the connection graph (Fig. 7 on the left) we can see that some images have a strong interconnection, while some other images have a weak or null interconnection. That reflects the imaging modes, in fact: (i) elliptical edge nodes refer to S2 ascending images; (ii) circular edge nodes refer to S4 ascending images; (iii) triangular edge nodes refer to S1 descending images; and (iv) inverse triangular edge nodes refer to S6 descending images. In this case, traditional pairwise methods for image co-registration failed to process this data set because many images did not have any correspondences with the 'master' image. On the other hand, the simultaneous processing based on MIRA exploited the link between the ascending and the descending blocks of images (56DF vs. 57F0) along with the links between each image pair to calculate a global co-registration of the entire satellite time series.

3 Conclusions

The growing availability of digital images captured with different sensors is an amazing opportunity for several users involved in many real applications.

When the images must be turned into metric products, the use of automated, efficient and reliable registration procedures has a primary importance to prepare all the data for the following stages of processing. Such preliminary operation cannot be neglected when accurate deliverables must be produced, requiring robust solutions able to deal with huge datasets in a fully automated way.

The integration of the SfM strategy in Photogrammetry and Remote Sensing has led to novel processing algorithms, which have significantly changed the traditional workflow of several surveying applications based on digital images.

Acknowledgements The authors would like to thank the Italian Society of Photogrammetry and Topography (SIFET) for the UAV benchmarking data set 'Fornace Penna'.

References

Barazzetti L, Remondino F, Scaioni M (2009) Combined use of photogrammetric and computer vision techniques for fully automated and accurate 3D modeling of terrestrial objects. In: Videometrics, range imaging, and applications X, SPIE, paper no. 74470M

Barazzetti L, Binda ML, Scaioni M, Taranto P (2011) Photogrammetric survey of complex geometries with low-cost software: application to the 'G1' Temple in Myson, Vietnam. J Cult Herit 12:253–262

Barazzetti L, Gianinetto M, Scaioni M (2014) Automatic co-registration of satellite time series via Least Squares adjustment. Eur J Remote Sens 47:55–74

Barazzetti L, Previtali M, Roncoroni F (2018) Can we use low-cost 360 degree cameras to create accurate 3D models? Int Arch Photogramm Remote Sens Spat Inf Sci 42(2):69–75

Colomina I, Molina P (2014) Unmanned aerial systems for photogrammetry and remote sensing: a review. ISPRS J Photogramm Remote Sens 92:79–97

Eltner A, Kaiser A, Castillo C, Rock G, Neugirg F, Abellán A (2016) Image-based surface reconstruction in geomorphometry-merits, limits and developments. Earth Surf Dyn 4:359–389

Gianinetto M, Monno V, Barazzetti L, Dini L, Daraio MG, Rota Nodari F (2016) Subpixel geocoding of COSMO-SkyMed and Sentinel-1 time series imaged with different geometry. In: International geoscience and remote sensing symposium (IGARSS 2016), Beijing, China, pp 5007–5010

Granshaw SI (2018a) RPV, UAV, UAS, RPAS … or just drone? Photogramm Rec 33:160–170

Granshaw SI (2018b) Structure from motion: origins and originality. Photogramm Rec 33:6–10

Hartmann W, Havlena M, Schindler K (2016) Recent developments in large-scale tie-point matching. ISPRS J Photogramm Remote Sens 115:47–62

Luhmann T, Robson S, Kyle S, Boehm J (2014) Close-range photogrammetry and 3D imaging. Walter de Gruyter, p 684

Luhmann T, Fraser CS, Maas H-G (2016) Sensor modelling and camera calibration for close range photogrammetry. ISPRS J Photogramm Remote Sens 115:37–46

Pepe M, Fregonese L, Scaioni M (2018) Planning airborne photogrammetry and remote-sensing missions with modern platforms and sensors. Eur J Remote Sens 51:412–435

Piras M, Di Pietra V, Visintini D (2017) 3D Modeling of industrial heritage building using COTSs system: test, limits and performances. Int Arch Photogramm Remote Sens Spat Inf Sci **42**(2/W6), 281–288

Remondino F, Spera MG, Nocerino E, Menna F, Nex F (2014) State of the art in high density image matching. Photogramm Rec 29:144–166

Scaioni M, Barazzetti L, Gianinetto M (2018) Multi-image robust alignment of medium-resolution satellite imagery. Remote Sens **12**(24)

Westoby MJ, Brasington J, Glasser NF, Hambrey MJ, Reynolds J (2012) 'Structure-from-Motion' photogrammetry: a low-cost, effective tool for Geoscience applications. Geomorphology 179:300–314

Yordanov V, Mostafavi A, Scaioni M (2019) Distance-training for image-based 3D modelling of archeological sites in remote regions. Int Arch Photogramm Remote Sens Spat Inf Sci **42**(2/W11):1165–1172

Geo-Referenced Procedure to Estimate the Urban Energy Demand Profiles Towards Smart Energy District Scenarios

Simone Ferrari, Federica Zagarella and Paola Caputo

Abstract To effectively reduce cities' environmental impact, current policies boost building stocks transition towards smart energy systems (i.e. distributed energy generation, renewables integration, energy storage, connection to district heating and cooling networks). Smart energy systems feature complex-related energy fluctuations, since they include the intermittent and unpredictable behaviour of renewable energies and the variable energy demand of buildings. However, the technical literature underlines the need for accurate methods adoptable in several urban contexts for detailed energy demand assessment. In this framework, a method to estimate the energy demand profiles of urban buildings has been developed with particular regard to the Italian contexts. The method includes a geo-referenced procedure to assess the volumetric consistency of a building stock by age, characterizing different technological solutions, and by the mostly diffuse urban use categories, i.e. residential and common tertiary (office), affecting different usage profiles, thanks to data available for the national territory. Specifically, spatial datasets on buildings from the Topographic Database, currently under standardization based on the European (INSPIRE) directive, and from the National Institution of Statistics (Istat) were used. In order to determine current hourly energy profiles, a set of dynamic energy simulations is foreseen, based on simplified reference buildings. Hence, the energy behaviour of selected building portions, representative of different heat exchanges boundary conditions, can be assessed. The derived hourly energy profiles per built volume can, therefore, be consistently associated with the considered building stock to obtain the overall hourly energy demand. Moreover, by assigning the upgraded technological properties to the reference buildings, it is possible to replicate the procedure and derive the variation of energy profiles for the defined retrofit scenario. The method has been tested on the city of Milan and is validated on a yearly basis.

S. Ferrari (✉) · F. Zagarella · P. Caputo
Architecture, Built Environment and Construction Engineering—ABC Department, Politecnico di Milano, Milan, Italy
e-mail: simone.ferrari@polimi.it

B. Daniotti et al. (eds.), *Digital Transformation of the Design, Construction and Management Processes of the Built Environment*, Research for Development,
https://doi.org/10.1007/978-3-030-33570-0_33

Keywords Smart energy district planning · Geographic information system (GIS) · Building stock geodatabase · Urban energy profiles estimation · Topographic database · National institution of statistics (Istat) census

1 Introduction

The recently issued Agenda 2030 included the improvement of the energy sector and cities as goals to achieve global sustainable development (UN 2015). In fact, cities are responsible for 70% in the global energy demand (UNEP 2015), which is expected to increase in step with urban population. Hence, given the opportunity for a further exploitation of fossil resources, substantial changes in the energy production mix, largely integrating renewable energy sources (RES), are promoted by worldwide policies.

Italy is currently one of the lowest energy intensity European countries, having boosted in past decade the adoption of energy efficiency measures and the integration of RES with mandatory requirements and considerable tax subsidies. In spite of this, its energy mix is not one of the most RES based in the European Union (EU), despite its large potential. Therefore, within the National Energy Strategy (MiSE 2017), a large increment of RES share and a reduction in final energy consumption is predicted, thanks to several indicated actions, e.g. distributed energy generation, replacement in buildings of conventional systems with heat pumps and biomass ones, installation of photovoltaic panels on buildings surfaces, district heating and cooling networks in densely populated areas, demand response programmes, energy storage at building and district levels.

On a broader scale, the transition of urban areas towards Smart Energy Districts implies a quite complex management of the mismatch between energy demand and supply. In fact, the renewable energy availability is quite unpredictable and variable both on daily and annual bases. Moreover, the demand side is variable, considering an existing plenty of different uses and the increasing role of active consumers. To properly manage these energy systems, by minimizing the related inefficiencies and costs, an accurate knowledge of their dynamic behaviour is fundamental.

Indeed, for supporting urban energy planners in this task, several different computational tools have been developed. Focusing on tools for smart energy planning on an urban/districts scale, the more detailed ones require the hourly based profiles of energy demand to be inserted for the studied context (Ferrari et al. 2019a). At the same time, having detailed energy profiles is a challenging task, since the availability of actual energy profiles for any context is not widespread (Fig. 1). In technical literature, this has led either to adopting approximate energy data, which could not be adequate for this purpose, or to defining robust estimation methods, which usually refer to 'engineering' and 'statistical' categories and the choice of which can be driven by the level of detail of held and desired data (Ferrari et al. 2019a, b).

With this in mind, within a Ph.D. research at the ABC Department (Zagarella 2019), a method has been defined for the accurate estimation of the energy demand

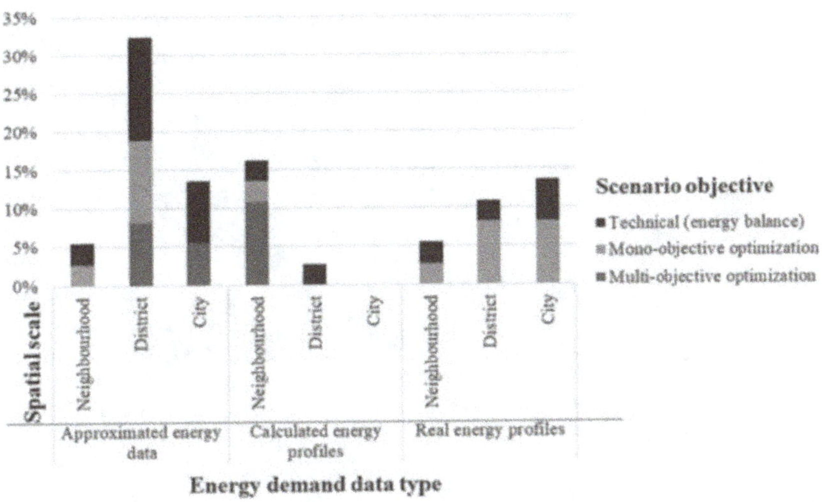

Fig. 1 Surveyed studies by spatial scale, energy scenario objective and type of energy demand data

profiles of an existing building stock, with reference to spatial data largely available from public administration across the Italian territory.

2 Digitized Spatial Data on the Built Environment

The EU highlighted the relevance of data availability to boost environmental policies as well as their ease of access, harmonisation and interoperability. Thus, with the Directive 2007/2/EC, the EU established general rules for the Infrastructure for Spatial Information in the European Community (INSPIRE). In Italy, the digitization process of public data had already begun back in 2005 (D.Lgs. 82/2005) but carried out only a limited number of concrete initiatives (Pasquinelli and Guzzetti 2006). In 2010, the INSPIRE Directive was implemented with the D.Lgs. 32/2010. From the national inventory of spatial data (Repertorio Nazionale dei Dati Territoriali or RNDT), which collects spatial metadata available at public administrations, a plenty of different types of data emerges with regards to the built environment (Fig. 2), although not all of it is uniformly collected and diffused across the territory.

Among the potentially available spatial datasets on buildings, a fundamental one is the so-called Topographic Database (TDb), introduced by the D.M. 10/11/2011 and currently under standardization to European (INSPIRE) and National (Intesa GIS) specifications. The TDb is made up of geo-referenced and updatable data, hierarchically structured into 'Layers', 'Themes' and 'Classes'. The latter are the mandatory basic unit of a TDb and comprise geometries associated with tabular data. Within the Layer 'Real estate and human settlements' and the Theme 'Built Environment', the following Classes are foreseen: the 'Building', the 'Volumetric Unit' (a building portion with same height), the 'Building Group' (a sum of buildings

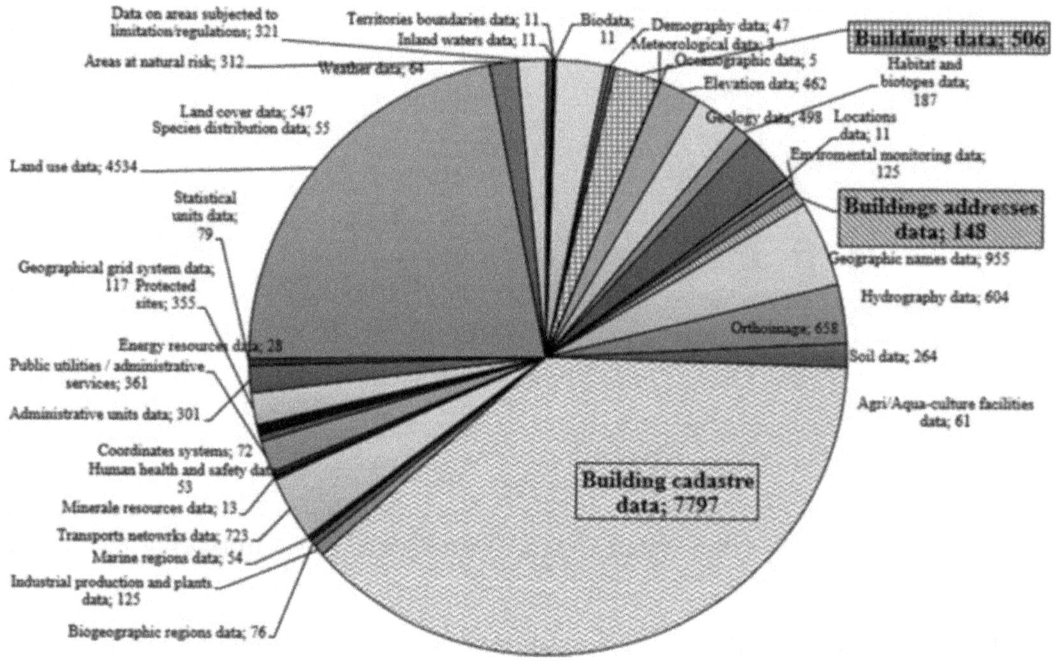

Fig. 2 Digitized spatial datasets classified by INSPIRE theme reported in the RNDT

having only external walls), the 'Roof Element', the 'Architectural Detail', and the 'Minor Building' (unstable volume attached to a building).

Another well-known and proved database of spatial data on buildings in Italy is the one coming from the 15th General Census of Population and Houses (GCPH), charged to the National Institution of Statistics, (Istat 2011). Since 2001 the 10 years census has been made of the population survey, which is carried out by means of questionnaires filled by respondents, and the survey on the existing buildings, carried out on-site by Istat technicians. Regarding the former, data was collected through a short form, given to most people and including the minimum set of data required by the European Statistical Office, and a longer form, sent to a sample of people.[1] Information from the longer form could be truly interesting for modelling energy use of the built environment (e.g. installed systems for space heating, space cooling, DHW energy and RES-based systems for electricity) although their access is quite restricted.

Public data was disseminated for defined spatial units (e.g. Census Unit,[2] Census Area,[3] etc.) by means of two datasets: the 'Spatial Bases', which only consist of

[1]In cities with less than 20 thousand inhabitants on 1 January 2008, the sample corresponds to the whole residing population; in cities with a population above the set threshold and in chief towns it corresponds to one-third of the people, selected among the so-called Census Areas.

[2]The Census Unit is the smallest urban area containing groups of buildings.

[3]The Census Area is a group of contiguous Census Units, with a population of 13–18 thousand the boundary of which was defined based on administrative divisions, infrastructures and natural elements.

the geographic representation of the statistical units, and the 'Statistical Variables', which contain a selection of data on population, dwellings and buildings.

3 Methodology

A method has been defined for supporting bodies involved in the energy planning at urban/district level. As hereinafter described, it has been intended to provide a method that can be both easily updated, in order to consider building energy retrofit scenarios as well, and widespread adopted, thanks to largely available spatial data. In detail, a geodatabase of an urban building stock, characterized by related energy features, is implemented in a Geographic Information System (GIS), namely in the QGIS tool with the embedded Python console (Fig. 3).

Buildings stock characterization was driven by the following considerations: the buildings' energy demand is largely influenced by building geometry, technological solutions and usage profiles, hence a procedure to calculate the building stock volume, differed by mostly diffuse urban use categories, i.e. residential and common tertiary (office), construction periods (i.e. from old traditional buildings to recent buildings complying with energy-saving requirements) and different portions of buildings, i.e. referring to different heat exchange boundary conditions (thermal zones) was defined.

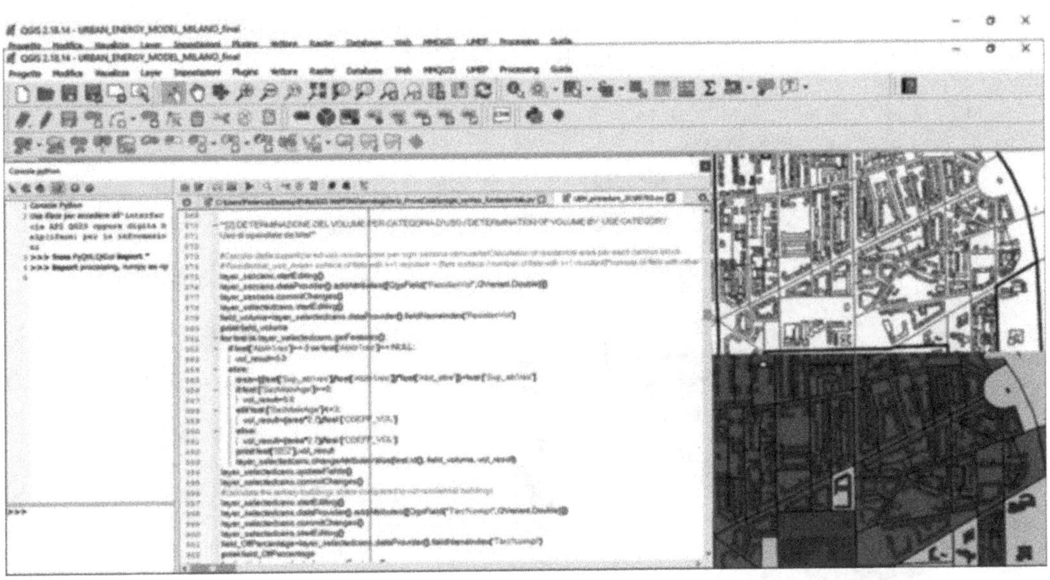

Fig. 3 QGIS tool with Python console embedded

First, a construction period for all buildings included in each Census Unit is determined taking into account the relative consistency in terms of building units,[4] which leads to the buildings' volume magnitude, based on the data reported in the GCPH.

Second, the characterization of buildings by thermal zones has been accomplished by prior calculation of the built volume based on TDb geometric data and its adjustment for accounting for only conditioned volumes (i.e. excluding staircases, small service volumes, etc.) and by grouping contiguous Volumetric Units (i.e. for considering only external walls in assessing the heat exchanges of the buildings' volumes). Then, for a previously defined simplified building energy model (Ferrari and Zagarella 2016), different thermal zones were classified with reference to the main boundary conditions of heat exchange: the ones placed on the intermediate storey (i.e. floors do not exchange heat), the ones on the first conditioned storey (i.e. floor exchanging heat with the unconditioned basement) and the ones on the last conditioned storey (i.e. ceiling bordering the sky). For each storey, thermal zones were additionally classified into the ones in angular positions (i.e. with two vertical walls towards outside) and the ones in intermediate positions (i.e. with just one external wall). Hence, the resulting six thermal zones of the simplified building model (Fig. 4) are consistently associated with the buildings of the considered stock.

Third, to characterize the building stock by use category, for each Census Unit, the total residential built volume is calculated based on the overall residential building units' number and floor surface data from the GCPH. Then, the not-residential volume is deduced from the total built volume minus the residential one; from the former, the common tertiary (office) volume is, in turn, deduced proportionally to the related office buildings number (data coming from the GCPH).

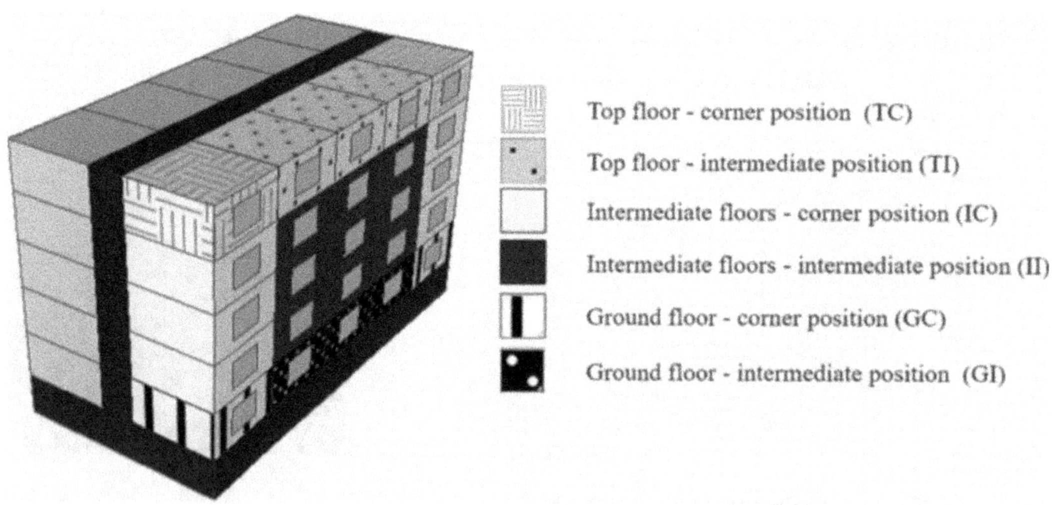

Fig. 4 Simplified building energy model and classification of thermal zones

[4]A building unit is either defined as a real estate unit (residential, office, commercial, etc.) accessed from a building collective distribution area or a residential estate unit accessed from outside and with its own address.

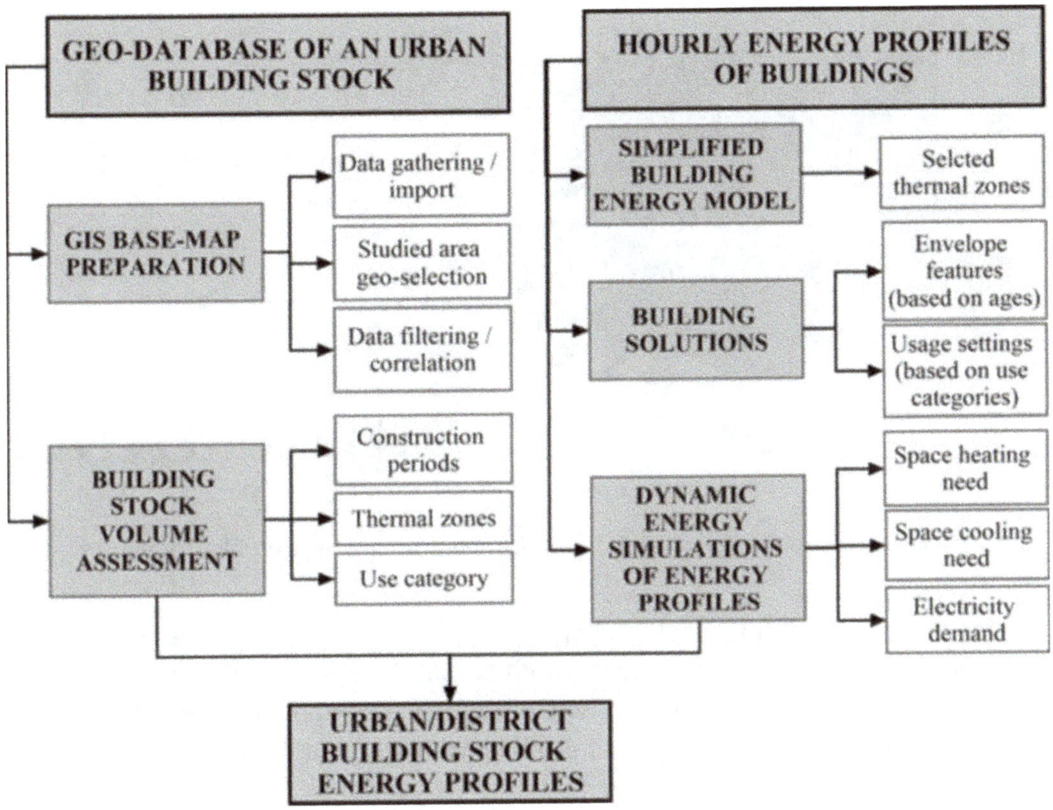

Fig. 5 Flowchart of the defined method

Once this is carried out, based on a set of alternative inputs for characterizing the building energy model into different building solutions, defined for covering the main construction periods and use categories, the hourly based energy profiles of space heating, space cooling and electricity for the selected thermal zones can be obtained through dynamic energy simulations.

Finally, the resulting energy profiles can be associated with the considered building stock volume to obtain the overall urban/district area energy profiles.

Figure 5 outlines the details of the defined method.

4 Case Study Application

The defined method to obtain the buildings' energy demand profiles was tested in the city of Milan. First, the building stock has been characterized by construction periods, thermal zones and use categories according to the GIS-based procedure. From results visualized in Fig. 6, it can be noted that in the city centre most of buildings are old, large (prevalence of intermediate thermal zones) and for office use, and vice versa in the outskirts as is what happens in reality.

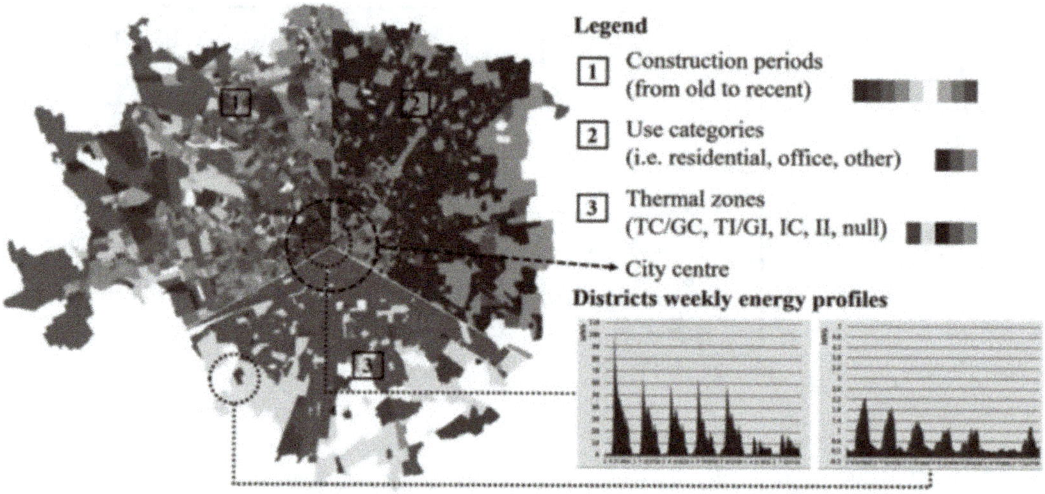

Fig. 6 Visualization of characterized building stock and district energy profiles for the case study

Once the urban building stock is characterized, a set of buildings energy solutions was alternatively associated with the simplified building energy model, by adopting different envelope characteristics based on age (Ferrari and Zanotto 2016a) and internal heat loads and air change rate profiles depending on the use categories (ISO 17772:2017).

The space heating activation period was set based on the D.P.R. 412/1993 and, according to (Ferrari and Zanotto 2016b), in residential cases, both the adoption of the adaptive approach for the set-point temperatures and a free cooling strategy for ventilation were considered during non-heating season. Hence, by means of dynamic energy simulations with TRNSYS (Klein et al. 2014), the energy profiles of space heating and cooling needs and of electricity demand were determined and converted into profiles of energy per built volume. These last ones have been associated with the considered built volume in GIS (an example of results for two districts is reported in Fig. 6).

Finally, due to a current lack in measured urban energy profiles, the energy demand calculated for the case study were validated on a yearly base, by comparing the results obtained for the overall residential building stock with the yearly urban residential energy consumptions derived from the regional energy balance (SIRENA 2012), assuming typical thermal systems' seasonal efficiencies from (UNI/TS 11300:2014 and D.M. 26/6/2015), and the energy consumption for Domestic Hot Water from (Ferrari and Zagarella 2015).[5]

Conversely, since tertiary buildings are usually equipped with quite complex HVAC systems, a reliable estimation of their energy performance, to be validated with yearly based registered consumption, will be possible only in further research development.

[5]Since tertiary buildings are usually equipped with quite complex HVAC systems, a reliable estimation of their energy performance, to be validated with yearly based registered consumption, will be possible only in further research development.

5 Conclusions

The assessment of hourly energy demand of the existing building stock, as well as the prediction of its variation due to energy efficiency measures, are fundamental activities for planning strategies of distributed energy generation, district heating and/or cooling networks, renewables integration, energy storage, etc., towards smart energy districts.

To support bodies in planning smart energy scenarios at urban/district level, a method has been defined to estimate the energy demand profiles of urban buildings. The method, which includes dynamic energy simulations of alternative solutions assigned to a simplified reference building model, foresees a GIS-based procedure useful to characterize the considered built volume by age, implying different technological solutions, and by the mostly diffuse urban use categories, i.e. residential and common tertiary (office), affecting different usage profiles. This procedure is based on spatial datasets which are widely available across the national territory (i.e. the Topographic Database, currently under standardization according to the INSPIRE directive, and the Istat population and houses census).

The method was applied to the case study of Milan building stock and accomplished elaborations produced valid results, as compared with available yearly based statistical energy data (i.e. the regional energy balance SIRENA).

The defined method is also suitable to assess the variation of the energy demand profiles due to buildings retrofit policies. In fact, by assigning the upgraded technological properties to the reference buildings, it is possible to replicate the procedure and derive the building stock profiles of the defined retrofit scenario.

Acknowledgements The authors thank the Annex 75 "Cost-effective Building Renovation at District Level Combining Energy Efficiency & Renewables" working group of the International Energy Agency - Energy in Buildings and Communities Programme - for having stimulated some deepenings of the present research.

References

D.Lgs. 27 gennaio 2010, n. 32. Attuazione della direttiva 2007/2/CE, che istituisce un'infrastruttura per l'informazione territoriale nella Comunità europea (INSPIRE), Rome (in Italian)

D.Lgs. 7 marzo 2005, n. 82. Codice dell' amministrazione digitale, Rome (in Italian)

D.P.R. n. 412/1993, Regolamento recante norme per la progettazione, l'installazione e la manutenzione degli impianti termici degli edifici, ai fini del contenimento dei consumi di energia, in attuazione dell'art. 4, comma 4 della legge 9 gennaio 1991, n. 10. (aggiornata dal D.P.R.551/99) (in Italian)

Ente Nazionale Italiano di Unificazione (UNI). UNI/TS 11300-2:2014. Prestazioni energetiche degli edifici Parte 2: Determinazione del fabbisogno di energia primaria e dei rendimenti per la climatizzazione invernale, per la produzione di acqua calda sanitaria, per la ventilazione e per l'illuminazione in edifici non residenziali

EU (2007) Directive 2007/2/EC of the European Parliament and of the Council of 14 March 2007 establishing an Infrastructure for Spatial Information in the European Community (INSPIRE), Brussels

Ferrari S, Zagarella F (2015) Costs assessment for building renovation cost-optimal analysis. Energy Proc 78:2378–2384

Ferrari S, Zagarella F (2016) Assessing buildings hourly energy needs for urban energy planning in southern European context. Proc Eng 161:783–791

Ferrari S, Zagarella F, Caputo P, Bonomolo M (2019a) Assessment of tools for urban energy planning. Energy 176:544–551

Ferrari S, Zagarella F, Caputo P, D'Amico A (2019b) Results of a literature review on methods for estimating buildings energy demand at district level. Energy 175:1130–1137

Ferrari S, Zanotto V (2016a) Defining representative building energy models. In: Building energy performance assessment in southern Europe, SpringerBriefs in Applied Sciences and Technology, pp 61–77

Ferrari S, Zanotto V (2016b) Energy performance analysis of typical buildings. In: Building energy performance assessment in southern Europe. SpringerBriefs in Applied Sciences and Technology, pp 79–98

International Organization for Standardization (ISO). ISO 17772-1:2017. Energy performance of buildings—Indoor environmental quality—Part 1: indoor environmental input parameters for the design and assessment of energy performance of buildings

Intesa Stato Regioni Enti Locali Sistemi Informativi Territoriali. Specifiche per la Realizzazione dei Database Topografici di Interesse Generale—Il catalogo degli oggetti—Revisione delle Specifiche di contenuto 1n 1007_1 e 1n 1007_2, 2006 (in Italian)

Istat (2011) Spatial data of the 15th general census of population and houses. https://www.istat.it/it/archivio/104317 (in Italian). Accessed 16 Apr 2019

Klein SA, Beckman WA, Mitchell JW, et al (2014) TRNSYS—A transient system simulation program user manual. The Solar Energy Laboratory—University of Wisconsin, Madison

Ministero dello Sviluppo Economico (MiSE), Ministero dell'Ambiente e della Tutela del Territorio e del Mare (2017) Decreto Ministeriale 10 novembre 2017, Rome (in Italian)

Ministero per la Pubblica Amministrazione e l'Innovazione. Decreto 10 novembre 2011. Regole tecniche per la definizione del contenuto del Repertorio nazionale dei dati territoriali, nonché delle modalità di prima costituzione e di aggiornamento dello stesso (GU n. 48 del 27/02/2012—Suppl. Ordinario n. 37) (in Italian)

MiSE, Ministro dell'Ambiente e della Tutela del Territorio e del Mare, Ministro delle Infrastrutture e dei Trasporti, Ministro per la Semplificazione e la Pubblica Amministrazione, Decreto Ministeriale 26/06/2015. Applicazione delle metodologie di calcolo delle prestazioni energetiche e definizione delle prescrizioni e dei requisiti minimi degli edifici (in Italian)

Pasquinelli A, Guzzetti F (2006) Knowledge for intelligence: discussing the state and the role of building data in Italy. In: International conference on smart data and smart cities, 30th UDMS, Split, Croatia

QGIS Development Team, QGIS Geographic Information System (2019) Open Source Geospatial Foundation Project. http://qgis.osgeo.org. Accessed 16 Apr 2019

Regional energy balance SIRENA20 (2012). http://www.energialombardia.eu/sirena20. Accessed 16 Apr 2019

Repertorio Nazionale dei Dati Territoriali (RNDT). http://geodati.gov.it/geoportale/. Accessed 08 Sep 2018

United Nations (UN) (2015) Transforming our world: the 2030 agenda for sustainable development. Report no. A/RES/70/1, New York

United Nations Environment Programme (UNEP) (2015) District energy in cities—Unlocking the potential of energy efficiency and renewable energy, Nairobi

Zagarella F (2019) Estimating the buildings hourly energy demand for smart energy district planning, Ph.D. thesis, Politecnico di Milano, Milan

Advanced Digital Technologies for the Conservation and Valorisation of the UNESCO Sacri Monti

Cinzia Tommasi, Cristiana Achille, Daniele Fanzini and Francesco Fassi

Abstract Information technology touches all the main activities that orbit Cultural Heritage, including management, communication, monitoring and conservation. In particular, advanced digital tools can help the process of preservation, fostering a participatory process connecting diverse experts with various skills and educational backgrounds, and empowering the maintenance activities. The practical example of this statement is its application to the Sacri Monti site. Thanks to its inclusion on the UNESCO World Heritage list, the needs of the site are highlighted in the Periodic Report, a tool useful for monitoring the existing condition of the site. The research presented here attempts to develop a solution for the conservation issue, using advanced technology as a tool for managing the safeguarding activities, creating a product that actively involves users and stakeholders, contributing to the valorisation process.

Keywords UNESCO site · Valorisation · Digital · Communication · Survey · 3D model · Participatory process · Co-design

1 Introduction

The valorisation of Cultural Heritage is an entire process within a structured and territorial vision. The strategic lines dealing with the conservation and preservation of buildings are several. The site is one of the main actions that help this process, and it should be the result of an interdisciplinary vision that connects public institutions, various experts and local communities (Laing 2018). In this framework, the first obstacle to overcome is the communication between many actors, and the use of advanced digital technologies allows for better management of the information at every stage of the project (Della Torre 2014). In order to overcome this issue, valorisation (MiBACT 2019) consists of the activities aimed to promote awareness

C. Tommasi (✉) · C. Achille · D. Fanzini · F. Fassi
Architecture, Built Environment and Construction Engineering—ABC Department, Politecnico di Milano, Milan, Italy
e-mail: cinzia.tommasi@polimi.it

B. Daniotti et al. (eds.), *Digital Transformation of the Design, Construction and Management Processes of the Built Environment*, Research for Development,
https://doi.org/10.1007/978-3-030-33570-0_34

of national heritage, and to ensure the best conditions of use and access to that heritage by all members of the public, to stimulate cultural development. Improving the diffusion and the involvement of local communities in heritage awareness can foster identities, strengthen communities and support business and sustainable development.

This premise considers valorisation as knowledge awareness and improvement, and the first step to trigger this kind of process is to collect tangible data and intangible information about the application field, finding a way to use and combine it in a sharable and accessible system. Current research in the field of Cultural Heritage concentrates on the investigation of advanced conservation strategies that also include the possibility of sharing and use of the relative structured information and 3D models (Benatti et al. 2014; Apollonio et al. 2017). One of the methods adopted to achieve this goal is to use a structured web platform (Fassi and Parri 2012). This kind of online system can handle 3D data models, created to contain and support punctual information. They simultaneously respond to the need for storing different types of data (coming from heterogeneous sources) and sharing information between different expert users.

2 Sacri Monti Application Field

The research focused on the nine Sacri Monti of Piedmont and Lombardy (Northern Italy), which are holy paths in a beautiful, isolated landscape (Fig. 1). The chapels and the church along the route contain artistic scenes (statues, paintings, etc.) that evoke the theme of the pathway. They have been included in the UNESCO World Heritage list since 2003 (UNESCO 2003). As part of this global institution, they have to maintain the parameters and criteria[1] (UNESCO 1972; UNESCO 2005) to stay on the list. The Italian Lex n° 77 of 2006 (and subsequent updates) identifies the UNESCO sites as points of Italian excellence, representing the country on an international level. It recognises the management plans required from the organisation as tools to ensure the conservation and valorisation of the sites, defining intervention priorities and the implementation methods and the actions for finding public and private resources. The Sacri Monti circuit is subject to the drafting of a ‘periodic report’, one of the core conservation monitoring mechanisms of the World Heritage Convention.

Every 6 years, the States Parties are invited to submit to the World Heritage Committee a Periodic Report on the application of the World Heritage Convention in their territory.[2]

The Report (UNESCO 2014) is composed of two sections, in the form of a questionnaire, compiled by the World Heritage Site Manager (Sect. 2) and the National

[1] Operational Guidelines for Implementation of the World Heritage Convention.

[2] Unesco Periodic reporting https://whc.unesco.org.

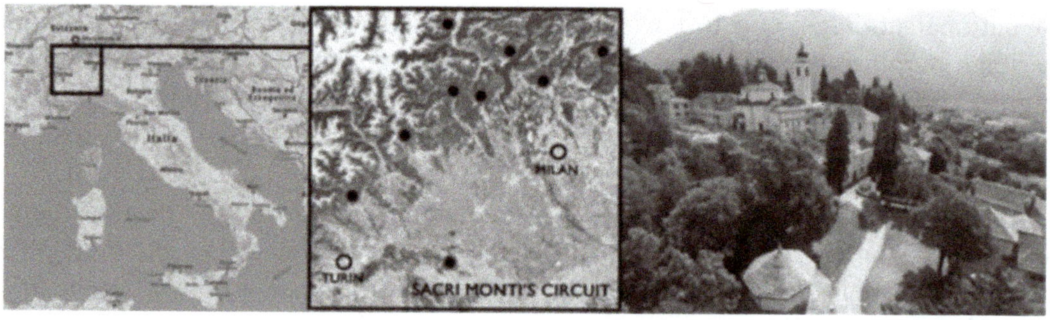

Fig. 1 The Sacri Monti circuit geolocalisation

Focal Point (Sect. 1 and validating Sect. 2): the first lists the State Party and the institutions and groups involved in the preparation of the Report; the second provides information about the primary data of the property and its management.

Thus, the document contains a series of criteria through which to assess the 'health status' of the Sacri Monti and as such, is a good starting point to define the needs of the site.

The categories represented in the Periodic Report Second Cycle (2012–2015 for Europe), highlight these main requirements:

- Annual conservation work/action plan.
- Education and awareness programmes.
- Local community and stakeholder engagement.
- Tourism monitoring.

The management needs of the sites required a monitoring programme for improving the understanding of Outstanding Universal Value[3] and knowledge of the areas. The research activity in progress aims to respond to these needs, contributing to the fulfilment of the Third Cycle of the Periodic Reporting (2019–2024). In particular, this chapter focuses on the management system, proposing a structured methodology for planning and storing conservation activities and supporting interdisciplinary work.

The main steps of the project are outlined below (Tommasi et al. 2019):

- Knowledge phase: Experimentation and integration of different systems and technologies of the 3D survey for a multi-scale representation (architectural and landscape scale);
- Modelling phase: Definition of the workflow from the data (point cloud, bidimensional drawings DWG, etc.) to accurately building the 3D model to support related information; creation of support for the information collected;

[3] Outstanding Universal Value (UNESCO 2005, Operational Guidelines) Par. 49. Outstanding universal value means cultural and/or natural significance, which is so exceptional as to transcend national boundaries and to be of common importance for present and future generations of all humanity. As such, the permanent protection of this heritage is of the highest importance to the international community as a whole. The Committee defines the criteria for the registration of properties on the World Heritage List.

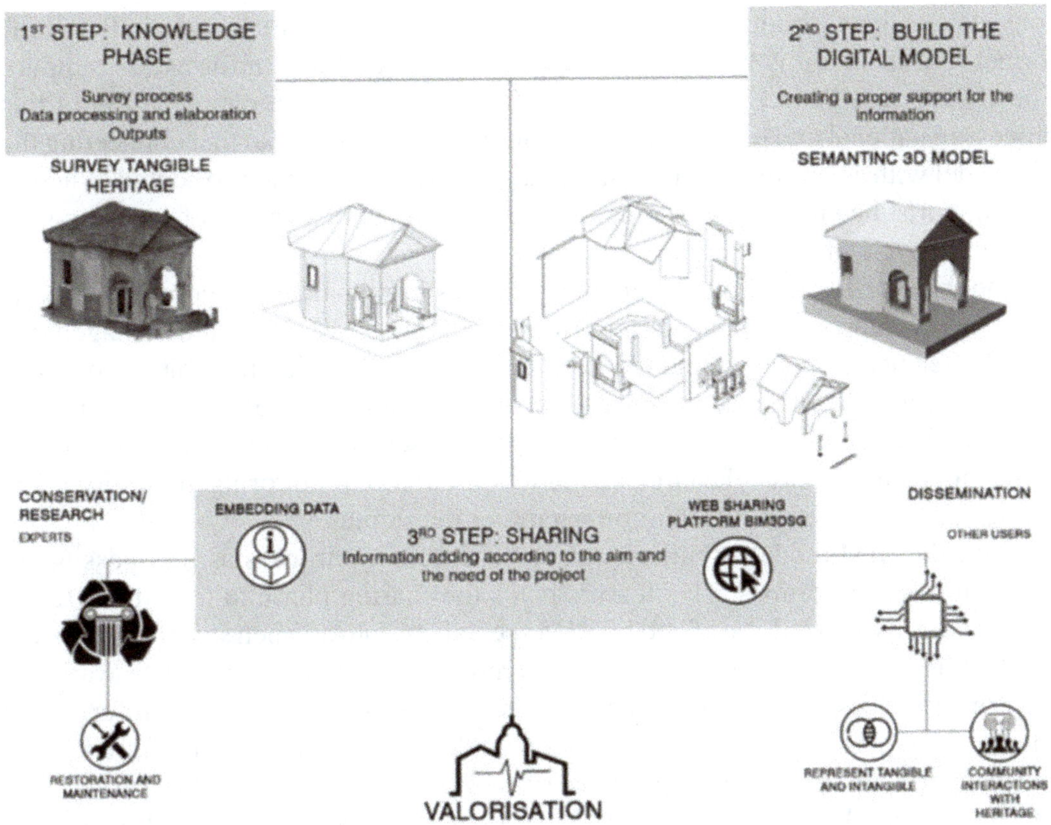

Fig. 2 Workflow from survey data to modelling, web sharing access and final visualisation

- Sharing Phase: Methods and modalities for access to and web sharing of the 3D multi-scale reconstructions and further information from experts with different skills (Fig. 2).

3 Data Management Through an Online Information System for Conservation

The first topic addressed is the elaboration of the existing survey data for building 3D models of chapels used as pilots; in this case, the data inserted came from the Ente di Gestione dei Sacri Monti[4] and the summer school Laboratory of Places 2018 by the Politecnico di Milano (Achille et al. 2018). Starting with the survey data,[5] which

[4]From 2012, the Regional Lex n° 19/2009, merged the seven Sacri Monti of Piedemont in the Ente di Gestione dei Sacri Monti, which is a regional management body.

[5]During the Laboratory of Places 2018 six chapels of Sacro Monte di Domodossola were surveyed, the data acquired and inserted in the web system includes: general point cloud of the site coming from Leica Pegasus Backpack, point cloud coming from Leica C10 TLS, photogrammetric point cloud obtained from the integration of UAV and terrestrial photogrammetry. The clouds are georeferenced

consists mostly of bi-dimensional drawings in DWG format and images in RASTER format, the first step is to build a 3D model, which provides information support. The choice of BIM technique[6] is due to: (i) the increasing implementation of this process in national and international regulations; (ii) the potential that connecting the 3D model with an entire information system offers; (iii) the object-oriented structure that supports the building of a semantic model; (iv) the possibility of managing the whole life cycle of the building.

The criterion for building the model was its description in all the constitutive parts necessary to support the punctual information about conservation and management. The Politecnico's operators and the Ente di Gestione identified the objects to model and the level of detail needed to ensure correct georeferencing of all the information required. Therefore, it was necessary to know and select the data before assignment to the model. In this case, the information was about roof monitoring, humidity, static damage, and restoration information regarding the statues.

Once the BIM model is built, it can serve different purposes as it is possible to export it in many formats. This feature helps the sharing phase to transition from a BIM model that can only be edited by BIM specialists, into an open system, accessible to all kinds of specialist. In assessing the 3D model and the monitoring data available, the primary necessities are: to manage and easily visualise a structured information database, to share the models and their content between experts with different skills and backgrounds and to manage conservation activities, enhance the survey data and the output. The tangible result for these specific needs is the BIM3DSG web platform[7] which allows for:

- Loading, visualisation and use of 3D models inside an ordinary web browser, selecting the level of detail desired.
- Monitoring the conservation status of Sacri Monti.
- Managing the information system.
- Using the system in portable devices.
- Sharing data among experts with different skill sets and backgrounds.

The benefit of this system is the possibility to shape it according to the needs of the application field. Thanks to teamwork and the bottom-up approach, the result is accurate and tailored to the real needs of the chapels. Within the environment, the three-dimensional objects modelled become categories of real elements, to which different kinds of information is assigned, according to typology: information that

---

using the Leica Pegasus Backpack data as reference. Modelling data: the BIM model of Chapels, georeferenced as point clouds, and divided into all the elements and categories required from the restorers and workers of Enti Gestione Sacri Monti. Conservation information: structural and humidity analyses, decay, 2D drawings, photos, and more.

[6]The models were elaborated with the software Autodesk Revit, as the software guarantees complete compatibility with the DWG drawings of the Ente (coming from Autodesk AutoCAD, same software house), and a very good management of the point clouds, making the modelling phase much easier.

[7]The system was developed starting from the prototype BIM3DSG (3DSurvey - Patent Pending MI2014A002016), an online platform created ad hoc and through the collaboration between the Politecnico di Milano and the final users/consumers.

belongs to the objects, information that changes after a 3D model update and information that varies according to different time thresholds (Fig. 3).

The first operation is the creation of a structured Database (DB), which represents 'the skeleton' of the user interface. Once the fields to be listed were identified, it was necessary to organise them at an IT level, following the BIM3DSG logic of

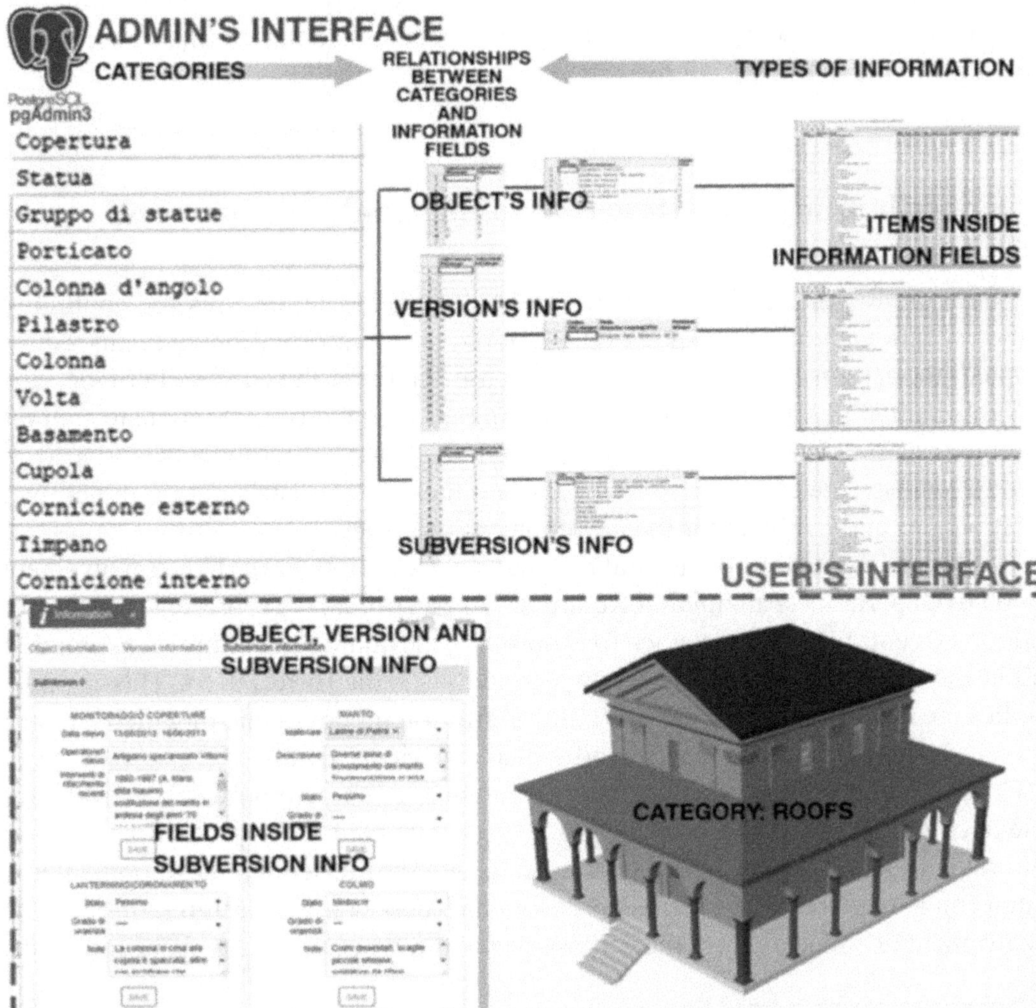

Fig. 3 Structure of the admin system: each 3D element is assigned to a category, and each category has structured information attached. In the lower part, the visualisation of the structure of the database in the user interface

structure and relationships between pieces of information[8] (Rechichi et al. 2016). Thus, it was possible to link in a dynamic way the models and the data about the state of conservation and the interventions on the Chapels. This structure is presented in PostgreSQL, a 'free database management system' developed through PgAdmin software.

Moreover, the system allows various data to be saved and stored as digital documents, 3D models, videos, images, etc., and these may be consulted within the web environment. In particular, each model was attached to the original report in the format provided (mainly PDF format).

4 Contribution of the Research to the Co-design Process

The Periodic Reporting Questionnaire is the document that ensures the continuity of the exceptional universal value of the CH, providing information on the measures adopted, the actions taken and the state of conservation of the cultural heritage. The previous paragraphs have illustrated the usefulness of having a 3D model to support the conservation and monitoring activities, but the same technologies could also be used to increase the awareness of citizens and their active participation in defining the strategies and methods for the development of the area.

The 'experimentation' of cultural resource consumption through the involvement of creative operators (Fanzini and Rotaru 2015; Fanzini 2017) represents a sector of undoubted centrality and relevance for improving the quality of cultural consumption and at the same time triggering a new economy (Della Torre 2015). This formula implies the transition from a passive cultural fruition model to one in which the user actively participates, co-producing individual experiences.

Citizens' involvement is increasingly acknowledged as an important success factor able to contribute to all phases of the processes connected to the protection of world cultural heritage, from the initial start-up stages to the subsequent development and actual implementation. The terms co-design and user-centred design (stressing the involvement of endusers from the early stages of the process—fuzzy front end), have

---

[8]The system's architecture includes:

- The categories that represent the 3D components division, and refer to classes of real elements;
- The information that is divided into three classes: object info: the data that does not change in time, version info: the 3D model changes over time, and subversion info: the 3D model does not change over time, but the information related to it is updated, allowing operations to be traced and inspections already made or in progress in the buildings, handling their life-cycle and planning the events over time.

Each class of information has different macro fields, and each macro field has specific "boxes" inside, defining the name and typology of fields (text, date, multiple choice, and more).

The classes of information are connected to the categories through relationships that relate the number of the macro field to the number of categories. The structure of the database has a user interface inside the web browser, e.g. Mozilla firefox.

been introduced to describe this particular endeavour. The evolution of user involvement models, from the participatory formulas to the real co-design, has changed the role of professional designers, and also that of the project users, establishing new creative domains.

As noted by Jannack et al. (2015), the real challenge for the future co-design environment is to enable communication and collaboration between large numbers of experienced citizens and professionals, ensuring a more creative, targeted and secure design process for projects of public interest. In other words, there is a need for collaborative tools and environments that allow creative professionals to capitalise on the ingenuity of crowds, to follow opinions, feelings, values and exploit widespread intelligence. The Building Information Modelling (BIM) turned out to be an effective expert instrument in similar cases. The intermediation of information exchange from the computer model of the real-life building enables a better understanding of the planned intervention, opening the design process to different stakeholders not necessarily familiar with specialised representation conventions. As part of the proposal, BIM is, therefore, the enabling technology that could be used to connect different information, from the conscious and compelling expression of the client's needs, to the broad and operational sharing of the project, also fostering the possibility to access and use diffuse data (Fig. 4).

In addition to the responsible use of technologies, special attention is needed for the configuration of methods and processes for the creative involvement of users. In other words, computer skills have to be combined with the in-depth knowledge of collaborative design practices that can specifically involve the original contribution of users.

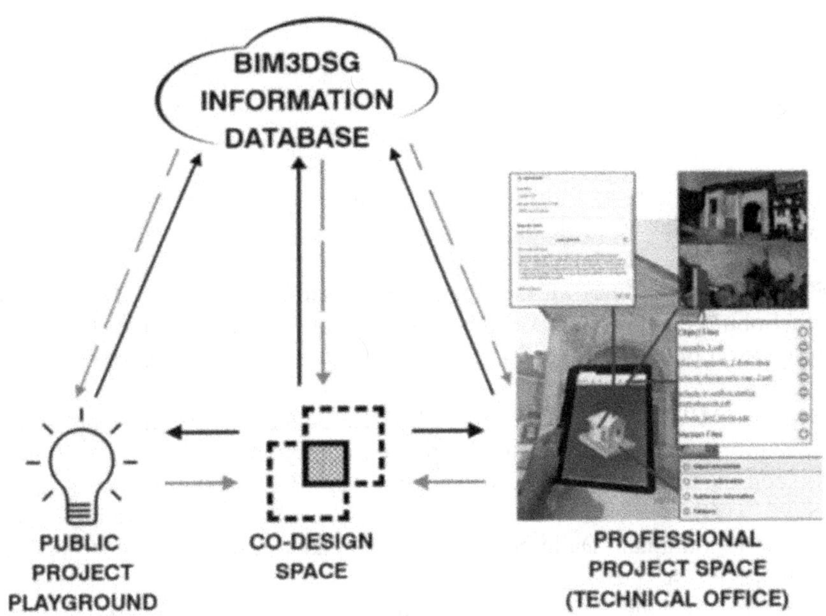

Fig. 4 Re-elaboration of the co-design environment (Jannack et al. 2015)

In this sense, the BIM 3D model could contribute to the realisation of interactive systems that facilitate public participation by anticipating problems and allowing creative interaction between experts, local communities and the various stakeholders.

5 Conclusion

In conclusion, the research aims to show how it is possible to build a valid tool for the conservation and monitoring activities of CH, following national and international legislation and the specific needs of the application field. In particular, it will broaden the criteria to build a semantic 3D model that will link a database of information, specifying the logic and the specifications behind it.

In the same way, it will show how to build a database collaborating with local institutions and experts, following the real needs of the site. The benefit is that this kind of tools gives multidisciplinary works and projects a shared environment for storing and viewing a variety of data. They can help both to implement conservation activities and be open to the public. The result will help to update a section of the UNESCO Periodic Report and will contribute to shaping 'standard' best practices to monitor and safeguard all similar case studies.

Acknowledgements The authors would like to thank the Ente di Gestione Sacri Monti, managers of the seven Sacri Monti of Piedmont. They are actively collaborating with the 3D Survey Group in the ABC Department of the Politecnico di Milano, providing DWG and PDF materials and are involved in the decision-making process of the valorisation strategy.

References

Achille C, Fassi F, Mandelli A, Fiorillo F (2018) Surveying cultural heritage: summer school for conservation activities. Appl Geomatics 10(4):579–592. https://doi.org/10.1007/s12518-018-0225-3

Apollonio FI, Basilissi V, Callieri M, Dellepiane M, Gaiani M, Ponchio F, Rizzo F, Rubino AR, Scopigno R, Sobrà G (2017) A 3D-centered information system for the documentation of a complex restoration intervention. J Cult Herit 29:89–99. https://doi.org/10.1016/j.culher.2017.07.010

Benatti E, Borgarino MP, Della Torre S (2014) PLANET beni architettonici. Uno strumento per la conservazione programmata del patrimonio storico-architettonico. In: ICT per il miglioramento del processo consevrativo. Ed Nardini, Milano, Italy, pp 13–31

Della Torre S (2014) Oltre il restauro, oltre la manutenzione. Keynote lecture in La strategia della Conservazione programmata. Dalla progettazione alle attività di valutazione degli impatti. Ed. Nardni, Milano, Italy, pp 1–10

Della Torre S (2015) Lezioni imparate sul capo dei distretti culturali. In: Il Capitale Culturale: studies on the value of cultural heritage 03, 61–73. https://doi.org/10.13138/2039-2362/1175

Ente di Gestione dei Sacri Monti. https://www.sacri-monti.com/. Accessed 13 May 2019

Fanzini D (2017) Tecnologie e processi per il progetto del paesaggio. Reti e modelli distrettuali, 1st edn, Maggioli Editore, Santarcangelo di Romagna, IT

Fanzini D, Rotaru I (2015) Processi inclusivi e project anticipation per la rigenerazione delle città e dei territori. Techne 10:102–109. https://doi.org/10.13128/Techne-17506

Fassi F, Parri S (2012) Complex Architecture in 3D: from survey to web. Int J Herit Digit Era 1:379–398

Jannack A, Munster S, Noenning JR (2015) Enabling massive participation: blueprint for a collaborative urban design environment. In: Proceedings of IFKAD 2015, publisher: international forum on knowledge asset dynamics, editors: Giovanni Schiuma, 2363–2380

ICOMOS International Council on Monuments and Sites. Retrieved from: https://www.icomos.org/. Accessed 18 March 2019

Laing R (2018) Digital participation and collaboration in architectural design, 1st edn. Routledge, Oxon, UK

Ministero per I Beni e le Attività Culturali – MiBACT. http://www.beniculturali.it/mibac/export/MiBAC/index.html. Accessed 14 March 2019

Rechichi F, Mandelli A, Achille C, Fassi F (2016) Sharing high-resolution models and information on web: The web module of BIM3DSG system. In: International archives of the photogrammetry, remote sensing and spatial information sciences, vol XL-B5, pp 703–710. https://doi.org/10.5194/isprs-archives-XLI-B5-703-2016

Regione Piemonte e Lombardia (2015) Sacri Monti of Piedemont and Lombardy. Ed. Sagep, Genova, Italy

SACHER Smart Architecture for Cultural Heritage in Emilia-Romagna (POR FESR 2014–2020). http://www.sacherproject.com/. Accessed 2 April 2019

SICaRweb – Sistema informativo per I cantieri di restauro. http://sicar.beniculturali.it/. Accessed 2 April 2019

Tommasi C, Fiorillo F, Jiménez Fernández-Palacios B, Achille C (2019) Access and web-sharing of 3D digital documentation of environmental and architectural heritage. In: International archives of the photogrammetry, remote sensing and spatial information sciences, vol XLII-2/W9, pp 707–714. https://doi.org/10.5194/isprs-archives-xlii-2-w9-707-2019

Standards and Laws

Codice dei Beni Culturali e del paesaggio (2004, January 22) Decreto Legislativo 22 gennaio 2004, n. 42. Retrieved from https://www.beniculturali.it/mibac/multimedia/MiBAC/documents/1226395624032_Codice2004.pdf

Lex n°77/2006—Misure speciali di tutela e fruizione dei siti e degli elementi italiani di interesse culturale, paesaggistico, e ambientale, inseriti nella lista del patrimonio mondiale, posti sotto la tutela UNESCO

UNESCO (1972, November 21) Convention Concerning the Protection of the World Cultural and Natural Heritage. Retrieved from https://whc.unesco.org/en/conventiontext/

UNESCO (2003) Document of Sacri Monti's Nomination. Retrieved from https://whc.unesco.org/list/1068/docs/

UNESCO (2005, revised version 2017) The Operational Guidelines for the Implementation of the World Heritage Convention. Retrieved from https://whc.unesco.org/en/guidelines/

UNESCO (2014) Periodic Report, Second Cycle. Retrieved from https://whc.unesco.org/list/1068/docs/

Survey and Scan to BIM Model for the Knowledge of Built Heritage and the Management of Conservation Activities

Raffaella Brumana, Daniela Oreni, Luigi Barazzetti, Branka Cuca, Mattia Previtali and Fabrizio Banfi

Abstract Surveying a historic building means to measure, to detect and to analyse its geometries, its structural elements, the connections still existing between the different parts, in order to define its state of conservation, to make structural analysis and finally to plan a proper project of conservation, consolidation and reuse. The survey represents the first necessary moment for building's knowledge investigation. Nowadays, the wide use of tools and accurate surveying techniques makes it possible to achieve an adequate level of accuracy of information related to the buildings; BIM tools offer a great potential, in terms of both planning and evaluation of the entire knowledge and conservation process of an historical building, and in terms of its management and future maintenance. In particular, the BIM technologies allow the communication between data coming from different software, allowing a greater exchange of information between many actors. In recent years, the generative process of Building Information Modelling (BIM) oriented to the digitization of built heritage has been supported by the development of new commands modelling able to integrate the output data produced by laser scanner surveys (point clouds) in major modelling applications. Structural elements, such as vaulted historical systems, arches, decorations, architectural ornaments and wall partitions with variable cross sections, require higher levels of detail (LOD) and information (LOI) compared to the digitalization process of new buildings. Therefore, the structure of a BIM model aimed at representing existing and historical artefacts (HBIM) requires the definition of a new digital process capable of converting traditional techniques used for the management of new buildings to those suitable for creation of digital versions of historical buildings that are unique of their kind. The aim of this paper is to present the results of the ongoing researches and activities carried out on survey and HBIM model of historical buildings.

R. Brumana · D. Oreni (✉) · L. Barazzetti · B. Cuca · M. Previtali · F. Banfi
Architecture, Built Environment and Construction Engineering—ABC Department, Politecnico di Milano, Milan, Italy
e-mail: daniela.oreni@polimi.it

B. Daniotti et al. (eds.), *Digital Transformation of the Design, Construction and Management Processes of the Built Environment*, Research for Development,
https://doi.org/10.1007/978-3-030-33570-0_35

1 Introduction

This paper tries to make a synthesis of the different lessons learnt, related both to the positive and critical aspects concerning Historic BIM (HBIM) feasibility, sustainability and usefulness, and to the challenges of using HBIM for restoration and preservation activities. The theoretical and practical approach adopted, overcame the current BIM logic, based on sequential Level of Detail (LoD) that is typical of new buildings from simplex to complex (Volk et al. 2014; AEC (CAN) BIM Protocol; American Institute of Architects 2013), from the preliminary to the executive design (Nuovo Codice Appalti 2017). This choice was made in favour of the maximum precision and articulated description and representation of each component of the existing building, in order to derive the conservation project. The challenge was to obtain a cost-effective HBIM able to embody the complexity of each element of an existing historically important building (i.e. walls, pillars, vaults, beams, timbers). A Non-Uniform Rational Basis-Splines (NURBS) based parametric generative modelling process is here proposed in order to get sustainable rich modelling, able to match the related information, moving HBIM towards the actors. On the lesson learnt from this experience and many others, the process of updating the current codification criteria (UNI11337-2017) started with a draft proposal, stimulating a debate for the future of HBIM adoption.

2 The Main Case Study: The Basilica of Collemaggio in L'Aquila

The Basilica of Collemaggio was significantly damaged by the earthquake of Richter Magnitude 5.9 that had struck the town of L'Aquila (Central Italy) on 6 April 2009: the dome, the transept and triumphal arches collapsed with their pillars, and great damage occurred to the apses, to the pillars of the arched walls of the nave, and to the longitudinal north front with the 'Holy Door'. The Basilica is a Romanesque masterpiece characterized by a dense, fascinating history and different construction phases, which began in 1270 (Bartolomucci 2004). An important event called 'Festa della Perdonanza' is celebrated every year on 28 and 29 August, with the procession ceremony transferring the original Papal bull established by Pope Celestino V from the Municipality to the Holy Door of the Basilica. The result is an extraordinary, unique mix of tangible and intangible community values to be preserved and transferred to the future generations. Thus, the restoration project *'Ripartire da Collemaggio' (Restarting from Collemaggio)* has been undertaken by ENIservizi with the aim of giving new hope to the L'Aquila citizens. The Superintendence Office carried on the restoration project, with the scientific support of the University of L'Aquila and the Sapienza University of Rome, under the coordination of the Politecnico di Milano. The challenge was to improve on one side the use of BIM tools for existing buildings (Oreni et al. 2013; Brumana et al. 2017), able to combine all the complexity of the geometrical shapes, and on the other the use of these tools for processes of built heritage conservation (Della Torre 2015). The HBIM has been carried on by integrating the surveying (Oreni et al. 2014) with the information coming from the historical and archive research and with the construction technology analysis. The aim was to support the following activities: (i) the design project, preservation and decision-making, (ii) the management of critical issues, regarding the conservation of material authenticity and construction techniques and (iii) the need to guarantee safety also in case of other potentially stronger earthquakes. Given these main goals of the project, the production of HBIM of the Basilica needed to take into account the complex morphology of the different ancient structures and their transformations along the centuries.

3 From 3D Geometric Survey to HBIM: Generative Modelling

The surveys of the Basilica (Barazzetti et al. 2015a, b) has been carried out to check the geometry of the whole damaged Basilica, the out of plumbs, the structural state of the art, in order to support the preservation plan. A geodetic network has been carried out to strengthen the clouds registry (Laser scanner Faro Focus ©), and the 3D photogrammetric image block processing, covering all the vertical walls and the vaulted system.

The first draft release of 2D digital drawings (plans and vertical sections) was delivered starting from the information collected during on-site surveying campaigns (May 2013); the first HBIM drafted version was delivered in support of the preliminary steps of the design project (end of 2013) and to the definitive-executive steps (2014–15).

In the case of Collemaggio Basilica, the precondition was the reliability of the HBIM from the point of view of geometry and interoperability: every element of the Basilica was be linked and related to the different kinds of information, and to be managed by various actors (Barazzetti et al. 2016). Therefore, the HBIM has been addressed to manage the geometric richness and morphology acquired by the point clouds in order to support the following modelling phases. The modelling procedure developed had to be able to gather all the richness of information contained in the point clouds (Fig. 1).

Under the BIM tool technological aspect, parametric BIM models are generally managed for new construction processes; even the latest developments of parametric applications do not provide advanced tools, being time consuming for model generation. Therefore, different methods of modelling and representation of complex architectonical elements and managing of parametric BIM (Scans to NURBS) with multiple levels of details have been experimented with promising results, (Banfi 2016; Banfi et al. 2017).

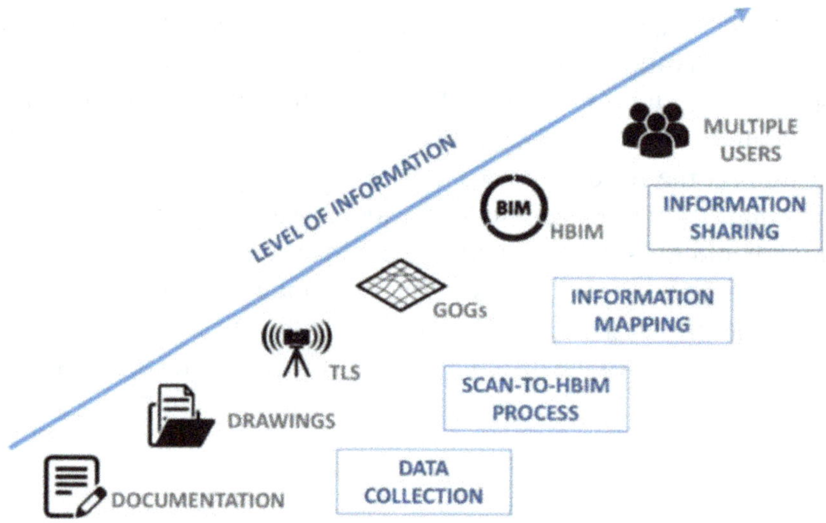

Fig. 1 The primary, secondary data sources and tools used for the generation of the HBIM model

A multi-slice-based wireframe model and/or cloud-based model have been adopted for each 3D parametric object, maintaining the morphologic richness and precision acquired by the surveying. A comprehensive overview of exiting procedures has been taken into account to define new processes and to boost the use of HBIM (Banfi 2016). The definition of the novel concept of 'Grade of Generation' (GOG 1–10) and Grade of Information (GOI), related to the object modelling, allows to enrich the logic of the traditional LOD sequence (100–500), defined by the AIA specifications (2013–2017), introducing as part of the process the method and functionalities adopted to generate the model objects themselves (Fig. 2). Starting from the cloud scans, the objects' models were generated within the CAD environment and within other modelling environments (such as MC Nell Rhinoceros ©), in order to be managed further within the BIM tools (Banfi 2017).

In Fig. 3, the generative process from the first phase of processing is represented by the creation of accurate 3D parametric objects following the GOG 9 and GOG 10 generative modelling process and protocols implemented (Banfi 2019).

The study carried out on NURBS and geometric primitives led to improvements in the level of automation and accuracy of each element. Thanks to the combination of various advanced scan to BIM modelling requirements, and the proper use of generative profiles, the level of automation has passed from semi-automatic to automatic, avoiding long lead times and the relatively high costs involved in the digital 2D/3D representation. The generation of proper geometric primitives and the management of the generative digital process led to the creation of transmissible three-dimensional elements (Fig. 2). Clustering exchange formats (.dgw,.sat,.stp,.pts) made it possible to obtain and transfer 3D slices, profiles, control edges and complex shapes from a versatile 3D NURBS modeller (MC Neel Rhinoceros ©) to the most used comprehensive BIM application (Autodesk Revit ©).

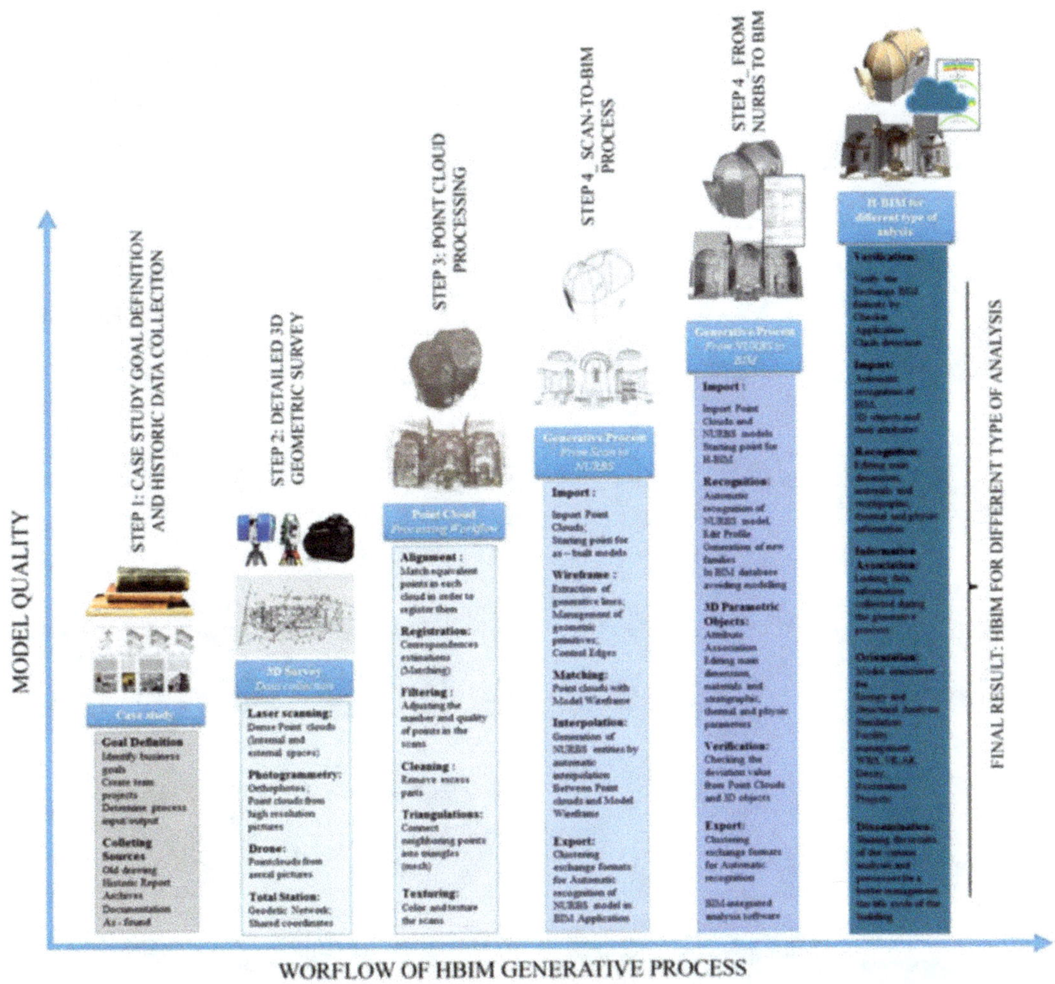

Fig. 2 The generative process applied to the HBIM of Basilica of Collemaggio. *Source* after Banfi (2019) holistic generative modelling process for HBIM

4 HBIM as Building Site Tool to Share Different Data Among Actors

The HBIM model was conceived to support operators, architects, structural engineers, and their activities: economic computation, construction site management, diagnostic analysis, design, construction tender and restoration itself. The adopted approach allows the HBIM to be updated and adapted by the different BIM actors in order to support the different phases: (i) models that provide a comparison and impact assessment of the simulation scenario of different solutions for decisions making about the crashed dome; (ii) conservation plan of the overall restoration design, progressively including the diagnostic analysis within the BIM (material analysis, surface decay analysis and preservation project) (Fig. 4); (iii)Finite Element Analysis (FEA) in support of the structural analysis, addressing the intervention on the damaged pillars, the walls and vaulted system; (Barazzetti et al. 2014; Crespi et al. 2016)

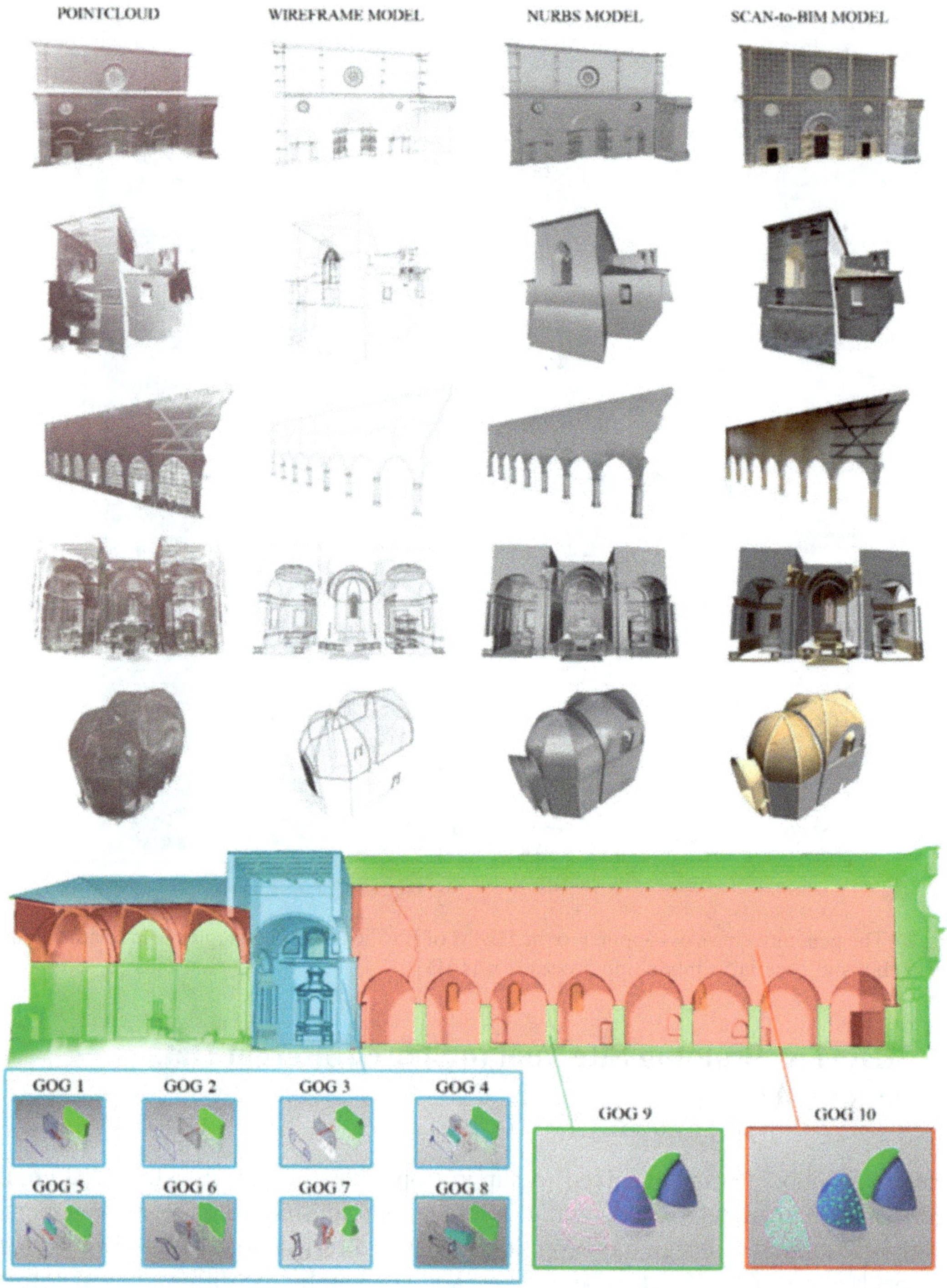

Fig. 3 The different GOGs applied during the generative process. *Source* Banfi (2019) holistic generative modelling process for HBIM

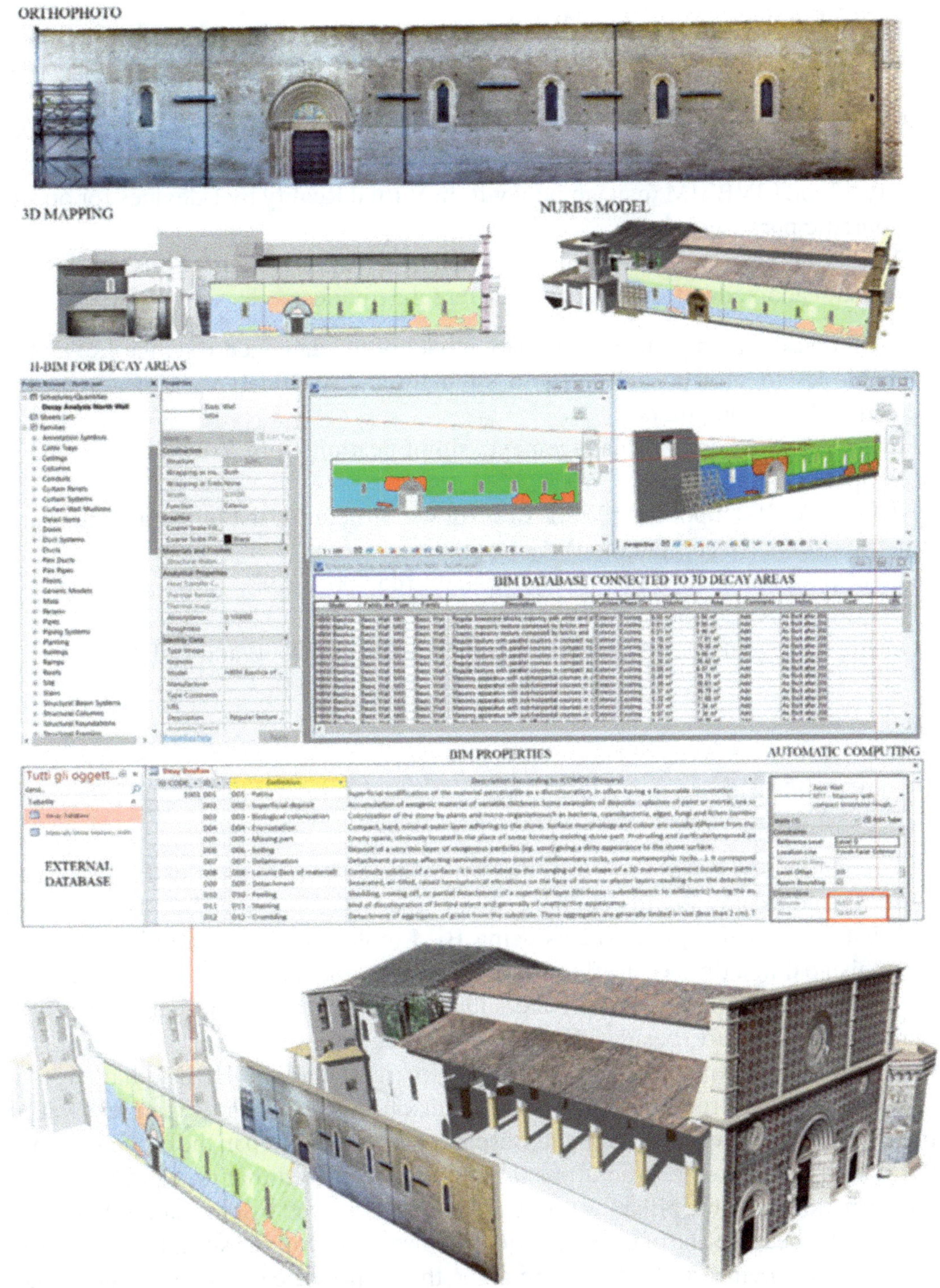

Fig. 4 The utility of HBIM for 3D mapping and the generation of specific BIM database related to the new type of information such as decay areas and stratigraphic information

(iv) Work Breakdown Structure (WBS) of the restoration activities and the computational phase; (Brumana et al. 2018) (v) Construction Site Management (Co.Si.M) carried out on the HBIM adding machinery, construction site structures, simulation of the scaffoldings to manage the construction site and to simulate partitions in order to make the Basilica partially usable during the last phases of restoration; (vi) Virtual Reality/Cloud LIVEBIM management within Virtual Reality by I-devices for on site and touristic purposes.

During the Scan to HBIM process, different Level of Detail (LODs) have been defined and adopted in function of the different activities:

- HBIM conservation plan: 3D complex model of the damaged Basilica LoD500;
- HBIM design project and scenario simulation: LoD300 (i.e. solutions on the crashed dome and pillars);
- HBIM construction site management: simplification from LoD500 to LoD300 (complex to simplex);
- HBIM to WBS (supporting Work Breakdown Structures, list of the activities and metric-estimative computation): LoD500;
- HBIM preservation and maintenance plan: LoD 500;
- HBIM to FEA: LoD500;
- On site management and HBIM as-built updating (LoD500 and LoD500–300).

5 Conclusions

They remain a few open questions: did we manage the paradigm of HBIMs complexity linked to preservation challenges? In other words, can we boost the level of preservation of challenging aspects using BIM tools and interoperability? How to reach this goal in a time-cost-effective manner, thus encouraging a more systematic use of BIM by the public and private subjects? Which lacks and gaps have to be overcome in the future, besides the BIM modelling issues?

After different experiences, it is possible to argue that one HBIM challenges is represented by the structural analysis and decision-making towards the preservation aims. In theory, this result certainly represents an important step in the process linking HBIM complexity to the structural analysis domain. Given that the demolition and reconstruction of all the pillars was out of options requested by the Superintendence, this issue has become one of the main points of discussion among specialists, restorers and structural engineers. Authors here argue that in the case of single elements data management (such as the case of the pillars preservation), the use of HBIM has heavily contributed to support the preservation process. Structural intervention able to embody all the richness is gained from the HBIM model and it needs to be further investigated. During the restoration project and the on-site intervention regarding the Basilica of Collemaggio, many difficult compromises have been undertaken, between conservation and safety issues, causing also difficult choices when it was necessary to partially sacrifice functional authenticity of structural elements. In those cases, in order to enhance the level of preservation and to adopt softer measures of connection and reinforcement among structural elements (i.e. pillars, arches, walls, vaults, domes, trusses facades and wooden roof interventions), taking into account the tradition of their construction technologies, the HBIM model was fundamental tool for simulation of different scenarios to support better informed decision-making processes.

Acknowledgements The work on the Basilica of Collemaggio was supported by ENIservizi. The authors would like to thank P. Strada (ENIservizi), A. Garofalo (Soprintendenza ai Beni Architettonici e Paesaggistici per l'Abruzzo); Prof. A. Franchi, P. Crespi for the images taken during the restoration and BIM-FEA and M.A. Trani for the COSIM ones.

References

AEC (CAN) BIM Protocol (2014) Implementing Canadian BIM standards for the architectural, engineering and construction industry based on international collaboration

AIA (2013–2017) BIM forum, LOD specification. BIMForum 2013–2017. The level of development specification: for building information models, version: 2013–2017

American Institute of Architects (2013) AIA Document G202-2013: project building information modeling protocol form2

Banfi F (2016) Building information modelling – A novel parametric modeling approach based on 3D surveys of historic architecture. In: Ioannides M, et al. (eds) Digital heritage. Progress in cultural heritage: documentation, preservation, and protection. EuroMed 2016. Lecture notes in computer science, vol 10058. Springer, Cham

Banfi F (2017) Bim orientation: grades of generation and information for different type of analysis and management process. In: International archives of the photogrammetry, remote sensing and spatial information sciences, vol XLII-2-W5, pp 57–64

Banfi F, Fai S, Brumana R (2017) BIM automation: advanced modeling generative process for complex structures. In: ISPRS international conference on annals of the photogrammetry, remote sensing and spatial information sciences, vol IV-2-W2, pp 9–16

Banfi F (2019) Holistic generative modeling process for HBIM (Doctoral dissertation, Italy)

Barazzetti L, Brumana R, Oreni D, Previtali M, Roncoroni F (2014) UAV-based orthophoto generation in urban area: the Basilica of Santa Maria di Collemaggio in L'Aquila. In: Murgante B et al (eds) ICCSA 2014, part IV, LNCS 8582, 2014. © Springer International Publishing, Switzerland 2014, pp 1–13

Barazzetti L, Banfi F, Brumana R, Previtali M (2015a) Creation of parametric BIM objects from point clouds using. Photogram Rec 30(152):339–362

Barazzetti L, Banfi F, Brumana R, Gusmeroli G, Previtali M, Schiantarelli G (2015b) Cloud-to-BIM-to-FEM: structural simulation with accurate historic BIM from laser scans. Simul Model Pract Theory 57:71–87

Barazzetti L, Banfi F, Brumana R (2016) HBIM in the cloud. In: Digital heritage. progress in cultural heritage: documentation, preservation, and protection, EuroMed 2016, Springer, pp 104–115 (Werner Weber Award)

Bartolomucci C (2004) Santa Maria di Collemaggio. Interpretazione critica e problemi di conservazione. Palombi Editore, Roma

Brumana R, Della Torre S, Oreni D, Previtali M, Cantini L, Barazzetti L, Franchi A, Banfi F (2017) HBIM challenge among the paradigm of complexity, tools and preservation: the Basilica di Collemaggio 8 years after the earthquake (L'Aquila). In: International archives of the photogrammetry, remote sensing and spatial information sciences, vol XLII-2-W5, pp 97–104

Brumana R, Della Torre S, Previtali M, Barazzetti L, Cantini L, Oreni D, Banfi F (2018) Generative HBIM-modeling to embody complexity. Surveying, preservation, site intervention. The Basilica di Collemaggio (L'Aquila). Appl Geomatics 10(4):545–567

Crespi PG, Franchi A, Giordano N, Scamardo MA, Ronca P (2016) Structural analysis of stone masonry columns of the Basilica S. Maria di Collemaggio. Eng Struct 129:81–90

Della Torre S (2015) Shaping tools for built heritage conservation: from architectural design to program and management. learning from Distretti culturali. In: Van Balen K et al (eds) Community involvement in heritage (reflections on cultural heritage theories and practices), pp 93–102

Della Torre S (2014) ICT per il miglioramento del processo conservativo. Nardini Editore, Firenze

G.U. 29/01/2008, n. 24 (and 2011 updates) Direttiva per la valutazione e la riduzione del rischio sismico del patrimonio culturale con riferimento alle norme tecniche per le costruzioni

Oreni D, Brumana R, Della Torre S, Banfi F, Barazzetti L, Previtali M (2014) Survey turned into HBIM: the restoration and the work involved concerning the Basilica di Collemaggio after the earthquake (L'Aquila). In: ISPRS annals of the photogrammetry, remote sensing and spatial information sciences, vol II/5, pp 267–273

Oreni D, Brumana R, Georgopoulos A, Cuca B (2013) HBIM for conservation and management of built heritage: towards a library of vaults and wooden bean floors. In: Grussenmeyer P (ed) ISPRS annals of photogrammetry, remote sensing and spatial information sciences, vol II-5/W1, Copernicus Publications, pp 215–221

Piegl LA, Tiller W (1997) The NURBS book. Springer

Volk R, Stengel J, Schultmann F (2014) Corrigendum to "Building information modeling (BIM) for existing buildings—literature review and future needs". Autom Constr 38:109–127. Automation in Construction, Vol 43, July 2014, 204 p

Nuovo Codice Appalti, L. 21/06/2017, n.96., found at https://www.gazzettaufficiale.it/eli/gu/2017/06/23/144/so/31/sg/pdf last accessed in June 2019

www.ingramcontent.com/pod-product-compliance
Lightning Source LLC
LaVergne TN
LVHW082112200926
848484LV00008B/1943